NORMAL MODES AND LOCALIZATION IN NONLINEAR SYSTEMS

Normal Modes and Localization in Nonlinear Systems

Edited by

Alexander F. Vakakis

Reprinted from *Nonlinear Dynamics*,
Volume 25 Nos. 1–3, 2001

KLUWER ACADEMIC PUBLISHERS
DORDRECHT / BOSTON / LONDON

A C.I.P. Catalogue record for this book is available from the Library of Congress

ISBN: 0-7923-7010-4

Published by Kluwer Academic Publishers,
P.O. Box 17, 3300 AA Dordrecht, The Netherlands

Sold and distributed in North, Central and South America
by Kluwer Academic Publishers,
101 Philip Drive, Norwell, MA 02061, U.S.A.

In all other countries, sold and distributed
by Kluwer Academic Publishers,
P.O. Box 322, 3300 AH Dordrecht, The Netherlands.

Printed on acid-free paper

Printed in the Netherlands

Contents

Nonlinear Dynamics **25:** 1, 2001.

Preface

Traditionally, the concept of 'normal mode' is associated with linear vibration theory and is closely related to the principle of linear superposition. 'Nonlinear normal modes' (NNMs), first defined by Lyapunov and then studied by Rosenberg and others, were conceived as nonlinear extensions of the linear normal modes; that is, as free, periodic, and synchronous nonlinear oscillations of unforced discrete or continuous oscillators.

Although the principle of linear superposition does not generally apply to nonlinear oscillation problems, NNMs can still provide a useful framework for studying numerous problems in nonlinear dynamics, such as bifurcations that give rise to NNMs with no linear analogs; symmetry-breaking localized nonlinear periodic motions and nonlinear motion confinement; forced resonances and 'jump' phenomena; NNM-based order-reduction schemes for discretizing the dynamics of continuous oscillators; and nonlinear system identification.

Localization phenomena are encountered in numerous applications in engineering and applied mathematics, such as in elastic buckling, in shear banding, in the free and forced dynamics of extended symmetric oscillators, and in the solutions of certain nonlinear partial-differential equations (solitons and solitary waves). As such, studies of localization phenomena can find applicability in a wide range of applications.

This special issue devoted to 'Nonlinear Normal Modes and Localization in Dynamical Systems' presents a collection of original papers on NNMs and/or localization. Of particular interest to the reader will be the numerous mathematical techniques and methodologies presented in these works, as well as the diverse applications of NNMs and localization to problems in different areas of engineering and applied mathematics.

Alexander F. Vakakis
University or Illinois at Urbana/Champaign
Urbana, Illinois

Nonlinear Dynamics **25:** 3–31, 2001.

Invariant Manifolds, Nonclassical Normal Modes, and Proper Orthogonal Modes in the Dynamics of the Flexible Spherical Pendulum

IOANNIS T. GEORGIOU[*,**] and IRA B. SCHWARTZ
Naval Research Laboratory, Special Project in Nonlinear Science, Code 6700.3, Plasma Physics Division, Washington, DC 20375, U.S.A.

(Received: 19 January 1999; accepted: 10 August 1999)

Abstract. It is shown that the flexible spherical pendulum undergoes purely slow motions with master and slaved components. The family of slow motions is realized as a three-dimensional invariant manifold in phase space. This manifold is computed analytically by applying the method of geometric singular perturbations. This manifold is nonlinear and for all energy and angular momentum levels is characterized precisely by three PO (proper orthogonal) modes. Its submanifold of zero angular momentum is a two-dimensional invariant manifold; it is the geometric realization of a nonclassical nonlinear normal mode. This normal mode is characterized precisely by two PO modes. The slaved slow dynamics are characterized precisely by a single PO mode. The stability of the slow invariant manifold as well as interactions between fast and slow dynamics are considered.

Keywords: Coupled flexible-body/rigid-body dynamics, reduced slow dynamics, invariant manifolds, nonlinear normal modes, proper orthogonal modes.

1. Introduction

In many sectors of engineering, one encounters machines composed of flexible components that perform rigid-body motions. For example, cables in marine lifting machines, tethers in tethered vehicles, disks in computers, and shafts in propulsion systems are flexible structural elements that perform rotational and/or translational motions. Due to nonlinearities, large amplitude interactions between rigid-body dynamics and flexible-body structural dynamics are possible. To prevent the interaction dynamics in applications, the flexibility is designed to be sufficiently low. However, it is not always possible to achieve high rigidity in the flexible components. Moreover, in new designs the introduction of low density new engineering materials and saving of weight in systems made of conventional materials increase the flexibility of the components in structural systems. Therefore, we need to understand the mechanisms causing interaction between flexible-body dynamics and rigid-body dynamics.

In this work we analyze the dynamics of the conservative flexible spherical pendulum. It is a major structural/mechanical element in many machines. The ideal spherical pendulum has been studied extensively [1–13]. As an infinite-dimensional dynamical system, the flexible spherical pendulum is a challenging problem. Our approach is geometric and computational [22, 23]. No assumptions are made regarding the nonlinearities and the magnitude of motions.

---

[*] Research Scientist, Science Applications International Corporation, McLean, VA 22102, U.S.A.
[**] Currently with the Department of Naval Architecture and Marine Engineering, National Technical University of Athens, Zografos 157 10, Greece.

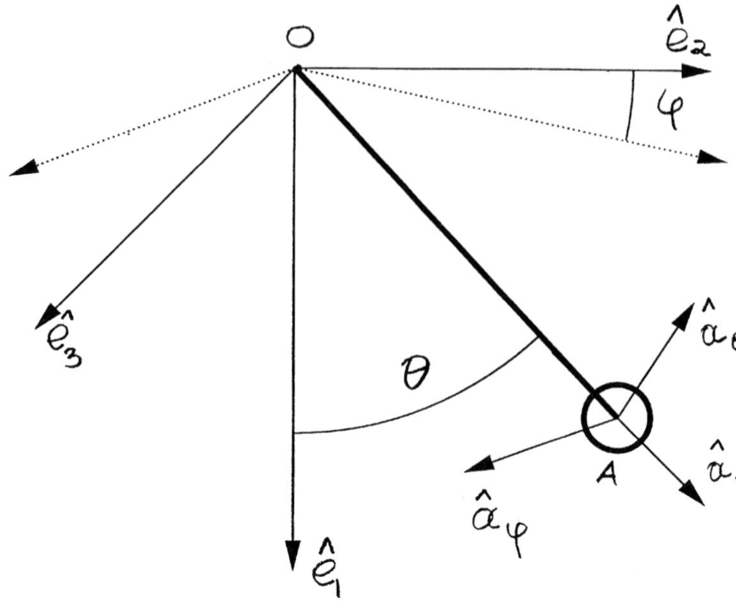

Figure 1. Schematic of the flexible spherical pendulum configuration.

We determine the reduced dynamics by computing analytically a slow invariant manifolds of motion. We apply the proper orthogonal decomposition method to compute independently the spatio-temporal characteristics, that is, shapes and amplitudes of PO modes, for the dynamics on the slow invariant manifold.

The work contributes in showing formally that the flexible spherical pendulum undergoes slow dynamics related to the dynamics of the ideal spherical pendulum. Moreover the work contributes in showing that the spatio-temporal natural characteristics of invariant manifolds of motion of infinite-dimensional conservative systems can be characterized by PO modes. Possibly this work is the first attempt to apply the POD method to identify the spatio-temporal characteristics of the reduced dynamics of infinite-dimensional Hamiltonian dynamical systems in solid continuum mechanics.

2. Equations of Motion

Figure 1 depicts a rigid-body/flexible-body configuration, referred to as the flexible spherical pendulum, in a gravitational field of constant acceleration g. The end O is inertially fixed. It consists of a rigid mass M_r attached to a cable of mass M_f. It has unstretched uniform cross-section A_0 and unstretched length L_0; it is made of linear elastic material of Young modulus E and mass density ρ. It is assumed that the cable undergoes only radial deformations.

Let the angle ϕ denote the orientation of the plane of pendulation defined by the unit vector $\hat{\mathbf{e}}_1$ and the body-fixed unit vector $\hat{\mathbf{a}}_r$. Let the angle θ denote the orientation of the cable in the pendulation plane. Let $u(x,t)\hat{\mathbf{a}}_r$ denote the radial deformation of the cable.

We introduce the normalized independent variables: $\xi = x/L_0$, $\tau = \omega_r t$, and the normalized cable radial displacement: $U = u/L$. Moreover, the following frequency and mass ratios are introduced:

$$\mu = \frac{\omega_r}{\omega_f}, \quad \beta = \frac{M_r}{M_f}, \tag{1}$$

where $\omega_r = \sqrt{g/L_0}$, $\omega_f = \pi/L_0 \times \sqrt{E/\rho}$ are the pendulation linear frequency and the fundamental frequency of the flexible cable.

The static stable and unstable equilibrium configurations are given by

$$C: \quad 8\hat{U}(\xi) = \mu^2\pi^2(2(1+\beta)\xi - \xi^2), \quad \hat{\theta} = 0, \quad \text{for all } \hat{\phi}, \tag{2a}$$

$$S_{\pm 1}: \quad 8\hat{U}(\xi) = \mu^2\pi^2(2(1+\beta)\xi - \xi^2), \quad \hat{\theta} = \pm\pi, \quad \text{for all } \hat{\phi}. \tag{2b}$$

Let $V(\xi, \tau) = U(\xi, \tau) - \hat{U}(\xi)$ denote the displacement of the cable with respect to the stable equilibrium configuration C. Moreover, let V_A, $\dot{V}_A$, $\ddot{V}_A$ denote respectively the relative displacement, velocity, and acceleration of the cable end A.

The motion of the flexible spherical pendulum is described by the following system of coupled equations:

$$\ddot{\theta} = -\frac{8(\sin(\theta) + 2\dot{\theta}\dot{V}_A)}{8 + \mu^2\pi^2(1+2\beta) + 8V_A} + \frac{P_\phi^2 \cos(\theta)}{\sin^3(\theta)}, \tag{3a}$$

$$\dot{\phi} = \frac{P_\phi}{\sin^2(\theta)}, \quad \dot{P}_\phi = -\frac{16 P_\phi \dot{V}_A}{8 + \mu^2\pi^2(1+2\beta) + 8V_A}, \tag{3b}$$

$$\begin{aligned} \mu^2\pi^2\ddot{V}(\xi, \tau) &= 4\, V_{\xi\xi}(\xi, \tau) - \mu^2\pi^2(1 - \cos(\theta)) \\ &\quad + \mu^2\pi^2(\xi + \hat{U}(\xi) + V(\xi, \tau))(\dot{\theta}^2 + P_\phi^2/\sin^3(\theta)), \end{aligned} \tag{3c}$$

$$V(\xi = 0, \tau) = 0, \quad V_\xi(\xi = 1, \tau) + \mu^2\pi^2\beta(1 - T_A) = 0, \tag{3d}$$

where the term $P_\phi \equiv \dot{\phi}\sin^2(\theta)$ denotes a 'generalized' angular momentum. Moreover, the term T_A in the boundary conditions (3d), given in Appendix A, denotes the reaction force exerted by the pendulum mass on the cable end A. Details on the derivation of the equations of motion are given in Appendix A.

The coupling between the rigid-body dynamics, Equations (3a) and (3b), and the flexible-body dynamics (3c) is parametric nonlinear. It appears in the equations of the rigid-body and the boundary conditions for the flexible-body. The equations of motion depend smoothly on the coupling parameters (μ, β).

The dynamics of flexible-body/rigid-body configurations may be governed by a few degrees-of-freedom. A degree-of-freedom is an oscillator. Mathematically it is realized as a generalized coordinate. In phase space, it is realized as a two-dimensional invariant manifold resulting from its position and momentum description. The objective of this work is to show that under certain conditions the dynamics of the flexible pendulum are dominated by low-dimensional dynamics.

3. POD Analysis

The spatio-temporal dynamic behavior of a physical system or a mathematical model can be obtained in the form of a high-resolution database, experimental for the physical system and numerical for the mathematical model. The issue is to extract from the database the essential spatio-temporal characteristics of the dynamics. The method of proper orthogonal decomposition (POD) computes optimum degrees-of-freedom [30]. We will apply the POD to characterize the spatio-temporal characteristics of reduced dynamics which we will compute analytically.

In [22], the POD method is combined with the method of geometric singular perturbations to study systematically coupled systems in continuum mechanics. Moreover, in [24], the POD method has been developed further to study the finite element dynamics of highly nonlinear PDEs in continuum mechanics.

We present the basics of the POD method. We consider the coupled position vector field,

$$\mathcal{P}(\xi, \tau) = [V(\xi, \tau), \theta(\tau), \sin(\phi(\tau))]^T, \tag{4}$$

where $\xi \in [0, 1]$ and $\tau \in [T_1, T_2]$. Its mean value, defined by

$$\langle \mathcal{P} \rangle(\xi) \equiv \frac{1}{T_2 - T_1} \int_{T_1}^{T_2} \mathcal{P}(\xi, \tau)\, \mathrm{d}\tau, \tag{5}$$

is assumed to be zero; otherwise, we consider the demeaned vector field. The temporal autocorrelation function of $\mathcal{P}(\xi, \tau)$ is defined by

$$C(\tau, s) \equiv \frac{1}{T_2 - T_1} \left\{ \int_0^1 V(\xi, \tau) V(\xi, s)\, \mathrm{d}\xi + \theta(\tau)\theta(s) + \sin(\phi(\tau)) \sin(\phi(s)) \right\}. \tag{6}$$

The autocorrelation energy of the position field is defined by

$$E \equiv \int_{T_1}^{T_2} C(\tau, \tau)\, \mathrm{d}\tau. \tag{7}$$

Since the function $C(\tau, s)$ is symmetric, one may introduce naturally the eigenproblem

$$\int_{T_1}^{T_2} C(\tau, s) \alpha_m(s) ds = \lambda_m \alpha_m(\tau). \tag{8}$$

The eigenvalues λ_m, $m = 1, 2, \ldots, M$, are fundamental quantities: It can be shown that

$$E = \sum_{m=1}^{M} \lambda_m = 1. \tag{9}$$

Thus the eigenvalue λ_m represents a fraction of the autocorrelation energy E, which is normalized at 1.

Now the projection of the vector field onto the eigenfunctions $\alpha_m(\tau)$, that is,

$$\frac{1}{T_2 - T_1} \int_{T_1}^{T_2} \alpha_m(\tau) \left[V(\xi, \tau), \theta(\tau), \sin(\phi(\tau))\right] \mathrm{d}\tau \tag{10}$$

defines the spatial functions: $\Phi_m(\xi) \equiv \sqrt{\lambda_m}[\phi_m^f(\xi), \phi_m^{1r}, \phi_m^{2r}]$ satisfying the orthonormality condition:

$$\langle \Phi_m(\xi), \Phi_n(\xi) \rangle \equiv \sqrt{\lambda_m}\sqrt{\lambda_n} \left\{ \int_0^1 \phi_m^f(\xi)\phi_n^f(\xi)\,\mathrm{d}\xi + \phi_m^{r1}\phi_n^{r1} + \phi_m^{r2}\phi_n^{r2} \right\} = \delta_{mn}. \tag{11}$$

They are the shapes of Proper Orthogonal (PO) modes.

The sequence $\{\lambda_m, \alpha_m(\tau), \Phi_m(\xi)\}_{m=1}^{m=M}$ is a fundamental quantity of the vector field since it expands it as follows:

$$[V(\xi,\tau), \theta(\tau), \sin(\phi(\tau))] = (T_2 - T_1) \sum_{m=1}^{M} \alpha_m(\tau)\sqrt{\lambda_m}\Phi_m(\xi), \tag{12}$$

and the expansion is optimum. The sequence $\mathcal{S} = \{\lambda_m\}_{m=1}^{m=M}$ is called the POD spectrum.

Assuming the existence of a low-dimensional invariant manifold governing the reduced dynamics, we will apply the POD method to identify the spatio-temporal characteristics of motions residing on the invariant manifold of motion. The fundamental manifold which governs the reduced dynamics as well as the interaction dynamics in coupled structures is the slow invariant manifold [15, 17, 22]. The objective of the next sections is to introduce and compute the slow invariant manifold.

3.1. Singular Perturbation Analysis

Our system consists of a PDE coupled to two ODEs. Following [22], we decompose the PDE (3c) along with its boundary conditions (3d) into a set of oscillators using Galerkin projection onto the linear modes:

$$\phi_m(\xi) = \sin(\frac{\pi\xi}{2\mu_m}), \qquad \mu_m \equiv \frac{\hat{\omega}_1}{\hat{\omega}_m} = \frac{1}{2m-1}, \qquad m = 1, 2, \ldots, \infty. \tag{13}$$

where $\hat{\omega}_m$ is the m-th natural frequency of the uncoupled cable.

The modal decomposition transforms the equations of motion into a set of coupled oscillators. To introduce invariant manifolds of motion, we need to consider the problem in a special phase space setting. In particular, the change of variables:

$$\Psi_1 = \theta, \quad \Psi_2 = \dot{\theta}, \quad \Psi_3 = \phi, \quad \Psi_4 = P_\phi, \quad \Psi_5 = \dot{\phi}, \tag{14}$$

and the hierarchical stretching of the modal amplitudes and velocities:

$$\eta_m = \mu^2\mu_m^2 Z_{2m-1}, \qquad \dot{\eta}_m = \mu\mu_m^2 Z_{2m}, \tag{15}$$

introduced originally in [22], casts the equations of coupled oscillators into the first set of ordinary differential equations:

$$\begin{aligned}
\dot{\Psi}_1 &= \Psi_2, \\
\dot{\Psi}_2 &= -\frac{8(\sin(\Psi_1) + 2\mu\Psi_2 Q_N(Z))}{8 + \mu^2\pi^2(1+2\beta) + 8\mu^2 P_N(Z)} + \frac{\Psi_4^2\cos(\Psi_1)}{\sin^3(\Psi_1)}, && \text{(16a)} \\
\dot{\Psi}_3 &= \frac{\Psi_4}{\sin^2(\Psi_1)}, \\
\dot{\Psi}_4 &= -\frac{16\mu Q_N(Z)\Psi_4}{8 + \mu^2\pi^2(1+2\beta) + 8\mu^2 P_N(Z)}, && \text{(16b)}
\end{aligned}$$

$$
\begin{aligned}
&\mu \dot{Z}_{2m-1} = Z_{2m}, \\
&(1+2N\beta)\,\mu\mu_m^2\,\dot{Z}_{2m} \\
&\quad = -\left[1+2\beta(N-1)-\mu^2\mu_m^2(1+2\beta(N-2))(\Psi_2^2+\Psi_4^2/\sin^3(\Psi_1))\right] Z_{2m-1} \\
&\qquad -2\beta \sum_{j=1,\, j\neq m}^{N} (-1)^{m+j+2}\left[1-2\mu^2\mu_j^2(\Psi_2^2+\Psi_4^2/\sin^3(\Psi_1))\right] Z_{2j-1} \\
&\qquad +(C_{1m}+\mu^2 C_{2m})\,(\Psi_2^2+\Psi_4^2/\sin^3(\Psi_1)) + D_m(1-\cos(\Psi_1)), \qquad (16c) \\
&\qquad m = 1, 2, \ldots, N \to \infty,
\end{aligned}
$$

where

$$
P_N(Z) \equiv \sum_{m=1}^{N} (-1)^{m+1}\mu_m^2 Z_{2m-1}, \qquad Q_N(Z) \equiv \sum_{m=1}^{N} (-1)^{m+1}\mu_m^2 Z_{2m}. \qquad (17)
$$

The various coefficients in Equation (16c) are given by

$$
\begin{aligned}
\pi^2 C_{1m} &\equiv (-1)^{m+1} 2\beta(\pi^2 - 8S_2(N)) + (-1)^{m+1} 8(1+2\beta N)\mu_m^2, \\
8\pi C_{2m} &\equiv (-1)^{m+1} 2\beta(\pi^3(1+2\beta) - 32\beta\pi S_2(N) - 32 S_3(N)) \\
&\quad + (-1)^{m+1} 32\beta\pi(1+2\beta N)\mu_m^2 + 32(1+2\beta N)\mu_m^3, \\
\pi D_m &\equiv (-1)^{m+1} 2\beta(-\pi + 4S_1(N)\mu_m) - 4(1+2\beta N)\mu_m. \qquad (18)
\end{aligned}
$$

Equations (16a) and (16b) describe the rigid-body dynamics, whereas Equation (16c), the oscillator representation of the original PDE, describes the flexible-body dynamics. Equation (16a) is the equation of motion in singular perturbation formulation because the frequency ratio μ appears on the left-hand side of Equation (16c), and can assume arbitrarily small values, including zero.

The formulation of the equations of motion of coupled mechanical systems as singular perturbation problems seems to be a basic step towards a global geometric analysis of the dynamics [15]. Recent developments and applications of geometric singular perturbations are given in [26].

To effectively apply symbolic computations to compute the slow dynamics, we rewrite the equations of motion in the compact form

$$
\begin{aligned}
\dot{\Psi} &= \mathbf{F}^0(\Psi) + \mathbf{F}^\mu(\Psi, Z; N), \qquad (19a) \\
\mu \dot{Z} &= \mathbf{B}^0(N)Z + \mathbf{G}^0(\Psi; N) + \mu^2\left[\mathbf{B}(\Psi; N)Z + \mathbf{G}(\Psi; N)\right], \quad N \to \infty, \qquad (19b)
\end{aligned}
$$

where $\Psi \equiv (\{\Psi_1, \Psi_2\}, \{\Psi_3, \Psi_4\})$, and $Z \equiv (\{Z_1, Z_2\}, \{Z_3, Z_4\}, \ldots \{Z_{2N-1}, Z_{2N}\})$. The vector-valued functions $\mathbf{F}^0$, $\mathbf{F}^\mu$, $\mathbf{B}^0$, $\mathbf{G}^0$, $\mathbf{B}$, $\mathbf{G}$ are determined from the right-hand side of Equation (16a). Moreover, the inverse of the matrix $\mathbf{B}_0$ exists for all orders of truncation N. The dimension of the phase space is $4 + 2N$.

3.2. The Ideal Spherical Pendulum

The singular perturbation formulation addresses the issues of dimension increase and dimension decrease. This can be seen clearly by considering the limiting behavior of Equations (19a). In particular, for fixed β, the limit of the rigid-body system (19a) as $\mu \to 0$, yields

$$\dot{\Psi} = \mathbf{F}^0(\Psi), \tag{20}$$

with detailed description

$$\begin{aligned}
\dot{\Psi}_1 &= \Psi_2, \\
\dot{\Psi}_2 &= -\sin(\Psi_1) + \frac{\Psi_4^2 \cos(\Psi_1)}{\sin^3(\Psi_1)}, \qquad (21a)\\
\dot{\Psi}_3 &= \frac{\Psi_4}{\sin^2(\Psi_1)}, \\
\dot{\Psi}_4 &= 0. \qquad (21b)
\end{aligned}$$

This is the equation of motion of the ideal spherical pendulum, that is, the spherical pendulum with perfectly rigid arm. The ideal spherical pendulum is completely integrable since we have the well known fact that its total energy and angular momentum given respectively by

$$2E = \Psi_2^2 + \Psi_4^2/\sin^2(\Psi_1) - \cos(\Psi_1), \quad \Psi_4 = \Psi_5 \sin^2(\Psi_1) \tag{22}$$

are constants of motion.

The interesting feature of the limiting process resides in the fact that Equation (19b) for the flexible-body dynamics is reduced into an algebraic equation, whose solution is

$$Z = \mathbf{H}^0(\Psi), \tag{23}$$

where the m-th pair-element $\{H^0_{2m-1}, H^0_{2m}\}$ of the $2N$-dimensional vector-valued function $\mathbf{H}^0$ is given by

$$\begin{aligned}
H^0_{2m-1}(\Psi;\mu_m) &= (-1)^{m+1} 2\beta \left(-1 + \cos(\Psi_1) + \Psi_2^2 + \frac{\Psi_4^2}{\sin^3(\Psi_1)}\right) \\
&\quad + (-1)^{m+1} \frac{8\mu_m^2}{\pi^2}\left(\Psi_2^2 + \frac{\Psi_4^2}{\sin^3(\Psi_1)}\right) - \frac{4\mu_m}{\pi}(1 - \cos(\Psi_1)), \qquad (24a)\\
H^0_{2m}(\Psi) &= 0. \qquad (24b)
\end{aligned}$$

This solution is defined for all $m \geq 1$. It does not depend on the order of truncation N. Note that $\mathbf{H}^0$ becomes singular at $\Psi_1 = 0$. However, this singularity is not physical but mathematical.

The graph of the solution $\mathbf{H}^0$, that is,

$$\mathcal{W}_0 = \left\{(Z, \Psi) : \quad Z = \mathbf{H}^0(\Psi)\right\} \tag{25}$$

defines a three-dimensional invariant manifold in the infinite-dimensional phase space. This limiting manifold is global and nonlinear.

The limiting dynamics evolve on the slow time scale and reside on the three-dimensional manifold $\mathcal{W}^0$. They are decomposed into the *master* slow dynamics (21a) and the *slaved* slow dynamics (23).

The issue here is *whether there exists an invariant manifold for sufficiently small* $\mu \neq 0$ *and* $N = \infty$. Physically, this would mean that the coupled system can experience dynamics that evolve on a slow time scale.

4. The Slaved Slow Dynamics

For conservative systems there are no theorems to deduce the existence of an invariant manifold in the neighborhood of $\mathcal{W}_0$ for $\mu \neq 0$ due to the lack of hyperbolicity transversely to this limiting slow invariant manifold. We will formally construct slow manifolds and conjecture the existence of an invariant slow manifold via numerical experiments and the method of proper orthogonal decompositions. The computations will be based on finite N. It will be shown how the results can be extended to $N = \infty$.

Physically, to say that the system possesses a slow invariant manifold means that the flexible-body dynamics are driven by low-dimensional dynamics parametrized by the rigid-body variable Ψ. Motivated by the limiting slaved slow dynamics, we assume that the slaved slow dynamics can be described by

$$Z = \mathbf{H}^{\mu}(\Psi) = \mathbf{H}^0(\Psi) + \mathbf{H}_{\mu}(\Psi). \tag{26}$$

The graph of the function $\mathbf{H}^{\mu}$ defines the three-dimensional manifold:

$$\mathcal{W}_{\mu} = \{(Z, \Psi): \quad Z = \mathbf{H}^{\mu}(\Psi)\}. \tag{27}$$

This manifold is called slow. To say that it is invariant means that the vector field in phase space defined by the equations of motion belongs to the tangent vector bundle [19] of the graph of the function. It is easy to show that the property of invariance gives rise to the slow manifold condition:

$$\mathbf{B}^0(\Psi)\mathbf{H}^0 + \mathbf{G}^0(\Psi) = \mathbf{L}_1^{\mu}(\Psi, \mathbf{H}^0, \mathbf{H}^{\mu}) + \mathbf{L}_2^{\mu}(\Psi, \mathbf{H}^0, \mathbf{H}^{\mu}) + \mathbf{L}_3^{\mu}(\Psi, \mathbf{H}^0, \mathbf{H}^{\mu}). \tag{28}$$

The symbol D_{Ψ} denotes the derivative with respect to Ψ. The terms on the left-hand side are respectively $O(\mu)$, $O(\mu^2)$, $O(\mu^3)$; they are defined by

$$\begin{aligned}
\mathbf{L}_1^{\mu}(\Psi, \mathbf{H}^0, \mathbf{H}^{\mu}) &\equiv \mu D_{\Psi}\mathbf{H}^0\mathbf{F}^0(\Psi) - \mathbf{B}^0(\Psi)\mathbf{H}^{\mu}, \\
\mathbf{L}_2^{\mu}(\Psi, \mathbf{H}^0, \mathbf{H}^{\mu}) &\equiv \mu\left[D_{\Psi}\mathbf{H}^{\mu}\mathbf{F}^0(\Psi) + D_{\Psi}\mathbf{H}^0\mathbf{F}^{\mu}(\Psi)\right] - \mu^2\left[\mathbf{B}(\Psi)\mathbf{H}^0 + \mathbf{G}(\Psi)\right], \\
\mathbf{L}_3^{\mu}(\Psi, \mathbf{H}^0, \mathbf{H}^{\mu}) &\equiv \mu D_{\Psi}\mathbf{H}^{\mu}\mathbf{F}^{\mu}(\Psi) - \mu^2\mathbf{B}(\Psi)\mathbf{H}^{\mu}.
\end{aligned}$$

The existence of $\mathbf{H}^{\mu}$ implies the existence of a slow invariant manifold.

The slow manifold condition is a highly nonlinear partial differential equation; it is solved approximately by assuming

$$\mathbf{H}^{\mu}(\Psi) = \mathbf{H}^0(\Psi) + \sum_{k=1}^{\infty} \mu^k \mathbf{H}^k(\Psi). \tag{29}$$

Approximations to the solution up to $O(\mu^3)$ are computed in Appendix B. The computations have been carried out by using the computer algebra code Maple. The symbolic computations reveal that the approximate slow manifold depends on the order of truncation N through convergent sequencies of the frequency ratios μ_m. Clearly, this allows the extension of the results to the infinite system, that is, $N = \infty$. *We have constructed through symbolic computations an approximate slow manifold for the infinite system.* We will verify that indeed this manifold is in the neighborhood of a slow invariant manifold.

The slow manifold $\mathcal{W}_\mu$ contains the submanifolds of physical significance

$$\mathcal{N}_\mu = \left\{(Z, \Psi): \quad Z = \mathbf{H}_S^\mu(\Psi_1)\right\}, \tag{30a}$$

$$\mathcal{V}_\mu = \left\{(Z, \Psi): \quad Z = \mathbf{H}_P^\mu(\Psi_1, \Psi_2)\right\}, \tag{30b}$$

$$\mathcal{M}_\mu = \left\{(Z, \Psi): \quad Z = \mathbf{H}_M^\mu(\Psi_1, \Psi_4)\right\}, \tag{30c}$$

where the functions $\mathbf{H}_S^\mu$, $\mathbf{H}_P^\mu$, and $\mathbf{H}_M^\mu$ are the restrictions of the solution $\mathbf{H}^\mu$ to the planes Ψ_1–Ψ_3, Ψ_1–Ψ_2, and Ψ_1–Ψ_4.

The submanifolds $\mathcal{N}_\mu$ and $\mathcal{M}_\mu$ are called the stagnation and angular momentum manifolds; they are not invariant. The submanifold $\mathcal{V}_\mu$ is invariant and is called the pendulation manifold since it supports motions with zero angular momentum. The corresponding submanifolds of the computed slow manifold will be used to select initial conditions in numerical experiments.

The stagnation submanifold is parametrized by the angle Ψ_1 or equivalently by the energy level. This submanifold provides initial conditions for slow motions with zero angular momentum.

5. The Slow Dynamics

In this section, approximations to the slow dynamics of the flexible spherical pendulum are computed. The restriction of the dynamics to the slow invariant manifold $\mathcal{W}_\mu$ is described by

$$\dot{\Psi} = \mathbf{F}^0(\Psi) + \mathbf{N}^\mu(\Psi), \tag{31a}$$

$$Z = \mathbf{H}^\mu(\Psi), \tag{31b}$$

where $\mathbf{N}^\mu(\Psi) \equiv \lim_{N\to\infty} \mathbf{F}^\mu(\Psi, \mathbf{H}^\mu(\Psi; N); N)$, and the term $\mathbf{H}^\mu$ is an infinite-dimensional vector-valued function whose mth component is denoted by $\{\mathcal{H}_{2m-1}^\mu, \mathcal{H}_{2m}^\mu\}$ and whose graph represents the m-th pair-element $\{\mathcal{W}_{2m-1}^\mu, \mathcal{W}_{2m}^\mu\}$ of the slow invariant manifold. The position and velocity of the cable during slow motions are given by

$$V(\xi, \tau) = \mathcal{H}_1^\mu(\xi, \Psi(\tau)) \equiv \mu^2 \sum_{m=1}^{\infty} \mu_m^2 \mathcal{H}_{2m-1}^\mu(\Psi(\tau))\phi_m(\xi),$$

$$\dot{V}(\xi, \tau) = \mathcal{H}_2^\mu(\xi, \Psi(\tau)) \equiv \mu \sum_{m=1}^{\infty} \mu_m^2 H_{2m}^\mu(\Psi(\tau))\phi_m(\xi). \tag{32}$$

The dynamics on the slow invariant manifold are decomposed into the *master* and *slaved* dynamics.

To confirm the fact that the ideal spherical pendulum predicts the slow dynamics and the computational procedure generates valid approximations to an invariant slow manifold, we need to compare the ideal pendulum to approximations to the master slow system. The $O(\mu^2)$ and $O(\mu^4)$ approximations to the master slow system (31a) have been computed in Appendix C.

The $O(\mu^2)$ approximation to the master slow system is found to be

$$\dot{\Psi} = \mathbf{F}^0(\Psi) + \mu^2\mathbf{N}_2(\Psi). \tag{33}$$

To compare it to the ideal spherical pendulum, we present its detailed description:

$$\begin{aligned}
\dot{\Psi}_1 &= \Psi_2, \\
\dot{\Psi}_2 &= -\left[1 + 2\mu^2 Q_1(\Psi_1, \Psi_2)\right]\sin(\Psi_1) \\
&\quad + \left[1 + 2\mu^2 Q_2(\Psi_1)\right]\frac{\Psi_4^2\cos(\Psi_1)}{\sin^3(\Psi_1)} + 2\mu^2 Q_3(\Psi_1)\Psi_2^2\Psi_4^2,
\end{aligned} \tag{34a}$$

$$\begin{aligned}
\dot{\Psi}_3 &= \frac{\Psi_4}{\sin^2(\Psi_1)}, \\
\dot{\Psi}_4 &= 2\mu^2\left[P_1(\Psi_1) + P_2(\Psi_1, \Psi_4)\right]\Psi_2\Psi_4,
\end{aligned} \tag{34b}$$

where the various functions of Ψ in the above equation have expressions

$$\begin{aligned}
16\pi^2 Q_1(\Psi_1, \Psi_2) &= 16\pi(2S_3 + \beta\pi S_2)(1 - \cos(\Psi_1)) \\
&\quad - 16(4\pi S_3 + 7\beta\pi^2 S_2 + 20S_4)\Psi_2^2 - \pi^4(1 + 2\beta), \\
\pi^2 Q_2(\Psi_1) &= (\beta\pi^2 S_2 + 4S_4)\tan(\Psi_1), \\
\pi^2\sin^4(\Psi_1)Q_3(\Psi_1) &= 2(\beta\pi^2 S_2 + 4S_4)\cos(\Psi_1)(3 - 2\sin(\Psi_1)), \\
\pi^2 P_1(\Psi_1) &= (6\beta\pi^2 S_2 + 4\pi S_3 + 16S_4)\sin(\Psi_1), \\
\pi^2\sin^4(\Psi_1)P_2(\Psi_1) &= 2(\beta\pi^2 S_2 + 4S_4)\cos(\Psi_1)(3 - 2\sin(\Psi_1))\Psi_4^2.
\end{aligned} \tag{35}$$

Clearly, the second-order master slow system (34a) is a regular perturbation in μ^2 of the ideal spherical pendulum, the zeroth-order master slow system. *Is the second-order system completely integrable?*

Since the issue of integrability is a delicate issue, we have computed the $O(\mu^4)$ slow master system:

$$\dot{\Psi} = \mathbf{F}^0(\Psi) + \mu^2\mathbf{N}_2(\Psi) + \mu^4\mathbf{N}_4(\Psi). \tag{36}$$

The functions $\mathbf{N}_2$ and $\mathbf{N}_4$ are given in Appendix C.

We resort to direct numerical experiments to study the master slow systems. Figure 2a shows that the master slow systems behave almost identically for motions with zero angular momentum. Figure 2b shows that the $O(\mu^4)$ master slow system (36) behaves qualitatively the same as the lower-order slow master systems. The quantitative difference vanishes if the coupling μ is reduced, see Figure 2c. Figure 2d reveals that the approximate slow master systems undergo unidirectional precession, just as the ideal pendulum, during motions with nonzero angular momentum.

The numerical experiment suggests that the approximate master slow systems are completely integrable. For $\Psi_4 = 0$, the master slow systems become perturbations in μ^2 of the

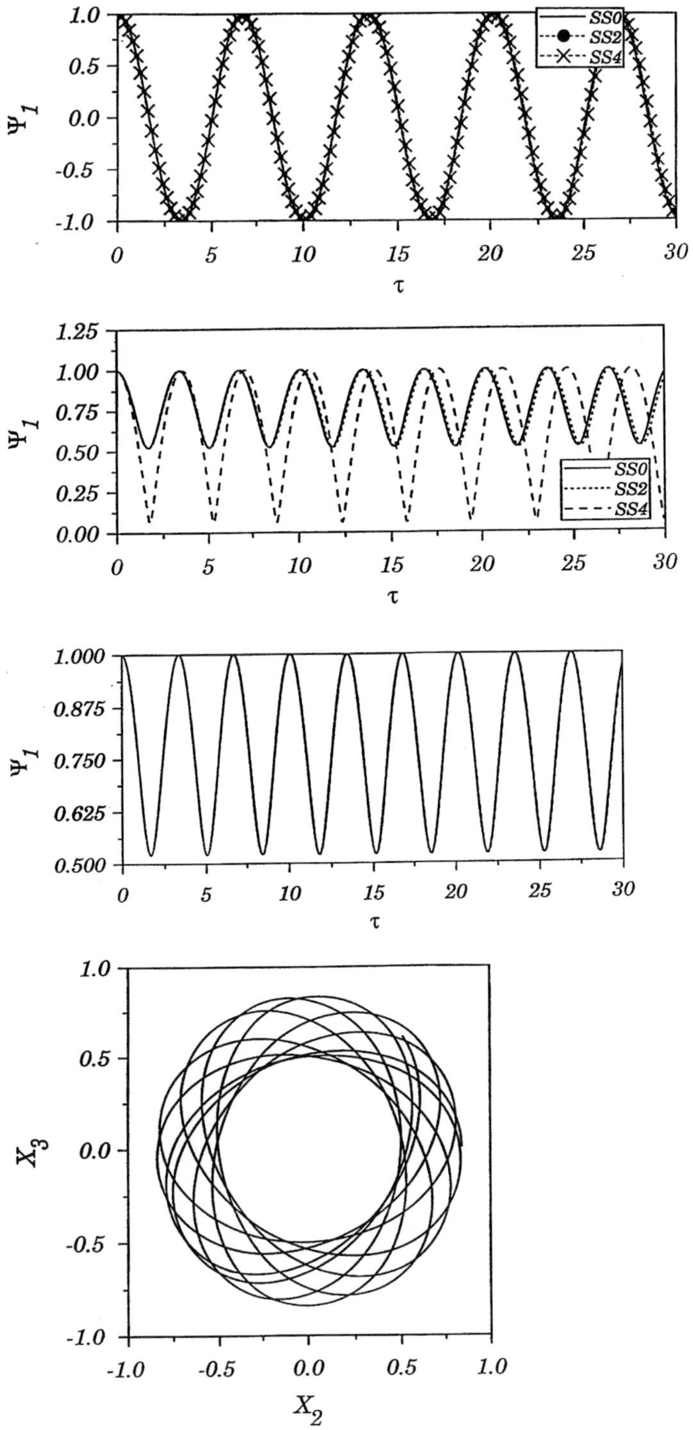

(a)

(b)

(c)

(d)

Figure 2. Time history of the pendulation angle Ψ_1 for $\beta = 10$: (a) pure pendulation motion, $\mu = 0.01$, (b) motion with angular momentum, $\mu = 0.01$, (c) motion with angular momentum, $\mu = 0.001$, (d) projection of the motion onto the X_2–X_3 plane, $\mu = 0.001$. $SS0$: ideal spherical pendulum, $SS2$: $O(\mu^2)$ master slow system, $SS4$: $O(\mu^4)$ master slow system.

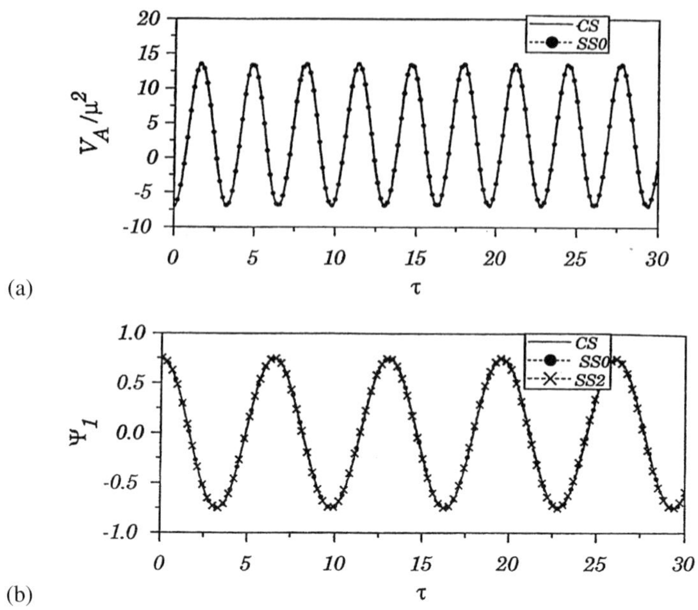

Figure 3. Time history of the displacement of the cable end A during a motion initiated on the stagnation submanifold: (a) slaved slow dynamics, (b) master slow dynamics. CS: complete system, $SS0$: zeroth order slow system, $SS2$: second-order slow system.

ideal planar pendulum. It has been shown analytically that these systems have a constant of motion that is a perturbation of the energy of the ideal planar pendulum [17, 23]. Thus, the new constants of motion of the master slow system should be $E_\mu = E + O(\mu^2)$ and $P_\phi^\mu = P_\phi + O(\mu^2)$, where E and P_ϕ are the constants of motion of the ideal spherical pendulum.

6. Exploring the Nature of the Slow Dynamics

In this section, we use the formally constructed slow manifold to perform numerical experiments in order to conjecture the existence of an invariant slow manifold, and explore the stability of the reduced slow dynamics. The numerical simulations, using the Gear method, are carried out for the system (16a) truncated at $N = 33$ with mass ratio fixed at $\beta = 10.0$. This is the case of a heavy payload. The frequency ratio μ is fixed at $\mu = 0.01$.

First we study pure pendulation motions initiated on the stagnation submanifold. Figure 3a shows that the time history of the displacement of the cable end A during a motion initiated on the stagnation submanifold $\mathcal{N}_\mu$, Equation (30a), with $\Psi_1(0) = 0.75$, is predicted exceptionally well by the zeroth-order slaved slow system (23). Figure 3b reveals that the zeroth-order master slow system, the ideal pendulum (21a), and the second-order master slow system (34a) predict exceptionally well the dynamics of the rigid-body variable Ψ.

The numerical experiment reveals that all motions initiated on the stagnation submanifold evolve on a purely slow time scale. This family of slow periodic motions forms a two-dimensional submanifold of the three-dimensional slow manifold $\mathcal{W}_\mu$. This submanifold is invariant and is approximated exceptionally well by the limiting pendulation manifold $\mathcal{V}_0$ and

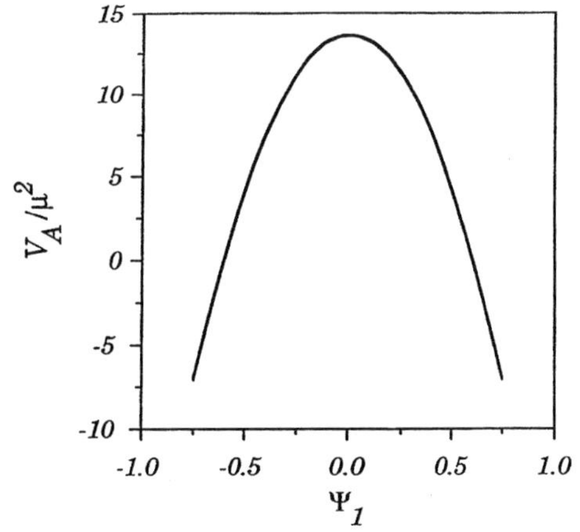

Figure 4. Projection of a motion on the two-dimensional slow invariant manifold $\mathcal{V}_\mu$.

the second-order computed pendulation manifold. This nonlinear manifold can be identified with a nonlinear normal mode since all motions residing on it are characterized by a number of common properties. Figure 4 shows the projection of a typical motion on the pendulation manifold $\mathcal{V}_\mu$ onto the configuration plane Ψ_1–V_A/μ^2. This motion does not satisfy the classical property of vibrations-in-unison which states that *the pendulum mass and a point of the cable reach their minimum and maximum values at the same time instance and that they cross zero simultaneously* [31]. During any motion on the pendulation slow invariant manifold, the cable displacement and angular displacement do not cross zero simultaneously. In fact, at zero pendulation $\Psi_1 = 0$, the cable attains its maximum displacement. Here we have the *case of a nonclassical nonlinear normal mode* [23].

Next we examine motions of the complete system initiated on the two-dimensional angular momentum submanifold $\mathcal{M}_\mu$. Figure 5a reveals that the motion of the cable during a typical motion initiated on the zeroth-order computed $\mathcal{M}_\mu$ is predicted exceptionally well by the computed slaved slow system. Figure 5b reveals that the generalized angular momentum Ψ_4 is a small perturbation of the conserved angular momentum of the ideal spherical pendulum. This quantity is predicted exceptionally well by the second-order master slow system. Clearly, this verifies the predictability of the formally constructed slow system. Thus, motions initiated on the angular momentum manifold evolve on the slow time scale of the ideal spherical pendulum. Figure 5c shows the projection of a slow motion onto the X_2–X_3 plane. We see that the dynamics on the slow invariant manifold experience the phenomenon of uniform unidirectional precession, predicted by the low-order master slow systems, see Figure 2d.

Now we examine the properties of motions with angular momentum on the slow invariant manifold $\mathcal{W}_\mu$. Figure 6a shows the projection of a slow motion of the complete system onto the configuration plane Ψ_1–V_A/μ^2. Figure 6b shows the projection of a set of slow motions with constant angular momentum and varying energy level. The zeroth-order slow system was used to generate the motions shown in Figure 6b. From these projections, we see that the motions on the three-dimensional slow invariant manifold share common properties. These properties can be used to define motions in-unison. Therefore the three-dimensional slow invariant manifold can be identified with a nonlinear normal hypermode since the dynamics are described by a two-degrees-of-freedom system. Recall that the normal mode of slow pendulations motion is two-dimensional.

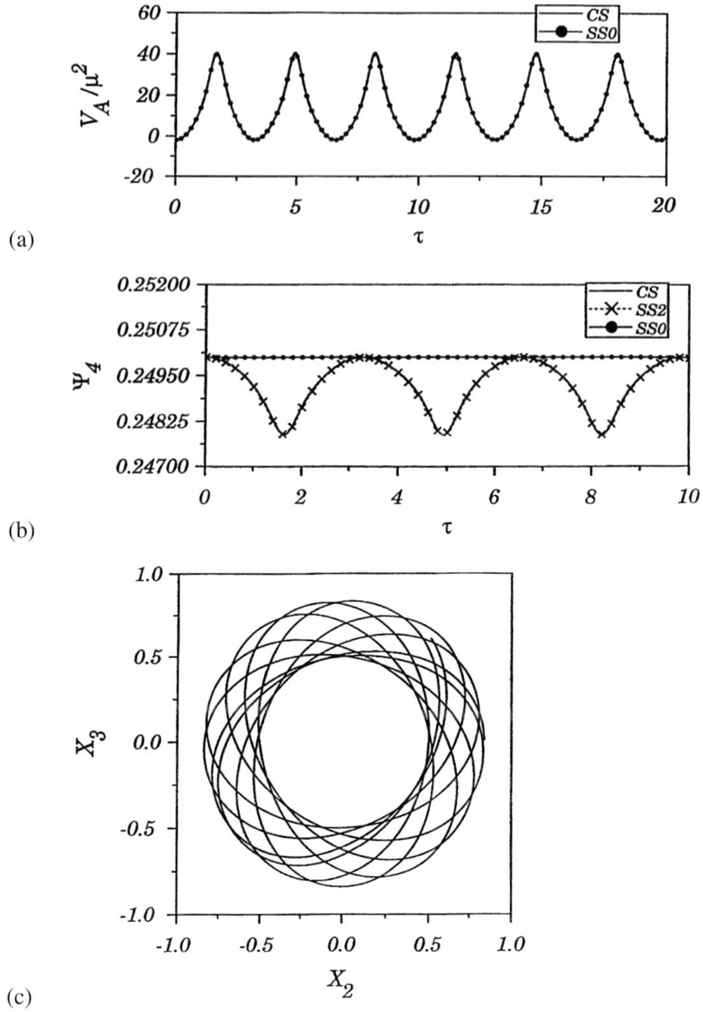

Figure 5. Time history of a motion initiated on the angular momentum slow manifold $\mathcal{M}_\mu$: (a) displacement of the cable end A, (b) generalized angular momentum, (c) projection onto the X_2–X_3 plane. CS: complete system, $SS0$: zeroth-order slow system, $SS2$: second-order slow system.

The numerical experiment reveals that the invariant manifold $\mathcal{W}_\mu$ has unstable regions. Figure 7a shows the behavior of the displacement of the cable end A during a typical motion initiated on the zeroth-order manifold $\mathcal{M}_0$. During a short initial time interval the motion is purely slow and is predicted by the zeroth-order slow system. After $\Psi_1 < \Psi_1^{\text{cr}}$, a fast oscillation is activated and is mixed with the slow oscillation. Note that the motion is underlined by a slow oscillation predicted by the slow system. The triggering and subsequent amplification of the fast oscillation, its frequency being close to the coupled natural frequency of the cable, indicates that the slow invariant manifold has developed transverse instabilities. This could be due to internal resonances between the cable slaved slow dynamics and the fast cable dynamics. Figure 7b shows the trajectory of the pendulum mass in the X_2–X_3 plane. The fast oscillation has not affected the uniformity of the unidirectional precession. This is only true for sufficiently small coupling μ, for fixed mass ratio β.

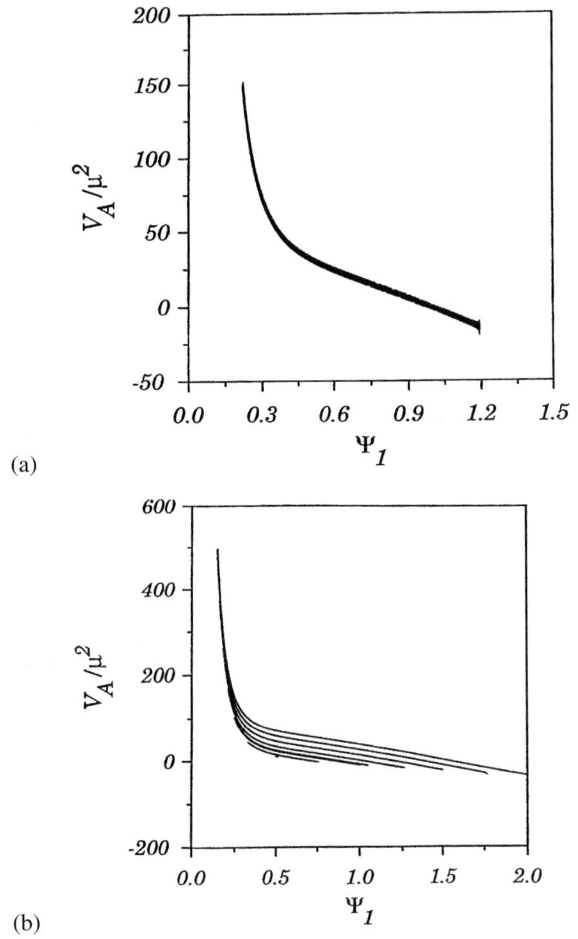

Figure 6. Projection of a motion on the three-dimensional slow invariant manifold $\mathcal{W}_\mu$: (a) a single motion of the complete system, (b) set of motions predicted by the zeroth-order slow system.

This numerical experiment reveals that a three-dimensional slow invariant manifold exists in the neighborhood of the limiting slow invariant manifold $\mathcal{W}^0$. For weak coupling μ, the slow invariant manifold is global since it approaches $\mathcal{W}_0$ as $\mu \to 0$.

7. Spatio-Temporal Analysis of the Slow Dynamics

In this section we apply the proper orthogonal decomposition method to compute the spatial and temporal characteristics of the slow reduced dynamics. The POD analysis will verify the existence of the slow invariant manifold.

As discussed in Section 3, we compute the following expansion:

$$\begin{aligned} [V(\xi,\tau), \theta(\tau), \sin(\phi(\tau))]^T &= (T_2 - T_1)\sum_{m=1}^{M} \alpha_m(\tau)\sqrt{\lambda_m}\Phi_m, \\ \Phi_m &\equiv \left[\phi_m^f(\xi), \phi_m^{1r}, \phi_m^{2r}\right]^T. \end{aligned} \tag{37}$$

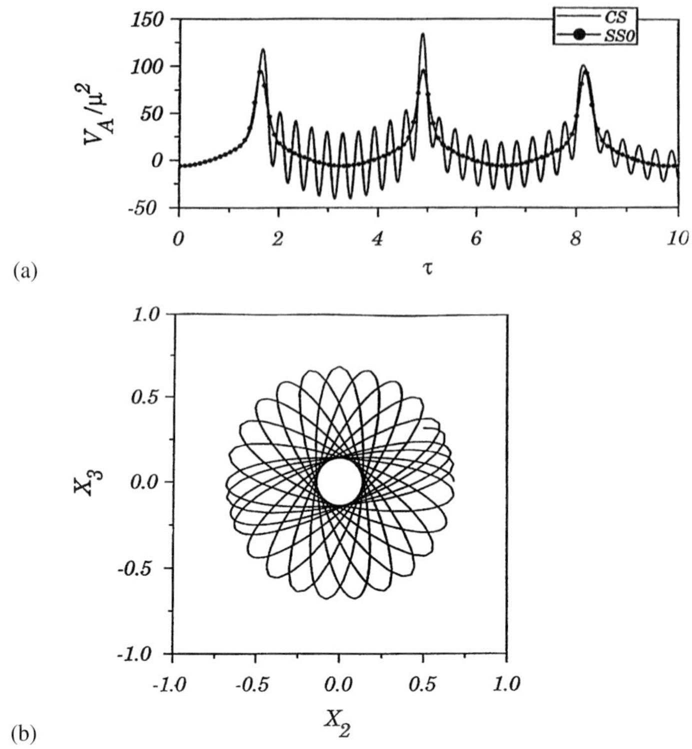

Figure 7. Time history of a motion initiated on the angular momentum slow manifold $\mathcal{M}_\mu$: (a) displacement of the cable end A, (b) projection of the motion onto the X_2–X_3 plane. CS: complete system, $SS0$: zeroth-order slow system.

The components $\phi_m^f(\xi)$, and ϕ_m^{1r}, ϕ_m^{2r} of the mode Φ_m denote flexible-body and rigid-body components, respectively.

The normalized flexible-body component of the spatial mode Φ_m is defined as follows:

$$\Phi_m^f(\xi) \equiv \frac{\phi_m^f(\xi)}{||\phi_m^f(\xi)||}, \qquad ||\phi_m^f(\xi)|| = \sqrt{\int_0^1 \phi_m^f(\xi)\phi_m^f(\xi)\,\mathrm{d}\xi}. \tag{38}$$

For any motion residing on the two-dimensional slow invariant submanifold $\mathcal{V}_\mu$, we find that $\lambda_1 + \lambda_2 = 1.00000$, where $\lambda_1 \gg \lambda_2$. Thus, the motion on the two-dimensional invariant manifold is characterized by two PO modes. In particular, the energy of the most energetic mode is contained in the rigid-body component ϕ_1^{1r}, whereas most of the energy of the least energetic mode is contained in the flexible-body component $\phi_2^f(\xi)$. It turns out that the two modes have identical flexible-body components, that is, $\Phi_1(\xi) \equiv \Phi_1^f(\xi) = \Phi_2^f(\xi)$. Now if we omit the dynamics of the rigid-body dynamics, the POD analysis reveals that all energy of the motion is contained in just one mode. The shape of the mode is identical to that of $\Phi_1(\xi)$.

Now for any motions on the stable region of the three-dimensional slow invariant manifold $\mathcal{W}_\mu$, the POD analysis gives exactly three PO modes: $\lambda_1+\lambda_2+\lambda_3 = 1.00000$. In particular, the first and second PO modes are dominated by the rigid-body components ϕ_2^{2r} and ϕ_1^{2r}, whereas the third PO mode, the least energetic, is dominated by the flexible-body component $\phi_3^f(\xi)$. Figure 8 shows the amplitudes of the three PO modes.

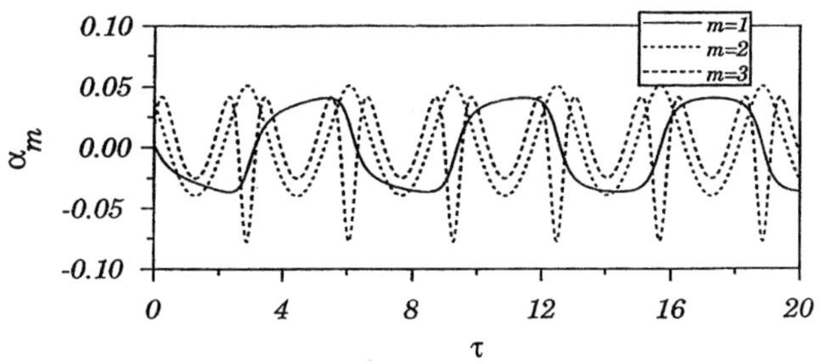

Figure 8. Time histories of the amplitudes of the three PO modes characterizing a motion on the three-dimensional slow invariant manifold $\mathcal{W}_\mu$.

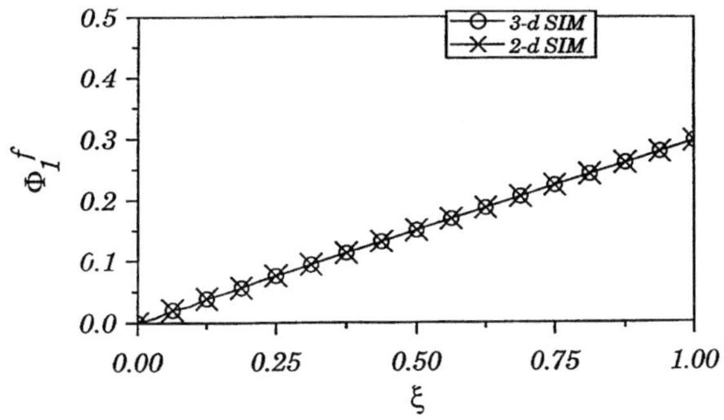

Figure 9. The shape of the flexible-body components of the PO modes characterizing the dynamics of a motion on the slow invariant manifold (SIM).

Regarding the spatial shapes of the PO modes, the computation reveals that the flexible-body components of the three PO modes are identical, that is, $\Phi_2(\xi) \equiv \Phi_1^f(\xi) = \Phi_2^f(\xi) = \Phi_3^f(\xi)$. If we omit the rigid-body dynamics, we obtain just one PO mode. Moreover, we find that $\Phi_1(\xi) = \Phi_2(\xi)$. Recall that $\Phi_1(\xi)$ denotes the PO mode of the slaved slow pendulation motions. *We conclude that during any slow motion the cable responds as one-degree-of-freedom*, that is, the position field of the cable admits the time-space factorization,

$$V(\xi, \tau) = \mathcal{H}_\mu^1(\xi, \Psi(\tau)) = \alpha(\Psi(\tau))\Phi(\xi). \tag{39}$$

Figure 9 presents the computed shape of the cable during slow motions.

The conclusion is that the POD analysis gives spatial and temporal characteristics that reflect the properties of the dynamics on the slow manifold, that is, the dynamics are governed by one-degree-of-freedom, for motions with zero angular momentum, or two-degrees-of-freedom system. For a motion on the pendulation slow invariant manifold, Figure 10a reveals a simple functional relationship between the amplitude of the mode dominated by the flexible-body dynamics and the amplitude of the mode dominated by the rigid-body dynamics. For a motion with angular momentum on the slow manifold, Figure 10b reveals a similar functional dependence of the PO mode dominated by the flexible-body dynamics on the PO mode dominated by the rigid-body dynamics. Figures 10a and 10b should be compared to Figures 4 and 6a showing a similar functional relationship between the flexible-body dynamics and the rigid-body dynamics for a motion on the slow invariant manifold. It is remarkable that the POD analysis reveals the fact that the slow motions consist of master and slaved components.

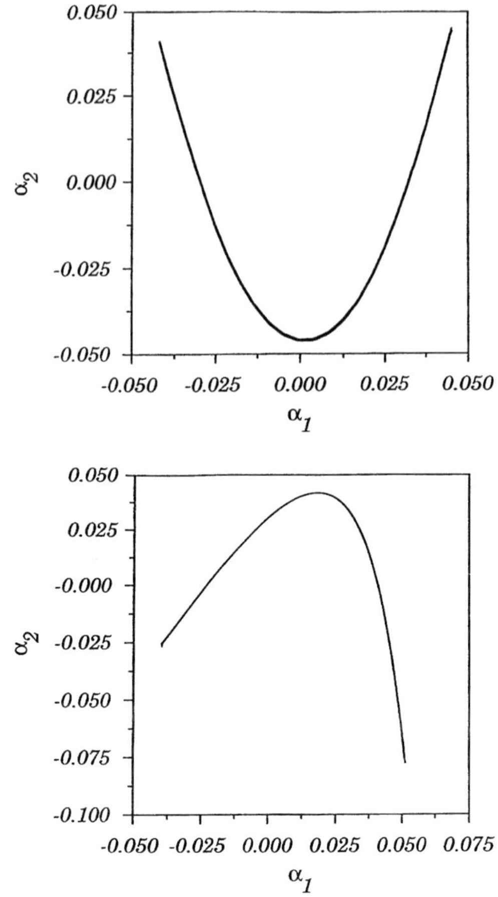

(a)

(b)

Figure 10. Cross-plots of the amplitudes of the POD modes. (a) Motion on the two-dimensional slow invariant manifold $\mathcal{V}_\mu$. (b) Motion on the three-dimensional slow invariant manifold $\mathcal{W}_\mu$.

The POD method does not produce mere mathematical expansions but expansions that reflect the natural spatio-temporal structure of the dynamics.

8. Interaction Dynamics

The reduced slow dynamics escape from the slow invariant manifold if the latter develops transversal instabilities. In this case we have interaction between the rigid-body slow dynamics and the cable fast dynamics. In the previous sections we have seen that the two-dimensional pendulation invariant manifold $\mathcal{V}_\mu$ is stable. However, the three-dimensional invariant manifold $\mathcal{W}_\mu$ has unstable regions.

If we apply the near identity transformation $Z = \mathbf{H}_\mu + Y$, and use the manifold condition (28), the system (19a) takes the form,

$$\dot{\Psi} = \mathbf{F}^0(\Psi) + \mathbf{F}^\mu(\Psi, \mathbf{H}^\mu(\Psi) + Y), \tag{40a}$$

$$\mu\dot{Y} = \left[\mathbf{B}^0 + \mu^2\mathbf{B}(\Psi)\right]Y - \underline{\mu D_\Psi \mathbf{H}^\mu(\Psi)\left[\mathbf{F}^\mu(\Psi, \mathbf{H}^\mu(\Psi) + Y) - \mathbf{F}^\mu(\Psi, \mathbf{H}^\mu(\Psi))\right]}. \tag{40b}$$

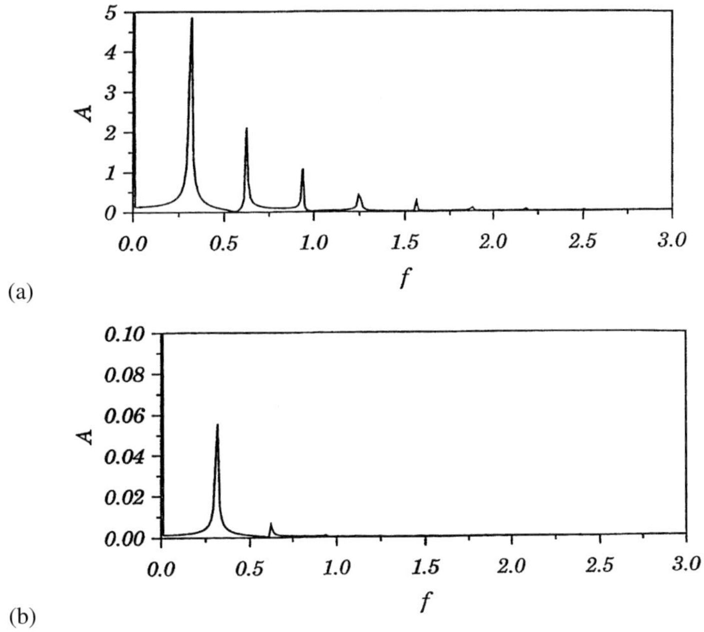

Figure 11. (a) Frequency spectrum of the cable displacement. (b) Frequency spectrum of the angular displacement.

The variable Y denotes the transverse distance of a state off the slow invariant manifold. If $Y = 0$, we recover the dynamics restricted to the slow invariant manifold.

The slaved slow dynamics can resonate with the fast dynamics since the underlined term is proportional to the gradient of the slaved slow dynamics. We compute the frequency spectrum of the dynamics of $\Psi_1(t)$ and $V_A(t)/\mu^2$ for a motion on the slow invariant manifold. Figure 11a reveals that the frequency spectrum of V_A/μ^2 contains the sequence $\{\omega, 2\omega, 3\omega, 4\omega, 5\omega, 6\omega\}$, where ω is the basic frequency of Ψ_1 whose frequency spectrum contains $\{\omega, 2\omega\}$, see Figure 11b. *The frequency spectrum of the slaved slow dynamics consists of multiples of the basic frequency of the master slow dynamics.* Therefore, the slaved slow dynamics evolve on a time scale faster than that of the master slow dynamics but at a time scale slower than that of the fast dynamics. Thus, it is possible to have internal resonances between the slaved slow dynamics and the fast dynamics. This can happen at an energy level far away from that of the static equilibrium. It has been predicted and confirmed experimentally in a cantilever beam/planar pendulum configuration [25].

By definition, we have interactions between the rigid-body dynamics and the flexible-body dynamics whenever a motion does not reside on the slow invariant manifold. Interesting interactions happen when a motion on the slow invariant manifold becomes unstable. Unstable slow motions escape naturally form the slow invariant manifold. Therefore, they intercact with the fast dynamics.

We analyze, as a function of μ, the frequency spectrum of the flexible-body dynamics during motions initiated on the second-order slow manifold. Figures 12a–12c show the time history of the cable displacement for $\mu = 0.02, 0.01, 0.0025$, respectively. We observe fast oscillations mixed with slow oscillations. The fast oscillations undergo amplitude modulations. This could indicate dynamics near heteroclinic of homoclinic connections due to transverse

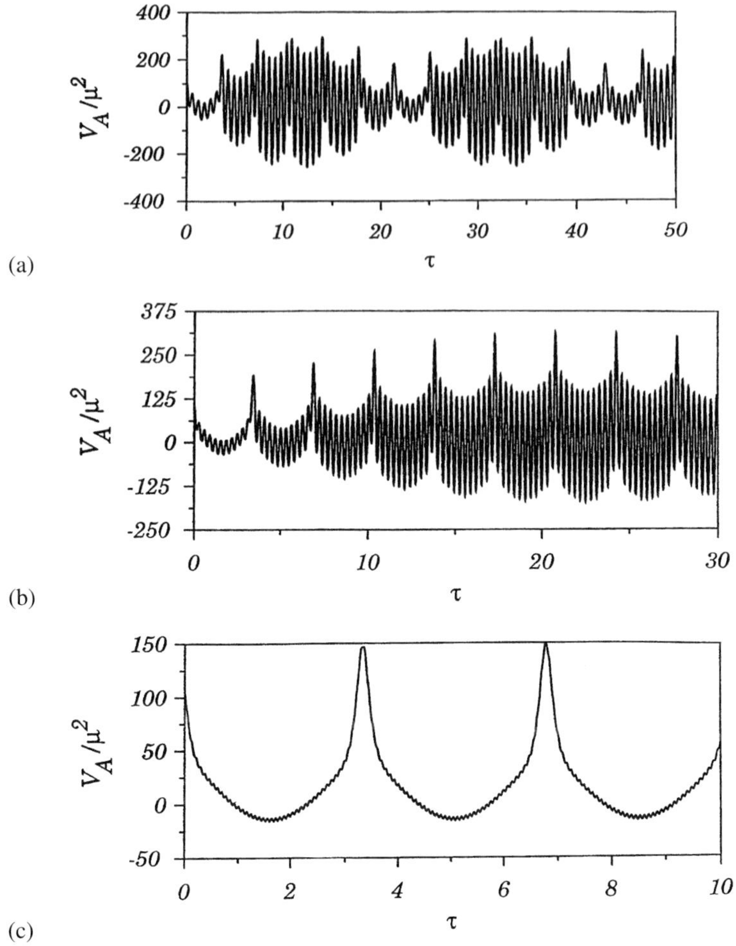

Figure 12. Time history of the cable displacement for a motion initiated at $\Psi_1(0) = 0.25$, $\Psi_2(0) = 0.50$, $\Psi_4(0) = 0.25$ on the angular momentum slow manifold, $\beta = 10$: (a) $\mu = 0.02$, (b) $\mu = 0.01$, (c) $\mu = 0.0025$.

instabilities of motions on the slow invariant manifold. As the strength of coupling is decreased the amplitude modulations of the fast oscillations decrease. Figure 12c shows a slow motion with a small fast oscillation. The amplitude of the fast oscillation does not grow as time progresses. This clearly indicates stability of the slow motion. This motion resides in the neighborhood of a stable region of the three-dimensional slow invariant manifold. Figures 13a–13c show the frequency spectra of the signals shown in Figures 12a–12c. Now we have the remarkable result that whenever we have mixing, thus interaction, of slow and fast dynamics, and this happens when the slow invariant manifold develops transverse instabilities, the fast frequency not only does fall in the frequency spectrum of the slaved slow dynamics, but also is locate close to a rational multiple of the basic frequency of the master slow dynamics. In particular, for the motion shown in Figure 12a, we have the relation: $f_f \approx 6 f_s$, see Figure 13a, where f_s and f_f are the basic slow and fast frequencies, respectively. And for the motion shown in Figure 12b, we have the relation: $f_f \approx 11 f_s$, see Figure 13b. The spectrum shown in Figure 13c does not contain the fast frequency which is $f_f \approx 13.90$. The connection

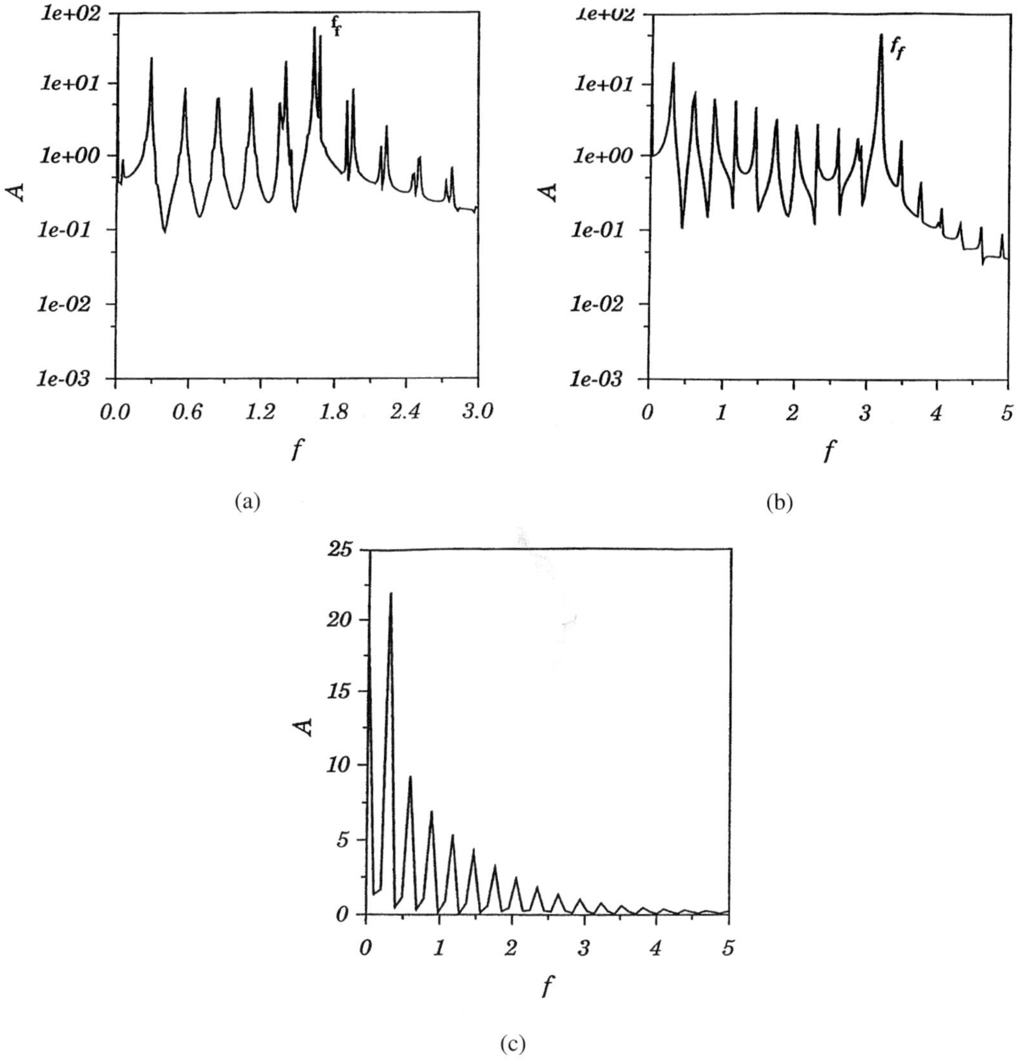

Figure 13. Corresponding frequency spectra for the motions shown in Figure 12.

of the interaction dynamics to the instability of the slow invariant manifold has been studied theoretically and experimentally in [25, 33] for a flexible-body/pendulum system.

The interaction can produce stochastic motions. Figure 14a show the trajectory of the pendulum mass onto the X_2–X_3 plane for motions initiate on the unstable regions of the slow invariant manifold for $\mu = 0.07715$. The initial energy of the motion is small. However, we see that the motion attains large amplitudes. The Poincaré map, defined by $\Psi_2 = 0$, $\Psi_1 > 0$, of this motion is shown is Figure 14b. Clearly, the motion is chaotic, or stochastic, since the system is conservative.

In the literature it has been reported that the ideal spherical pendulum undergoes motions during which the pendulation angle Ψ_1 remains constant. Such a motion traces a circle in the X_2–X_3 plane. Figure 15 reveals that the flexible spherical pendulum experiences motions with constant pendulation. These motion reside on the slow invariant manifold.

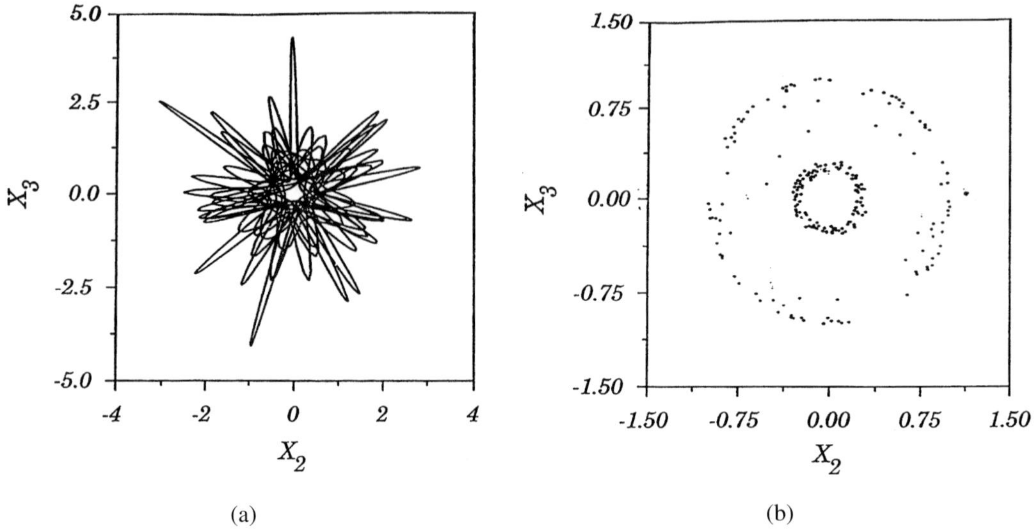

Figure 14. A stochastic motion, $\mu = 0.07715$, $\beta = 10.0$: (a) Projection onto the plane X_2–X_3, (b) Poincaré map.

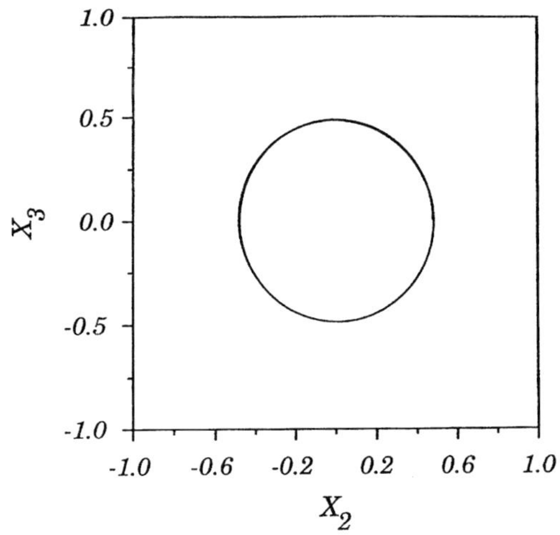

Figure 15. A whirling slow motion of the flexible pendulum with constant pendulation, $\mu = 0.0025$.

9. Conclusions and Discussion

We have studied some interesting aspects of the dynamics of the conservative spherical pendulum. In particular, we have computed the dominant dynamics for weak coupling. The dominant dynamics are slow and are realized as a three-dimensional manifold in phase space. We have computed formal approximations to the slow manifold. Numerical experiments and the method of proper orthogonal decomposition suggests that the slow manifold is a good approximation to a slow invariant manifold whose limit as the flexible-body dynamics become increasingly stiffer is computed exactly thanks to the singular perturbation approach to the problem.

The three-dimensional slow invariant manifold contains the two-dimensional pendulation submanifold. The two-dimensional slow invariant submanifold is the geometric realization of a nonclassical nonlinear normal mode. The master slow dynamics are described by a single nonlinear oscillator, which is a regular perturbation of the planar pendulum. The dynamics on the slow manifold is characterized by two POD modes for all energy levels. These two modes have flexible-body components that coincide and are identical to the shape of the single PO mode that characterizes the slaved slow dynamics.

The three-dimensional slow invariant manifold is the geometric realization of a mode of motion that involves two coupled degrees-of-freedom of equal natural linear frequencies. The master slow dynamics are described by a system that is a regular perturbation of the ideal spherical pendulum. Any motion on the slow invariant manifold is characterized by three PO modes. The flexible body components of these modes have identical shapes and coincide with the shape of the single PO mode characterizing the slaved slow dynamics.

The flexible cable is slaved to rigid-body slow motions. It behaves as a forced single degree-of-freedom during pure pendulation and whirling motions. It has been shown analytically and geometrically. It is verified with proper orthogonal projections. During slow motions the flexible pendulum behaves similarly to the ideal spherical pendulum. However, the slaved flexible dynamics, when destabilized, render the dynamics of the flexible pendulum very complex. Stochastic motions have been observed.

The slow invariant manifold of the weakly coupled conservative system is an important feature of the dynamics since it is perturbed to an attracting slow invariant manifold both for the dissipative and the forced dissipative systems.

Appendix A: Equations of Motion

To derive the equations of motion, an inertial frame $\{\hat{\mathbf{e}}_1, \hat{\mathbf{e}}_2, \hat{\mathbf{e}}_3\}$ is fixed at point O, and a body-fixed frame $\{\hat{\mathbf{a}}_r, \hat{\mathbf{a}}_\theta, \hat{\mathbf{a}}_\phi\}$ is fixed at the center A of the pendulum mass. The two reference frames are related by

$$\begin{aligned}
\hat{\mathbf{a}}_r &= \cos(\theta)\hat{\mathbf{e}}_1 + \sin(\theta)\cos(\phi)\hat{\mathbf{e}}_2 + \sin(\theta)\sin(\phi)\hat{\mathbf{e}}_3, && \text{(41a)}\\
\hat{\mathbf{a}}_\theta &= \sin(\theta)\hat{\mathbf{e}}_1 + \cos(\theta)\cos(\phi)\hat{\mathbf{e}}_2 + \cos(\theta)\sin(\phi)\hat{\mathbf{e}}_3, && \text{(41b)}\\
\hat{\mathbf{a}}_\phi &= -\sin(\phi)\hat{\mathbf{e}}_2 + \cos(\phi)\hat{\mathbf{e}}_3. && \text{(41c)}
\end{aligned}$$

All variables are normalized; and bold faced variables are vectors.

The position, velocity, acceleration, and angular velocity of a material point at the center of the cross-section of the cable are given by

$$\begin{aligned}
\mathbf{R}(\xi,\tau) &= (\xi + \hat{U}(\xi) + V(\xi,\tau))\hat{\mathbf{a}}_r, && \text{(42a)}\\
\dot{\mathbf{R}}(\xi,\tau) &= \dot{V}(\xi,\tau)\hat{\mathbf{a}}_r + (\xi + \hat{U}(\xi) + V(\xi,\tau))\dot{\hat{\mathbf{a}}}_r, && \text{(42b)}\\
\ddot{\mathbf{R}}(\xi,\tau) &= \ddot{V}(\xi,\tau)\hat{\mathbf{a}}_r + 2\dot{V}(\xi,\tau)\dot{\hat{\mathbf{a}}}_r + (\xi + \hat{U}(\xi) + V(\xi,\tau))\ddot{\hat{\mathbf{a}}}_r, && \text{(42c)}\\
\mathbf{\Omega}(\xi,\tau) &= \dot{\phi}\cos(\theta)\hat{\mathbf{a}}_r - \dot{\phi}\sin(\theta)\hat{\mathbf{a}}_\theta + \dot{\theta}\hat{\mathbf{a}}_\phi. && \text{(42d)}
\end{aligned}$$

The term $\hat{U}(\xi)$ denotes the static deformation when the system is at rest in its stable equilibrium configuration. Upon using the kinematic relations

$$\dot{\hat{\mathbf{a}}}_r = \mathbf{\Omega} \times \hat{\mathbf{a}}_r, \quad \ddot{\hat{\mathbf{a}}}_r = \dot{\mathbf{\Omega}} \times \hat{\mathbf{a}}_r + \mathbf{\Omega} \times \dot{\hat{\mathbf{a}}}_r, \tag{43}$$

we obtain the following expression for the acceleration:

$$\begin{aligned}\ddot{\mathbf{R}}(\xi,\tau) &= \left[\ddot{V}(\xi,\tau) - (\xi + \hat{U}(\xi) + V(\xi,\tau))(\dot{\theta}^2 + \dot{\phi}^2 \sin^2(\theta))\right]\hat{\mathbf{a}}_r \\ &+ \left[2\dot{V}(\xi,\tau)\dot{\theta} + (\xi + \hat{U}(\xi) + V(\xi,\tau))(\ddot{\theta} - \dot{\phi}^2 \sin(\theta)\cos(\theta))\right]\hat{\mathbf{a}}_\theta \\ &+ \left[2\dot{V}(\xi,\tau)\dot{\phi}\sin(\theta) + (\xi + \hat{U}(\xi) + V(\xi,\tau))(\ddot{\phi}\sin(\theta) + 2\dot{\phi}\dot{\theta}\cos(\theta))\right]\hat{\mathbf{a}}_\phi. \quad (44)\end{aligned}$$

Newton's Law yields the following vectorial equations of motion for the flexible pendulum:

$$\ddot{\mathbf{R}}_A(\tau) = -T_A\hat{\mathbf{a}}_r(\tau) + \hat{\mathbf{e}}_1, \quad (45a)$$

$$\mu^2\pi^2\ddot{\mathbf{R}}(\xi,\tau) = 4U_{\xi\xi}(\xi,\tau)\hat{\mathbf{a}}_r(\tau) + \mu^2\pi^2\hat{\mathbf{e}}_1, \quad (45b)$$

$$U(\xi = 0,\tau)\hat{\mathbf{a}}_r(\tau) = \mathbf{0}, \quad \left[U_\xi(\xi = 1,\tau) + \mu^2\pi^2\beta(1 - T_A)\right]\hat{\mathbf{a}}_r(\tau) = \mathbf{0}, \quad (45c)$$

We have assumed that the cable undergoes only radial deformations. The term T_A denotes the magnitude of the reaction force between the mass M_r and the cable end A. The projection of the equation of motion of the pendulum mass onto the unit vector $\hat{\mathbf{a}}_r$ yields

$$\begin{aligned}T_A &= -\ddot{V}_A + \cos(\theta) + (1 + \mu^2C(\beta) + V_A)(\dot{\theta}^2 + P_\phi^2/\sin^3(\theta)), \\ C(\beta) &\equiv \pi^2(1 + 2\beta)/8. \quad (46)\end{aligned}$$

By projecting the vectorial equation of motion onto the elements of the body-fixed frame, we obtain the equation of motion (3a).

The normalized length of the cable and the Cartesian coordinates of the rigid mass M_r, the payload, are given by

$$L_f(\tau) = 1 + \mu^2C(\beta) + V_A(\tau), \quad (47a)$$

$$(X_1, X_2, X_3) = L_f(\cos(\theta), \sin(\theta)\cos(\phi), \sin(\theta)\sin(\phi)). \quad (47b)$$

Appendix B: Computation of the Slaved Slow System

In this section, the computation of the slow manifold is presented. The solution to the slow manifold condition is approximated by the asymptotic series

$$\mathbf{H}_\mu(\Psi) = \mathbf{H}^0(\Psi) + \sum_{m=1}^{M}\mu^m\mathbf{H}_m(\Psi). \quad (48)$$

Moreover, we expand in asymptotic series all functions of μ. In particular, we have

$$\mathbf{F}^\mu(\Psi, \mathbf{H}_\mu) = \sum_{m=1}^{M}\mu^m\mathbf{F}^m(\Psi, \mathbf{H}^p), \quad p < m. \quad (49)$$

The first four terms expansion (49) have been computed. They are given below:

$$\begin{aligned}\mathbf{F}^1(\Psi, \mathbf{H}^p) &= (\mathbf{A}_2\mathbf{H}^0 \cdot \mathbf{e})\mathbf{F}_{11}(\Psi), \\ \mathbf{F}^2(\Psi, \mathbf{H}^p) &= \mathbf{F}_{21}(\Psi) + (\mathbf{A}_1\mathbf{H}^0 \cdot \mathbf{e})\mathbf{F}_{22}(\Psi) + (\mathbf{A}_2\mathbf{H}^1 \cdot \mathbf{e})\mathbf{F}_{11}(\Psi),\end{aligned}$$

$$\begin{aligned}
\mathbf{F}^3(\Psi, \mathbf{H}^p) &= (\mathbf{A}_2\mathbf{H}^0 \cdot \mathbf{e})\mathbf{F}_{31}(\Psi) + (\mathbf{A}_1\mathbf{H}^0 \cdot \mathbf{e})(\mathbf{A}_2\mathbf{H}^0 \cdot \mathbf{e})\mathbf{F}_{32}(\Psi) \\
&\quad + (\mathbf{A}_2\mathbf{H}^2 \cdot \mathbf{e})\mathbf{F}_{11}(\Psi) + (\mathbf{A}_1\mathbf{H}^1 \cdot \mathbf{e})\mathbf{F}_{22}(\Psi), \\
\mathbf{F}^4(\Psi, \mathbf{H}^p) &= \mathbf{F}_{41}(\Psi) + (\mathbf{A}_1\mathbf{H}^0 \cdot \mathbf{e})\mathbf{F}_{42}(\Psi) + (\mathbf{A}_1\mathbf{H}^0 \cdot \mathbf{e})^2\mathbf{F}_{43}(\Psi) \\
&\quad + (\mathbf{A}_2\mathbf{H}^3 \cdot \mathbf{e})\mathbf{F}_{11}(\Psi) + (\mathbf{A}_1\mathbf{H}^2 \cdot \mathbf{e})\mathbf{F}_{22}(\Psi) \\
&\quad + \left[(\mathbf{A}_1\mathbf{H}^0 \cdot \mathbf{e})(\mathbf{A}_2\mathbf{H}^0 \cdot \mathbf{e}) + (\mathbf{A}_1\mathbf{H}^1 \cdot \mathbf{e})(\mathbf{A}_2\mathbf{H}^0 \cdot \mathbf{e})\right]\mathbf{F}_{32}(\Psi).
\end{aligned}$$

The terms $\mathbf{F}_{km}$ are 4×1 vector-valued functions. The matrices $\mathbf{A}_1$ and $\mathbf{A}_2$ stem from the matrix representation of the cumulative coupling terms P_N and Q_N. In particular, we have

$$P_N(\mathbf{H}^\mu) = \mathbf{A}_1\mathbf{H}^\mu \cdot \mathbf{e}, = \mathbf{A}_1\mathbf{H}^0 \cdot \mathbf{e} + \sum \mu^m(\mathbf{A}_1\mathbf{H}^m \cdot \mathbf{e}), \tag{50}$$

$$Q_N(\mathbf{H}_\mu) = \mathbf{A}_2\mathbf{H}_\mu \cdot \mathbf{e} = \mathbf{A}_2\mathbf{H}^0 \cdot \mathbf{e} + \sum \mu^m(\mathbf{A}_2\mathbf{H}^m \cdot \mathbf{e}). \tag{51}$$

The nonzero elements of the above matrices are $(A_1)_{2m-1} = (-1)^{m+1}\mu_m^2$, and $(A_2)_{2m} = (-1)^{m+1}\mu_m^2$. Moreover, $\hat{\mathbf{e}} \equiv (1, 0, 0, \ldots, 0)$ is an $2N$-dimensional unit vector.

Upon substituting the asymptotic series (48) and (49) into the slow manifold condition (28), a recursive algorithm is obtained to determine the expansion terms $\mathbf{H}^m$. The $O(\mu^3)$ truncation of this algorithm yields the following equations for $\mathbf{H}^0$ and $\mathbf{H}^m$, $m = 1, 2, 3$:

$$\mathbf{B}^0\mathbf{H}^0 = -\mathbf{G}^0(\Psi), \tag{52a}$$

$$\mathbf{B}^0\mathbf{H}^1 = D_\Psi\mathbf{H}^0(\Psi)\mathbf{F}^0(\Psi), \tag{52b}$$

$$\begin{aligned}
\mathbf{B}^0\mathbf{H}^2 &= -\mathbf{G}(\Psi) - \mathbf{B}(\Psi)\mathbf{H}^0(\Psi) \\
&\quad + (\mathbf{A}_2\mathbf{H}^0(\Psi) \cdot \mathbf{e})D_\Psi\mathbf{H}^0(\Psi)\mathbf{F}_{11}(\Psi) + D_\Psi\mathbf{H}^1(\Psi)\mathbf{F}_0(\Psi),
\end{aligned} \tag{52c}$$

$$\begin{aligned}
\mathbf{B}^0\mathbf{H}^3 &= D_\Psi\mathbf{H}^0(\Psi)\left[\mathbf{F}_{21}(\Psi) + (\mathbf{A}_1\mathbf{H}^0(\Psi) \cdot \mathbf{e})\mathbf{F}_{22}(\Psi) + (\mathbf{A}_2\mathbf{H}^1(\Psi) \cdot \mathbf{e})\mathbf{F}_{11}(\Psi)\right] \\
&\quad + D_\Psi\mathbf{H}^2(\Psi)\mathbf{F}^0(\Psi) + D_\Psi\mathbf{H}^1(\Psi)(\mathbf{A}_2\mathbf{H}^0(\Psi) \cdot \mathbf{e})\mathbf{F}_{11}(\Psi) \\
&\quad - \mathbf{B}(\Psi)\mathbf{H}^1(\Psi).
\end{aligned} \tag{52d}$$

Equation (52a) has been solved symbolically for finite N by applying the computer algebra code Maple.

The solution to the first-order Equation (52b) is

$$\begin{aligned}
H^1_{2m-1}(\Psi) &= 0, \\
H^1_{2m}(\Psi; \mu_m) &= (-1)^{m+1}Q^1_1(\Psi) + Q^1_2(\Psi)\mu_m + (-1)^{m+1}Q^1_3(\Psi)\mu_m^2,
\end{aligned} \tag{53}$$

where

$$\begin{aligned}
Q^1_1(\Psi) &= -6\beta\Psi_2\sin(\Psi_1) - \frac{2\beta\Psi_2\Psi_4^2}{\sin^4(\Psi_1)}\cos(\Psi_1)(3 - 2\sin(\Psi_1)), \\
Q^1_2(\Psi) &= -\frac{4\Psi_2\sin(\Psi_1)}{\pi}, \\
Q^1_3(\Psi) &= -\frac{16\beta\Psi_2\sin(\Psi_1)}{\pi^2} - \frac{8\beta\Psi_2\Psi_4^2}{\pi^2\sin^4(\Psi_1)}\cos(\Psi_1)(3 - 2\sin(\Psi_1)).
\end{aligned} \tag{54}$$

The solution to the second-order Equation (52c) is

$$
\begin{aligned}
H^2_{2m}(\Psi_1;\mu_m,N) &= 0, \qquad (55a)\\
H^2_{2m-1}(\Psi_1;\mu_m,N) &= (-1)^{m+1}Q_2(\Psi)\mu_m^2 \\
&\quad + \left[Q_3(\Psi)+\hat{Q}_3(\Psi)\hat{H}_m\right]\mu_m^3 + (-1)^{m+1}Q_4(\Psi)\mu_m^4(-1)^{m+1} \\
&\quad \times \left[P_0(\Psi)+P_2(\Psi)S_2(N)+P_3(\Psi)S_3(N)+P_4(\Psi)S_4(N)\right]. \qquad (55b)
\end{aligned}
$$

The solution for the third-order Equation (52d) is

$$
H^3_{2m-1}(\Psi_1;\mu_m,N) = 0, \qquad (56a)
$$

$$
\begin{aligned}
H^3_{2m}(\Psi_1;\mu_m,N) &= (-1)^{m+1}\left[Q_0(\Psi;N)+Q_2(\Psi;N)\mu_m^2\right] \\
&\quad + (-1)^{m+1}\left[Q_3(\Psi)\mu_m^3+Q_4(\Psi)\mu_m^4+Q_5(\Psi)\mu_m^5+Q_6(\Psi)\mu_m^6\right], \qquad (56b)
\end{aligned}
$$

where

$$
\begin{aligned}
Q_0(\Psi;N) &= Q_{00}(\Psi)+Q_{02}(\Psi)(S_2(N)-\mu_m^2)-Q_{03}(\Psi)(S_3(N)-\mu_m^3) \\
&\quad + Q_{04}(\Psi)(S_4(N)-\mu_m^4)+(-1)^{m+1}\hat{H}_m\hat{Q}_{03}(\Psi), \\
Q_2(\Psi;N) &= Q_{20}(\Psi)+Q_{22}(\Psi)(S_2(N)-\mu_m^2)-Q_{23}(\Psi)(S_3(N)-\mu_m^3) \\
&\quad + Q_{24}(\Psi)(S_4(N)-\mu_m^4)+(-1)^{m+1}\hat{H}_m\hat{Q}_{23}(\Psi).
\end{aligned}
$$

The various functions of the variable Ψ in the above equations have the same structure. For example,

$$
Q_{20}(\Psi) = \sum_{m=0}^{5}\sum_{n=0}^{5} Q_{20mn}(\Psi_1)\Psi_2^{m-1}\Psi_4^{n-1}. \qquad (57)
$$

They have been computed symbolically and coded automatically in Fortran.

The symbolic computations reveal that the approximation terms $\mathbf{H}^p$, $p > 1$ depend explicitly on the order of truncation N through fast convergent sequencies of the frequency ratios μ_m. For instance, for $p = 2, 3$, we have

$$
S_2(N) \equiv \sum_{n=1}^{N}\mu_n^2, \quad S_3(N) \equiv \sum_{n=1}^{N}(-1)^{n+1}\mu_n^3, \quad S_4(N) \equiv \sum_{n=1}^{N}\mu_n^4. \qquad (58)
$$

Clearly, the computational results can be extended to the infinite system, that is, for $N \to \infty$. *Thus, we have constructed formal slow manifolds up to $O(\mu^3)$ to determine the slaved slow dynamics of the flexible spherical pendulum*. It remains to show that they are in the neighborhood of a slow invariant manifold.

Appendix C: Computation of the Master Slow System

The exact *master* slow system (31a) can be approximated as follows:

$$\dot{\Psi} = \mathbf{F}^0(\Psi) + \sum_{m=1}^{\infty} \mu^{2m} \mathbf{N}_{2m}(\Psi). \tag{59}$$

The approximation terms $\mathbf{N}_{2m}(\Psi)$ are 4×1 vector-valued functions, satisfying $(N_{2m})_1 = (N_{2m})_3 = 0$.

Upon using the analytic expressions for the approximations to the slow manifold, we find that the first two low-order correction terms are given by

$$\mathbf{N}_2(\Psi) = (\mathbf{A}_2\mathbf{H}^1(\Psi)\cdot\mathbf{e})\mathbf{F}_{11}(\Psi) + \mathbf{F}_{21}(\Psi) + (\mathbf{A}_1\mathbf{H}^0(\Psi)\cdot\mathbf{e})\mathbf{F}_{22}(\Psi), \tag{60}$$

$$\begin{aligned} \mathbf{N}_4(\Psi) = {} & (\mathbf{A}_2\mathbf{H}^3(\Psi)\cdot\mathbf{e})\mathbf{F}_{11}(\Psi) + (\mathbf{A}_1\mathbf{H}^2(\Psi)\cdot\mathbf{e})\mathbf{F}_{22}(\Psi) \\ & + (\mathbf{A}_2\mathbf{H}^1(\Psi)\cdot\mathbf{e})\mathbf{F}_{31}(\Psi) + (\mathbf{A}_1\mathbf{H}^0(\Psi)\cdot\mathbf{e})(\mathbf{A}_2\mathbf{H}^1(\Psi)\cdot\mathbf{e})\mathbf{F}_{32}(\Psi) \\ & + \mathbf{F}_{41}(\Psi) + (\mathbf{A}_1\mathbf{H}^0(\Psi)\cdot\mathbf{e})\mathbf{F}_{42}(\Psi). \end{aligned} \tag{61}$$

The nonzero elements of $\mathbf{N}_4$ are

$$\begin{aligned} N_{42}(\Psi) = {} & P_{20}(\Psi) + P_{21}(\Psi)(S_2S_3 - S_5) + P_{22}(\Psi)(S_3S_4 - S_7) \\ & + P_{23}(\Psi)(S_3^2 - S_6) + P_{24}(\Psi)(S_4^2 - S_8) + P_{25}(\Psi)(S_2^3 - S_6) \\ & + P_{26}(\Psi)S_3 + P_{27}(\Psi)S_2 + P_{28}(\Psi)S_4 + P_{29}(\Psi)(S_2^2 - S_4) \\ & + P_{210}(\Psi)S_5 + P_{211}(\Psi)S_6 + P_{212}(\Psi)S_7 + P_{213}(\Psi)S_8, \end{aligned} \tag{62}$$

$$\begin{aligned} N_{44}(\Psi) = {} & P_{41}(\Psi)(S_2S_3 - S_5) + P_{42}(\Psi)(S_3S_4 - S_7) \\ & + P_{43}(\Psi)(S_3^2 - S_6) + P_{44}(\Psi)(S_4^2 - S_8) + P_{45}(\Psi)(S_2^3 - S_6) \\ & + P_{46}(\Psi)S_3 + P_{47}(\Psi)S_2 + P_{48}(\Psi)S_4 + P_{49}(\Psi)(S_2^2 - S_4) \\ & + P_{410}(\Psi)S_5 + P_{411}(\Psi)S_6 + P_{412}(\Psi)S_7 + P_{413}(\Psi)S_8. \end{aligned} \tag{63}$$

The constants S_m are the limits of the infinite-sequencies of the frequency ratios μ_m

$$\begin{aligned} S_{2k-1} &= \sum_{m=1}^{\infty}(-1)^{m+1}\mu_m^{2k-1}, \quad k > 1, \\ S_{2k} &= \sum_{m=1}^{\infty}(-1)^{m+1}\mu_m^{2k}, \quad k \geq 1. \end{aligned} \tag{64}$$

We have computed approximations up to $Q(\mu^4)$ to the master slow system of the flexible spherical pendulum.

Acknowledgements

The numerical computations were part of the DOD High Performance Computing (HPC) project 'Dynamical Modeling and Control in Large Scale Coupled Structural and Electronic Systems'. The research is supported by the Office of Naval Research.

References

1. Miles, W. J., 'Stability of forced oscillations of a spherical pendulum', *Quarterly Applied Mathematics* **20**, 1962, 21–32.
2. Miles, W. J., 'Resonantly forced surface waves in a circular cylinder', *Journal of Fluid Mechanics* **149**, 1984, 15–31.
3. Kana, D. D., 'Distinguishing the transition to chaos in a spherical pendulum', *CHAOS* **5**(1), 1995, 298–310.
4. Kana, D. D., 'Validated spherical pendulum model for rotary liquid slosh', *AIAA Journal on Spacecraft and Rockets* **26**(3), 1989, 188–195.
5. Kana, D. D., 'A model for nonlinear rotary slosh in propellant tanks', *AIAA Journal on Spacecraft and Rockets* **24**(2), 1987, 169–177.
6. Irons, F. E., 'Concerning the nonlinear behavior of the forced spherical pendulum including the dowsing pendulum', *European Journal of Physics* **11**(2), 1990, 107–115.
7. Zou, M., 'Kolmogorov's condition for the square potential pendulum', *Physics Letters A* **166**, 1992, 321–329.
8. Bryant, P. J., 'Breakdown to chaotic motion of a forced, damped, spherical pendulum', *Physica D* **64**, 1993, 324–339.
9. Tritton, D. J., 'Ordered and chaotic motion of a forced spherical pendulum', *European Journal of Physics* **7**(3), 1986, 162–169.
10. Peters, D. P., 'Chaotic motion from support constraints of a nondriven rigid spherical pendulum', *Physical Review A* **38**(10), 1988, 5352–5359.
11. Chin, C., Nayfeh, A. H., and Mook, D. T., 'Dynamics and control of ship-mounted cranes', in *Proceedings of the 39th AIAA/ASME/ASCE/AHS/ASC Structures, Structural Dynamics, and Materials Conference*, Long Beach, CA, April 20–23, 1998, pp. 295–301.
12. Miles, J. W. and Zou, Q.-P., 'Parametric excitation of a detuned spherical pendulum', *Journal of Sound and Vibration* **164**(2), 1993, 237–250.
13. Krasnopolskaya, T. S. and Shvets, A. Yu., 'Chaotic oscillations of a spherical pendulum as an example of interaction with an energy source', *International Applied Mechanics* **28**(10), 669–674.
14. Numez-Yepez, H. N., Salas-Brito, A. L., and Vargas, C. A., 'Onset of chaos in an extensible pendulum', *Physics Letters A* **145**(2,3), 1990, 101–105.
15. Georgiou, I. T., Bajaj, A. K., and Corless, M., 'Slow and fast invariant manifolds, and normal modes in a two degree-of-freedom structural dynamical system with multiple equilibrium states', *International Journal of Non-Linear Mechanics* **33**(2), 1998, 275–300.
16. Kokotović, P. V., Khalil, H. K., and O'Reilly J., *Singular Perturbation Methods in Control: Analysis and Design*, Academic Press, London, 1986.
17. Georgiou, I. T. and Schwartz, I. B., 'The slow invariant manifold of a conservative pendulum-oscillator system', *International Journal of Bifurcation and Chaos* **6**(4), 1996, 673–692.
18. Lichtenberg, A. J. and Lieberman, M. A., *Regular and Chaotic Dynamics*, Springer-Verlag, New York, 1992.
19. Boothby, W. M., *An Introduction to Differential Manifolds and Riemanian Geometry*, Academic Press, London, 1985.
20. Taylor, A. E. and Lay, D. C., *Introduction to Functional Analysis*, second edition, Krieger, Malabar, FL, 1986.
21. Cusumano, J. P., Sharkady, M. T., and Kimple, B. W., 'Experimental measurements of dimensionality and spatial coherence in the dynamics of a flexible-beam impact oscillator', *Philosophical Transactions of the Royal Society of London A* **347**, 1994, 421–438.
22. Georgiou, I. T. and Schwartz, I. B., 'Dynamics of large scale coupled structural/mechanical systems: A singular perturbation/proper orthogonal decomposition approach', *Journal of Applied Mathematics (SIAM)* **59**(4), 1999, 1178–1207.

23. Georgiou, I. T. and Schwartz, I. B., 'The slow nonlinear normal mode and stochasticity in the dynamics of a conservative flexible rod/pendulum configuration', *Journal of Sound and Vibration* **220**(3), 1999, 383–411.
24. Georgiou, I. T. and Sansour, J., 'Analyzing the finite element dynamics of nonlinear in-plane rods by the method of proper orthogonal decomposition', in *Computational Mechanics, New Trends and Applications*, S. Idelsohn, E. Onate, and E. Dvorkin (eds.), CIMNE, Barcelona, Spain, 1998.
25. Georgiou, I. T., Schwartz, I. B., Emaci, E., and Vakakis, A. F., 'Interaction between slow and fast oscillations in an infinite degree-of-freedom linear system coupled to a nonlinear subsystem: Theory and experiment', *ASME Journal of Applied Mechanics* **66**, 1999, 448–459.
26. Jones, C., 'Geometric singular perturbation theory in dynamical systems', in Springer Lecture Notes in Mathematics, Vol. 1609, Springer-Verlag, New York, 1995, pp. 44–120.
27. Mira, C., 'Some historical aspects of nonlinear dynamics-possible trends for the future', *Journal of the Franklin Institute* **334 B**(5), 1997, 1075–1113.
28. Pontrjagin, L. S., 'Asymptotic behavior of solutions of differential equations when the higher derivatives contain a small parameter as a factor', *Izvestija Akademia Nauk (Ser. mat.)* **21**, 1975, 605–621 [in Russian].
29. Tikhonov, A. N., 'Systems of differential equations containing a small parameter with higher order derivatives', *Matematicheskii Sbornik* **31**(73), 1952, 575–584 [in Russian].
30. Sirovich, L., 'Turbulence and the dynamics of coherent structures, pt. I, Coherent structures', *Quartely of Applied Mathematics* **45**, 1987, 561–571.
31. Vakakis, A. F., Manevitch, L. I., Mikhlin, Yu. V., Pilipchuck, V. N., and Zevin, A. A., *Normal Modes and Localization in Nonlinear Systems*, Wiley Interscience, New York, 1996.
32. Schwartz, I. B. and Triandaf, I., 'Chaos and intermittent bursting in a reaction-diffusion process', *CHAOS* **6**, 1996, 229–237.
33. Schwartz, I. B. and Georgiou, I. T., 'Instant chaos and hysteresis in coupled linear-nonlinear oscillators', *Physics Letters A* **242**, 1998, 307–312.

Nonlinear Dynamics **25:** 33–48, 2001.

Normal Vibrations in Near-Conservative Self-Excited and Viscoelastic Nonlinear Systems

YU. V. MIKHLIN
Department of Applied Mathematics, Kharkov Polytechnical University, 21 Frunze Str., 61002 Kharkov, Ukraine

B. I. MORGUNOV[†]
Department of Mathematical Modelling, Moscow Institute of Electronics and Mathematics, 109028 Moscow, Russia

(Received: 14 January 1999; accepted: 2 February 2000)

Abstract. A perturbation methodology and power series are utilized to the analysis of nonlinear normal vibration modes in broad classes of finite-dimensional self-excited nonlinear systems close to conservative systems taking into account similar nonlinear normal modes. The analytical construction is presented for some concrete systems. Namely, two linearly connected Van der Pol oscillators with nonlinear elastic characteristics and a simplest two-degrees-of-freedom nonlinear model of plate vibrations in a gas flow are considered.

Periodical quasinormal solutions of integro-differential equations corresponding to viscoelastic mechanical systems are constructed using a convergent iteration process. One assumes that conservative systems appropriate for the dominant elastic interactions admit similar nonlinear normal modes.

Keywords: Self-excited nonlinear systems, nonlinear normal modes (NNMs), viscoelastic nonlinear systems, power series, iterations.

1. Introduction

Lyapunov proved that nonlinear finite-dimensional systems with an analytical first integral allow a one-parameter family of periodic solutions which tend towards linear normal vibration modes as amplitudes tend to zero. Natural frequencies of corresponding linearized systems must not be integrally related [1]. The Lyapunov's solutions possess all the properties of linear normal modes [2].

Nonlinear normal vibrations modes (NNMs) are a generalization of the normal (principal) vibrations of linear systems. In the normal mode, a finite-dimensional system behaves like a conservative one in having a single degree of freedom, and all position coordinates can be analytically parametrized by any one of them.

Rosenberg, in a series of papers [3–5] provided analytical methods for computing NNMs. He defined them as 'vibrations in unison' and introduced a broad class of essentially nonlinear conservative systems allowing for NNMs with rectilinear trajectories in a configuration space ('similar' NNMs).

For example, 'homogeneous systems' whose potential is an even homogeneous function of the variables belong to such a class. It is interesting to note that the number of the NNMs may exceed the number of degrees of freedom of the oscillator. In general, the NNMs trajectories are curvilinear instead of straight lines in linear systems.

[†] Deceased

For some particular cases, curvilinear trajectories were analyzed by Rosenberg and Kuo [5] and by Rand [6]. The power series method was proposed by Manevich and Mikhlin [7] for the construction of the curvilinear trajectories of NNMs.

Later, new results concerning NNMs of conservative systems were performed by several authors, including Vedenova et al. [8], Vakakis [9], Vakakis and Rand [10], Shaw and Pierre [11, 12], Nayfeh and Nayfeh [13], and Mikhlin [14].

In [8], nonlinear normal mode localization is studied in non linear systems with impact nonlinearities. In [9], for an oscillator with weak coupling stiffness, both localized and nonlocalized modes are detected using an asymptotic methodology. In [10], the NNMs and global dynamics of nonlinear systems are analyzed by means of Poincaré maps. In [11, 12], the authors reformulated the concept of NNMs for a general class of nonlinear discrete oscillators. The analysis is based on the computation of invariant manifolds of motion on which the NNM oscillations take place. In [13], a complex invariant manifold formulation is proposed. In [14], Padé approximations are used for an analysis of NNMs with large amplitudes.

Note that publications on the generalization of NNMs to nonconservative systems are not numerous.

It is well known that forced vibrations in nonlinear systems with one degree of freedom and small periodic disturbances are similar to those in an undisturbed conservative system (in the resonance region).

It should be noted that periodic solutions in nonautonomous systems close to Lyapunov systems were thoroughly investigated by Malkin [15].

The Rauscher method was used for analyses of normal vibrations in nonautonomous systems by Kinney and Rosenberg [16]. To solve the problem, one assumes that the external periodic force is proportional to some chosen positional coordinate raised to such a power that the resultant autonomous system is homogeneous.

NNMs in general finite-dimensional nonautonomous systems close to conservative systems allowing similar NNMs were considered by Mikhlin [17]. Rauscher's ideas and the power-series method for trajectories in a configurational space are used in the construction of resonance solutions.

Some other works on NNMs in nonautonomous systems, have been published by Yang and Rosenberg [18], Szemplinska-Stupnicka [19], and Vakakis and Caughey [20].

Of particular note is the work by Rand and Holmes [21], where periodic solutions of two weakly coupled Van der Pol oscillators are analyzed. Note that NNMs in a model of two strongly coupled Van der Pol oscillators are considered in this work.

Basic and new results on NNMs are presented in the book by Vakakis et al. [22] which describes quantitative and qualitative analyses of NNMs in conservative and nonautonomous systems, including localized modes, an analysis of stability, a generalization, and an analysis of NNMs in distributed systems.

Here, the perturbation methodology is utilized for the analysis of NNMs in broad classes of finite-dimensional self-excited nonlinear systems close to conservative systems allowing normal vibrations with rectilinear trajectories in a configuration space (Section 1). The analytical construction of NNMs is presented in Section 2 for some concrete systems. The first of them represents two linearly connected Van der Pol oscillators with nonlinear elastic characteristics. Another example refers to the problem of plate vibrations in the flow of gas. A simple, two-degrees-of-freedom nonlinear model is considered. In Section 3, periodical quasinormal solutions of integro-differential equations corresponding to viscoelastic mechanical systems are constructed. One assumes that elastic forces are dominant and conservative systems appropriate for the elastic interactions allow similar NNMs which are selected as generative solutions. A convergent iteration process is used here.

2. Normal Vibrations in Near-Conservative, Self-Excited Nonlinear Systems

2.1. Equations Describing the Trajectories of NNMs

We will demonstrate that the approaches for analysis of NNMs in conservative systems are applicable to near-conservative autonomous systems with small self-excited perturbations. We will also consider modes when all positional coordinates of the finite-dimensional, self-excited system are linked. In relation to these modes, the system behaves like a single-degree-of-freedom one. The periodic solutions close to NNMs in the generating conservative system could be called *nonlinear normal modes (NNMs) of the self-excited nonlinear system.*

Consider the following near-conservative system:

$$\ddot{x}_i + \Pi_{x_i}(x_1, x_2, \ldots, x_n) + \varepsilon g_i(x_1, \dot{x}_1, \ldots, x_n, \dot{x}_n) = 0, \tag{2.1}$$

where ε is a small parameter and the functions g_i may be nonlinear with respect to $\dot{x}_i$. The functions f_i and Π are assumed to be analytical in x_i, $\dot{x}_i$; a potential energy Π is subject to some limitations which will be shown later. The system may involve friction of any physical nature, such as viscous, dry, or turbulent. The conditions that ensure the occurrence of self-excited vibrations (limiting cycles) are discussed later.

Assume that a conservative system (c = 0) permits similar NNMs, i.e., normal modes with rectilinear trajectories in a configuration space: $x_{i0} = k_i x_{i0}$ ($i = 2, \ldots, n$; k_i are constants). Any terms may be regarded as a perturbation.

Consider one of these NNMs as a generating one. Rotating the coordinate axes so that the new x-axis is directed along the rectilinear trajectory, one obtains a generating solution in the new coordinates as

$$x_{i0}\ (i = 2, 3, \ldots, n), \qquad x_{i0} \equiv x = x(t). \tag{2.2}$$

Select some positional variable $x \equiv x_i$.

Let us find a solution that all phase coordinates are defined as single-valued and analytical functions of x:

$$x_i = x_i(x, \varepsilon), \qquad \dot{x} = \dot{x}(x, \varepsilon), \qquad \dot{x}_i = \dot{x}_i(x, \varepsilon). \tag{2.3}$$

Introducing a new independent variable x instead of t, one obtains from (2.1) the equations describing the trajectories of NNMs:

$$\begin{aligned} x_i''(\dot{x}(x))^2 \; &+ \; x_i'[-\Pi_x(x, x_2(x), \ldots, x_n(x)) - \varepsilon g_1(x, \dot{x}(x), x_2(x), \dot{x}_2(x), \ldots)] \\ &+ \; \Pi_{x_i}(x, x_2(x), \ldots, x_n(x)) + \varepsilon g_i(x, \dot{x}(x), x_2(x), \dot{x}_2(x), \ldots) = 0 \\ &\qquad (i = 2, 3, \ldots, n). \end{aligned} \tag{2.4}$$

Here and henceforth, a prime denotes differentiation with respect to x; it is clear that $\dot{x}_i(x, \varepsilon) = x_i'(x, \varepsilon)\dot{x}$.

Select the trajectory return points where all velocities are equal to zero. When $x = X_j$ $(j = 1, 2)$ and $\dot{x} = 0$, we obtain the additional conditions from (2.4):

$$\{x_i' [-\Pi_x(x, x_2(x), \dots, x_n(x)) - \varepsilon g_i(x, \dot{x}(x), x_2(x), \dot{x}_2(x) \dots)] \\ + \Pi_{x_i}(x, x_2(x), \dots, x_n(x)) + \varepsilon g_i(x, \dot{x}(x), x_2(x), \dot{x}_2(x) \dots)\}\big|_{x=X_j} = 0 \\ (i = 2, 3, \dots, n;\ j = 1, 2). \tag{2.5}$$

Here $(X, x_2(X), \dots, X_n(X))$ being the trajectory return points. Equations (2.4) in combination with conditions (2.5) allow a univalent determination of $x_i(x)$. Note that equations similar in form and additional boundary conditions were previously considered in the case of conservative systems [2–4, 22, 23].

In the zero approximation $(\varepsilon = 0)$ $\dot{x}_0^2 = 2[h_0 - \Pi(x, 0, \dots, 0)]$, where h_0 is the energy of the nonperturbed conservative system. We shall restrict our consideration of (2.4) to the first approximation in ε:

$$2x_{i1}''[h_0 - \Pi(x, 0, \dots, 0)] + x_{i1}'[-\Pi_x(x, 0, \dots, 0)] \\ + \sum_{k=2}^{n} x_{k1} \Pi_{x_i x_k}(x, 0, \dots, 0) + g_i(x, \dot{x}_0(x), 0, \dots, 0) = 0 \\ (i = 2, 3, \dots, n). \tag{2.6}$$

In this approximation, the boundary conditions are obtained from (2.4) at $\dot{x}_0 = 0$, $x = X_j$ $(j = 1, 2)$:

$$\Big\{x_{i1}'[-\Pi_x(x, 0, \dots, 0)] + \sum_{k=2}^{n} x_{k1} \Pi_{x_i x_k}(x, 0, \dots, 0) \\ + g(x, \dot{x}(x), 0, \dots, 0)\Big\}\Big|_{x=X_{1,2}} = 0 \\ (i = 2, 3, \dots, n). \tag{2.7}$$

The power series method may be applied to find $x_{i1}(x)$.

2.2. The Power Series Method

It is assumed, for example, that the generating conservative system is homogeneous, i.e. Π is an even homogeneous function of the power of $r+1$ in all the variables (r may take the values $r = 1, 3, 5, \dots$).

Since the unperturbed system is homogeneous, the matrix $B = \Pi_{x_i x_k}(x, 0, \dots, 0)$ may be written as $B = b_{ik} X^{r-1}$. Note that owing to the conservative nature of the system, $b_{ik} = b_{ki}$, and the symmetric matrix b_{ik} is reduced to a diagonal form by nondegenerate linear transformation of the positional coordinates [24]. The transformation can be chosen in order to preserve a generating solution in the form (2.1). Therefore, without loss of generality, one can assume that in (2.6) and (2.7) the function $\Pi_{x_i x_k}(x, 0, \dots, 0) = 0$ for $i \neq k$. Hence, the set (2.6),together with the boundary conditions (2.7) is 'split' in the variables x_{i1}.

Having substituted the series

$$x_{i1} = \sum_{j=0}^{\infty} a_{ij} x^j \tag{2.8}$$

into (2.6), one finds that the coefficients a_{ij} are interrelated by the following infinite set of recurrent relationships (it is written out coefficients at x^{r+j}):

$$\begin{aligned} &2h(r+j+2)(r+j+1)a_{i,r+j+2} - j(j+1)2\Pi(1,0,\ldots,0)a_{i,j+1} \\ &\quad - (j+1)\Pi_x(1,0,\ldots,0)a_{i,j+1} + \Pi_{x_i x_i}(1,0,\ldots,0)a_{i,j+1} = \phi_{i,r+j} \\ &\qquad (j = 0,1,2,\ldots), \end{aligned} \tag{2.9}$$

where $\phi_{i,r+j}$ denotes the corresponding terms in the Taylor-series expansions of the functions g_i. The recurrent relationships can be used to express all coefficients a_{ij} in terms of the leading coefficients a_{i0} and a_{i1} $(i = 2, 3, \ldots, n)$. Substituting the series $x_{i1}(x)$ into the boundary conditions (2.7), and in view of the recurrent relationships (2.9), one obtains the equations governing a_{i0}, a_{i1}:

$$\sum_{i=2}^{n} (R_{0i} a_{i0} + R_{1i} a_{i1}) = R_{2i} \quad (i = 2, 3, \ldots, n). \tag{2.10}$$

The determinant of the system may be represented as products of an infinite number of factors [2, 22, 23]

$$\begin{aligned} K_p &= |q_{ij}|, \\ q_{ij} &= \delta_i^j [p(p-1)2\Pi(1,0,\ldots,0) + p\Pi_x(1,0,\ldots,0) - \Pi_{x_i x_i}(1,0,\ldots,0)], \end{aligned} \tag{2.11}$$

where δ_i^j are Kronecker's delta, $p = 0, 1, 2, 3, \ldots$.

When the generating system is linear, conditions (2.11) mean an absence of internal resonances in the generating system.

The procedure presented here is similar to that applied in a conservative case [2, 22, 23], and it is possible to make use of analogous arguments. In particular, a convergence of the series (2.8) can be proved by L'Hospital's rule.

It follows from Poincaré's theorem on the small parameter series expansion [25] that there is a value of $\varepsilon_0 > 0$ such that, for all $|\varepsilon| < \varepsilon_0$, the series in powers of ε of the functions (2.3) converge in the domain; as $\varepsilon = 0$ this solution becomes the trivial generating solution $x_{j0} = 0$ $(j = 2, 3, \ldots, n)$ (see the details in [2, 22, 23]).

Note that the trajectories $x_j(x)$ can be derived not only in terms of power series in x, but also by the method of iteration [2, 23].

2.3. Single-Degree-of-Freedom Autonomous System and a Potentiality Condition

After having obtained a smooth NNM trajectory, the problem of finding a periodic solution reduces to the integration of the following single-degree-of-freedom autonomous system:

$$\ddot{x} + \Pi_x(x, x_2(x), \ldots, x_n(x)) + \varepsilon g_1(\dot{x}, x(x), x_2(x), \dot{x}_2(x), \ldots) = 0.$$

The solution $x(t)$ is determined as an inversion of the following quadrature:

$$t + \phi = \frac{1}{\sqrt{2}} \int_{x(0)}^{x} \frac{\mathrm{d}\xi}{(h - \Pi(\xi, x_2(\xi), \dots) - \varepsilon g_1(\xi, \dot{x}(\xi), x_2(\xi), \dots))^{1/2}}, \tag{2.12}$$

where h is the integration constant which has the meaning of an energy constant of the conditional single-degree-of-freedom system realizable along the NNM analytical trajectory; ϕ is a phase of the solution which can be taken to be equal to zero because the original system (2.1) is autonomous.

Next, the equation of phase trajectories is of the form ($\dot{x} = v$)

$$\frac{\mathrm{d}v}{\mathrm{d}x} = -\frac{\Pi_x(x, x_2(x), \dots, x_n(x)) + \varepsilon g_1(x, v, x_2(x), x_2'(x)v, \dots)}{v}.$$

In the general case, one has to employ approximate methods to find all phase trajectories and due allowance is to be made for the small values of ε. However, this equation allows exact integration if g_1 is a linear function with respect to $w = v^2$.

At the analytical closed phase trajectory of the limiting cycle, the system's behavior is similar to that of a conservative system with one degree of freedom. Therefore, the condition that the work of all forces over the period is equal to zero should hold (the condition can be called a *potentiality condition*):

$$\oint g_1(x, \dot{x}(x), x_2(x), \dot{x}_2(x), \dots)\, \mathrm{d}x = 0$$

or

$$\int_0^T g_1(x, \dot{x}(x), x_2(x), \dot{x}_2(x), \dots)\dot{x}\, \mathrm{d}t = 0. \tag{2.13}$$

The integral is taken over the period of vibrations. Equation (2.8) is used to find the value of the energy h_0 (and, accordingly, the amplitude values of X_j) for the generating solution of the conservative system, i.e. the limiting solution (at ε tending to zero) for the required NNMs of the self-excited system (2.1).

3. Self-Excited Systems: Examples

3.1. Two Linearly Connected Van der Pol Oscillators

As a simple example, let us consider two linearly connected Van der Pol oscillators with nonlinear elastic characteristics:

$$\begin{aligned} &\ddot{q}_1 + q_1 + \alpha q_1^3 + \beta(q_1 - q_2) - \varepsilon(1 - q_1^2)\dot{q}_1 = 0, \\ &\ddot{q}_2 + q_2 + \alpha q_2^3 + \beta(q_2 - q_1) - \varepsilon\gamma(1 - q_2^2)\dot{q}_2 = 0. \end{aligned} \tag{3.1}$$

Here ε is a small parameter. Two vibration modes may obtained in the zero approximation by ε: $q_2 = \pm q_1$. Let us introduce new coordinates x, y as follows:

$$q_1 + q_2 = x, \quad q_1 - q_2 = y.$$

In these coordinates, one of the NNMs, the so-called *in-phase* mode is governed by $y_0 = 0$, $x_0 = x_0(t)$ in the zeroth approximation by ε, and

$$\ddot{x}_0 = -x_0 - \frac{\alpha}{4}x_0^3, \quad \dot{x}_0^2 = 2\left(h - \frac{x_0^2}{2} - \frac{\alpha x_0^4}{16}\right), \tag{3.2}$$

where h is the system energy constant.

An equation describing the trajectory $y(x)$ in the first approximation with respect to ε has the form

$$y_1''\dot{x}_0^2 + y_1'\ddot{x}_0 + y_1\left(1 + 2\beta + \frac{3\alpha}{4}x_0^2\right) - \left(1 - \frac{x_0^2}{4}\right)\frac{\dot{x}_0}{2}(1-\gamma) = 0. \tag{3.3}$$

The equation should be solved together with the boundary conditions of the form (2.5) (at $\dot{x}_0 = 0$, $x_0 = Q$):

$$y_1'(Q)\ddot{x}_0(Q) + y_1(Q)\left(1 + 2\beta + \frac{3\alpha}{4}Q^2\right) = 0. \tag{3.4}$$

Substituting the solution $y(x)$ as a power series into (3.3), we obtain a set of recurrent equations in the expansion coefficients. (Note that $\dot{x}_0$ must be presented as a power series by x_0 too). One must also use the condition (3.4). Let $\alpha = 1.5$; $\beta = 1.5$; $\gamma = 0.5$, $\varepsilon = 0.2$. In the first approximation, one obtains the trajectory of the in-phase vibration mode:

$$y_1 \simeq -0.07203 + 0.01330x^2 - 0.000077x^4.$$

Here $x(0) = Q = 4$. The last value was obtained from the potentiality condition (2.8) in the first approximation with respect to ε in a quasiharmonic approximation.

The second of the NNMs, the so-called *out-of-phase* mode, is governed by $x_0 = 0$, $y_0 = y_0(t)$ in the zeroth approximation by ε; respectively,

$$\ddot{y}_0 = -(1+2\beta)y_0 - \frac{\alpha}{4}y_0^3; \quad \dot{y}_0^2 = 2\left(h - (1+2\beta)\frac{y_0^2}{2} - \frac{\alpha y_0^4}{16}\right). \tag{3.5}$$

The equation describing the trajectory $x(y)$ approximation with respect to ε has the form

$$x_1''\dot{y}_0^2 + x_1'\ddot{y}_0 + x_1\left(1 + \frac{3\alpha}{4}y_0^2\right) - \left(1 - \frac{y_0^2}{4}\right)\frac{\dot{y}_0}{2}(1-\gamma) = 0. \tag{3.6}$$

The equation should be solved together with the boundary conditions of the form (2.7) (at $\dot{y}_0 = 0$, $y_0 = Q$):

$$x_1'(Q)\ddot{y}_0(Q) + x_1(Q)\left(1 + \frac{3\alpha}{4}Q^2\right) = 0. \tag{3.7}$$

Substituting the solution $x(y)$ as a power series into (3.6) and (3.7), one obtains, in the first approximation, the following trajectory of the out-phase vibration mode:

$$x_1 \simeq -0.10179 + 0.00881y^2 - 0.000386y^4.$$

Here $y(0) = Q = 4$. The value was obtained from the potentiality condition as above.

There is good agreement between the approximation solution and the numerical calculations checked by a computer.

3.2. PLATE VIBRATIONS IN THE FLOW OF GAS

Another example refers to the problem of plate vibrations in the flow of gas. A simple, two-degrees-of-freedom model is chosen.

Next, q_1 is the vertical displacement of the bend of the center axis of the plate; q_2 is the angle of rotation around the axis.

Both the restoring force $P(q_1)$ and the restoring moment $M(q_2)$ are nonlinear:

$$P(q_1) = (d_1 + d_2 q_1^2)q_1, \quad M(q_2) = (d_3 + d_4 q_2^2)q_2,$$

where d_1, d_2, d_3, d_4 are coefficients depending on the plate's elastic characteristics.

The aerodynamic lift is

$$A(q_2) = \sigma v^2 \sin 2q_2 \simeq \sigma v^2 \left(2q_2 - \frac{4q_2^3}{3}\right),$$

where v is the flow velocity, and σ is dependent on the flow density and the airfoil aerodynamic chord.

Let ρ be the radius of inertia with respect to the bend (peck) center axis, l is the distance from the center of rigidity to the center of mass, and m is the mass of the airfoil per unit length. The equations of motion for the system are the following:

$$\begin{aligned} & m\ddot{q}_1 - ml\ddot{q}_2 + (d_1 + d_2 q_1^2)q_1 \\ & \quad = \sigma v^2 \left(2q_2 - \frac{4q_2^3}{3}\right) - ml\ddot{q}_1 + m(\rho^2 + l^2)\ddot{q}_2 + (d_3 + d_4 q_2^2)q_2 = 0. \end{aligned} \tag{3.8}$$

The analysis of the NNMs of the system (3.8) is based on the assumption that the values of d_3 and d_4 are close to those of d_1 and d_2 respectively, the sum $\rho^2 + l^2$ is close to unity and l^2 is much smaller than unity.

Make the coordinate transformation,

$$q_1 + q_2 = x\sqrt{2}, \quad q_1 - q_2 = y\sqrt{2}.$$

One obtains the following equations with respect to variables x, y in place of the system (3.8):

$$\begin{aligned} m\ddot{x} &+ d_1 x + 0.5d_2(x^3 + 3xy^2) \\ &= \varepsilon\big\{ml\ddot{x} + \sigma v^2[(x-y) - (x-y)^3/3] + 0.5m(1 - \rho^2 - l^2)(\ddot{x} - \ddot{y}) \\ &\quad + [(d_1 - d_3) + (d_2 - d_4)(x-y)^2/2](x-y)/2\big\}, \\ m\ddot{y} &+ d_1 y + 0.5d_2(y^3 + 3yx^2) \\ &= \varepsilon\big\{-ml\ddot{y} + \sigma v^2[(x-y) - (x-y)^3/3] - 0.5m(1 - \rho^2 - l^2)(\ddot{x} - \ddot{y}) \\ &\quad - [(d_1 - d_3) + (d_2 - d_4)(x-y)^2/2](x-y)/2\big\}, \end{aligned} \tag{3.9}$$

where ε is a formal small parameter characterizing the relative smallness of the right-hand sides of Equations (3.9). Consider the in-phase NNM close to the mode $q_1 = q_2$ or $x = x(t)$, $y = 0$ in the zeroth approximation with respect to ε. Make use of the relations (2.4) and (2.5).

The corresponding equation in the first approximation with respect to ε for the mode $y(x)$ is

$$2y''\left(h_0 - \frac{d_1}{2}x - \frac{d_2}{8}x\right) - y'\left(d_1 x + \frac{d_2}{2}x^3\right) + \left(d_1 + \frac{3}{2}d_2 x^2\right)y$$
$$= \sigma v^2\left(x - \frac{x^3}{3}\right) + \frac{(1-\rho^2-1)^2}{2}\left(d_1 x + \frac{d_2}{2}x^3\right)$$
$$- \left[(d_1 - d_3) + (d_2 - d_4)\frac{x^2}{2}\right]\frac{x}{2},$$

while the boundary conditions ($\dot{x} = 0$, $x = Q$) may be written as

$$\left\{y'\left(-d_1 x - \frac{d_2}{2}x^3\right) + \left(d_1 + \frac{3}{2}d_2 x^2\right)y - \sigma v^2\left(x - \frac{x^3}{3}\right)\right.$$
$$\left.+ \frac{(1-\rho^2-l^2)}{2}\left(d_1 x + \frac{d_2}{2}x^3\right) - \left[(d_1 - d_3) + (d_2 - d_4)\frac{x^2}{2}\right]\frac{x}{2}\right\}\Bigg|_{x=Q},$$

h_0 being the energy of the generating system.

Numerical calculations were performed for $m = 1$, $l = 0.2$, $d_1 = 2$, $d_2 = 1$, $d_3 = 2.3$, $d_4 = 1.1$, $\sigma = 0.5$, $\rho = 1$, and $v^2 = \{0; 0.2; 0.4\}$.

The following expressions were obtained for the NNM $y = y(x)$, $x(0) = Q = 1$ at $\varepsilon = 1$:

$$v^2 = 0: \quad y \simeq 0.0374x + 0.0018x^3 - 0.00014x^5,$$
$$v^2 = 0.2: \quad y \simeq 0.040x + 0.009x^3 - 0.0005x^5,$$
$$v^2 = 0.4: \quad y \simeq 0.125x + 0.022x^3 - 0.001x^5.$$

The potentiality condition (2.8) for the system at hand is identically satisfied for any vibration mode defined by an analytical function $y(x)$ (or $x(y)$), and therefore the NNMs are dependent on two free parameters, namely the vibration amplitude Q and the phase ϕ.

4. Iterative Computation of Quasinormal Vibrations in Nonlinear Viscoelastic Mechanical Systems

Here we will construct periodical solutions of integro-differential equations corresponding to some viscoelastic mechanical systems. One assumes that elastic forces are dominant. A conservative system corresponding to the elastic interactions is selected as generative and the NNMs of the system are selected as generative solutions.

4.1. Construction of Periodical Solutions in Two-Degrees-of-Freedom Nonlinear Model

Analysis of vibrations in nonlinear viscoelastic mechanical systems is associated with studies into integro-differential partial equations [26, 27].

By representing solutions of such equations as linear combinations of coordinate functions with time-dependent coefficients and by using the Bubnov–Galerkin method, sets of nonlinear integro-differential equations are obtained.

Restricting ourselves to two principal coordinate functions, we obtain the following set of equations with respect to the required coefficients of the linear combinations $x_1(t)$, $x_2(t)$:

$$\ddot{x}_i + \Pi_{x_i}(x_1, x_2) = \varepsilon g_i \quad (i = 1, 2), \tag{4.1}$$

where $\Pi = \Pi(x_1, x_2)$ is the potential energy that corresponds to the principal elastic interactions and is assumed to be an analytical function of its arguments; ε is a small parameter. Here and later on, the case of forced resonance will be called Case A, that of self-excited vibrations, Case B.

In Case A

$$g_i = g_{i1}(t, x_1, x_2, \dot{x}_1, \dot{x}_2) + \int\limits_{-\infty}^{t} R_i(t-s) g_{i2}(x_1(s), x_2(s))\, \mathrm{d}s,$$

while in Case B

$$g_i = g_{i1}(x_1, x_2, \dot{x}_1, \dot{x}_2) + \int\limits_{-\infty}^{t} R_i(t-s) g_{i2}(x_1(s), x_2(s))\, \mathrm{d}s,$$

Here, g_{i1}, g_{i2} are analytical functions which, in Case A, are continuous and periodical in t with a period of T; R_i are relaxation kernels that satisfy the following conditions:

$$\frac{\mathrm{d}R_i(\tau)}{\mathrm{d}\tau} < 0, \qquad \int\limits_{-\infty}^{\infty} R_i(\tau)\, \mathrm{d}\tau = 1.$$

We assume that a generating system (at $\varepsilon = 0$),

$$\ddot{x}_i + \Pi_{x_i}(x_1, x_2) = 0,$$

allows similar NNMs of the type

$$x_2 = k x_1, \tag{4.2}$$

wherein the constants k are defined by the algebraic equation

$$k \Pi_{x_i}(x_i, k x_i) = \Pi_{x_2}(x_1, k x_1).$$

Without loss of generality, we assume that after some rotation of the coordinate system (an axis of a new coordinate system is directed along the rectilinear trajectory of solution (4.2)), the generating system takes on the form

$$\ddot{x}_{i0} + \Pi_{x_1}(x_{10}, x_{10}) = 0, \quad x_{20} = 0,$$

i.e., $k = 0$ for the similar NNMs under condideration and the generating solution in Case A is

$$x_{10} = x_{10}(F, \Phi, t + \phi), \quad x_{20} = 0, \tag{4.3}$$

in Case B,

$$x_{10} = x_{10}(F, \Phi, t), \quad x_{20} = 0, \tag{4.4}$$

where F and Φ are two amplitudes of vibrations (two values of x_{10} that correspond to the cusp, $\dot{x}_{10} = 0$); ϕ is the generating solution phase which, in Case B, is arbitrary due to the system being autonomous and therefore is not explicitly introduced into the solution. The amplitudes F and Φ are related to the period of the generating solutions (4.3) and (4.4) by the equality

$$T(F, \Phi) = \sqrt{2} \int_{\Phi}^{F} \frac{\mathrm{d}x}{(h - \Pi(x, 0))^{1/2}},$$

where h is the energy of the generating solution,

$$h = \Pi(F, 0) = \Pi(\Phi, 0).$$

Without loss of generality, one assumes that F is greater or equal to Φ.

Let us look for a solution of the original perturbed system (4.1) in the form

$$x_1 = x_{10} + \xi, \quad x_2 = \eta.$$

In order to find ξ and η, we obtain from (4.1), selecting a linearized part of the system with respect to ξ and η:

$$\ddot{\xi} + p_1(x_{10})\xi = G_1, \quad \ddot{\eta} + p_2(x_{10})\eta = G_2, \tag{4.5}$$

where

$$p_1(x_{10}) = \Pi_{x_1 x_i}(x_{10}, 0), \quad p_2(x_{10}) = \Pi_{x_2 x_2}(x_{10}, 0);$$

$$G_1 = \varepsilon g_1 - \Pi_{x_1} - p_1(x_{10})\xi, \quad G_2 = \varepsilon g_2 - \Pi_{x_2} - p_2(x_{10})\eta,$$

here $g_1, g_2, \Pi_{x_1}, \Pi_{x_2}$ are calculated as $x_1 = x_{10} + \xi$, $x_2 = \eta$.

It is clear that the first equation of (4.5), linearized with respect to ξ and η, is a variational equation for the generating solution. Therefore, fundamental solutions of the linearized system can be obtained by the derivation of the generating solution with respect to the arbitrary parameters, namely the amplitude F and the phase ϕ [15]:

$$w_{11} = \frac{1}{\omega}\frac{\mathrm{d}x_{10}}{\mathrm{d}\phi} = \gamma \dot{x}_{10} \ (\gamma = \text{const.}), \quad w_{12} = \frac{1}{\omega'}\frac{\mathrm{d}x_{10}}{\mathrm{d}F},$$

where

$$\omega = \frac{2\pi}{T(F)}, \quad \omega' = \frac{\mathrm{d}\omega}{\mathrm{d}F}.$$

The first fundamental solution is periodical, while the second one involves a product of some periodical function by t.

A fundamental solutions of the second linearized equation in (4.5) may be obtained for certain classes of potential functions Π in the form of hypergeometric functions or Lamé functions [22, 23, 28]. In other cases, it may be constructed by a power series expansion in x_{10} or by other methods, see [2, 23]. Let us denote the fundamental solutions by w_{21}, w_{22}.

4.2. Iteration Process

In order to find periodical quasinormal solutions of Equations (4.5), we employ here the following iteration process:

$$\xi_{k+1} = \frac{1}{\Delta_1}\int_0^t G_{1k}^{\tau}(w_{11}^{\tau}w_{12}^{t} - w_{12}^{\tau}w_{11}^{t})\,\mathrm{d}\tau + \frac{1}{T\Delta_1}\left(w_{12} - w_{11}\frac{\dot{w}_{12}^{0}}{\dot{w}_{11}^{0}}\right)\int_0^t G_{1k}^{\tau}w_{12}^{\tau}\,\mathrm{d}\tau,$$

$$\eta_{k+1} = \frac{1}{\Delta_2}\int_0^t G_{2k}^{\tau}(w_{21}^{\tau}w_{22}^{t} - w_{22}^{\tau}w_{21}^{t})\,\mathrm{d}\tau + D_{1k}w_{21} + D_{2k}w_{22},$$

$$\dot{\xi}_{k+1} = \frac{1}{\Delta_1}\int_0^t G_{1k}^{\tau}(w_{11}^{\tau}\dot{w}_{12}^{t} - w_{12}^{\tau}\dot{w}_{11}^{t})\,\mathrm{d}\tau + \frac{1}{T\Delta_1}\left(\dot{w}_{12} - \dot{w}_{11}\frac{\dot{w}_{12}^{0}}{\dot{w}_{11}^{0}}\right)\int_0^t G_{1k}^{\tau}w_{12}^{\tau}\,\mathrm{d}\tau,$$

$$\dot{\eta}_{k+1} = \frac{1}{\Delta_2}\int_0^t G_{2k}^{\tau}(w_{21}^{\tau}\dot{w}_{22}^{t} - w_{22}^{\tau}\dot{w}_{21}^{t})\,\mathrm{d}\tau + D_{1k}\dot{w}_{21} + D_{2k}\dot{w}_{22}, \tag{4.6}$$

Here

$$\Delta_i = \begin{vmatrix} w_{i1} & w_{i2} \\ \dot{w}_{i1} & \dot{w}_{i2} \end{vmatrix} = \text{const.}; \qquad f_{ij}^{\theta}(z) = f_{ij}(z)|_{z=\theta};$$

$\xi_k^{(n)}$ $(n = 1, \ldots, 4)$ are substituted into the right-hand sides of the above relationships (4.6) together with ϕ_k (Case A) or F_k (Case B); D_{ik} are arbitrary constants.

Let us use the conditions of T-periodicity of the functions η_{k+1}, $\dot{\eta}_{k+1}$, to find the unknown constants D_{1k}, D_{2k} (the period T is given in Case A and unknown in Case B):

$$D_{1k}(w_{21}^{T} - w_{21}^{0}) + D_{2k}(w_{22}^{T} - w_{22}^{0}) = -\frac{w_{21}^{T}}{\Delta_2}\int_0^T G_{2k}^{\tau}w_{22}^{\tau}\,\mathrm{d}\tau + \frac{w_{22}^{T}}{\Delta_2}\int_0^T G_{2k}^{\tau}w_{21}^{\tau}\,\mathrm{d}\tau,$$

$$D_{1k}(\dot{w}_{21}^{T} - \dot{w}_{21}^{0}) + D_{2k}(\dot{w}_{22}^{T} - \dot{w}_{22}^{0}) = -\frac{\dot{w}_{21}^{T}}{\Delta_2}\int_0^T G_{2k}^{\tau}w_{22}^{\tau}\,\mathrm{d}\tau + \frac{\dot{w}_{22}^{T}}{\Delta_2}\int_0^T G_{2k}^{\tau}w_{21}^{\tau}\,\mathrm{d}\tau. \tag{4.7}$$

Besides: the periodicity conditions for ξ_{k+1} and $\dot{\xi}_{k+1}$ must be written in Case A:

$$\int_0^T G_{1k}^{\tau}w_{11}^{\tau}\,\mathrm{d}\tau = 0,$$

$$T = T(F, \Phi) = \sqrt{w}\int_{\Phi}^{F}\frac{\mathrm{d}x}{(\Pi(F,0) - \Pi(x,0))^{1/2}} \tag{4.8}$$

(the period T is known); $\Pi(F, 0) = \Pi(\Phi, 0)$.

From this set, the phase ϕ may be found in the corresponding approximation of the perturbation method, together with the amplitudes F, Φ of the generating solution.

In Case B, the periodicity conditions for ξ_{k+1} and $\dot{\xi}_{k+1}$ have a slightly different form:

$$\int_0^{T_k} G_{1k}^{\tau} w_{11}^{\tau}\,\mathrm{d}\tau = 0,$$

$$T = T_k = \sqrt{w}\int_{\Phi_k}^{F_k} \frac{\mathrm{d}x}{(\Pi(F_k, 0) - \Pi(x, 0))^{1/2}} \tag{4.9}$$

(the period T_k is not known); $\Pi(F_k, 0) = \Pi(\Phi_k, 0)$.

Equations (4.9) may evaluate the period T_k and the amplitudes F_k, Φ_k of generating solution in the corresponding approximation with respect to ε.

It is assumed that all variables and parameters vary over a limited region, all functions in (4.1) and their derivatives, inclusive of those of third-order, are bounded, satisfy the Lipschitz condition, and w_{ij}, $\dot{w}_{ij}$, Δ_{ij} $(i, j = 1, 2)$ are also bounded; the remaining conditions are given in [29, 30].

For the iterations of (4.6) to converge to the solution of the initial integro-differential equations,the following principal conditions are necessary: the variables ξ_{k+1}, η_{k+1}, $\dot{\xi}_{k+1}$, $\dot{\eta}_{k+1}$ should remain in a bounded region as should the phase ϕ_k and the amplitudes F and Φ in Case A or the period T_k and the amplitudes F_k, Φ_k in Case B. The operators that determine the transformation (4.6) should be contracting; the periodicity conditions (4.8) or (4.9) should be uniquely solvable with respect to ϕ_k, F, Φ or T_k, F_k, Φ_k; and the corresponding operators determining the phase, amplitude, and period of the vibrations should be contracting.

These requirements lead to a set of inequalities that make the small parameter ε bounded above [29, 30]; the inequalities are cumbersome and therefore are not shown here.

In the first approximation of the iteration process, the number of periodic solutions of the set of integro-differential equations is determined by the number of roots of Equations (4.8) in Case A or the number of roots F_k of Equations (4.9) in Case B. When ε tends towards zero, all these solutions tend towards NNMs of the generating conservative (elastic) system.

4.3. Resonance Nonlinear Transversal Vibrations of a Viscoelastic Rod

By way of example, let us consider resonance nonlinear transversal vibrations of a viscoelastic rod. Within the framework of the well-known Kirchhof hypothesis, the equations of motion are taken in the form

$$u_{tt} + EIu_{xxxx} - \frac{EI}{2F}\int_0^I (u_\xi)^2\,]rmd\xi u_{xx}$$

$$= \varepsilon\lambda \sin pt \sum_{k=1}^{\infty} a_k \sin\frac{k\pi x}{1} + \varepsilon \int_{-\infty}^{t} R(t-s)u^3(s)\,\mathrm{d}s,$$

where u is the transversal displacement; x is the longitudinal independent variable; E is the Young modulus; I is the moment of inertia, l is a distance between end points, $R(z)$ is the relaxation kernel, F is an area of the transversal cross-section of the rod, and ε is a small parameter. The rod ends are assumed to be restrained.

Setting the spatial wave formation in the form (m and n are integers)

$$u = x_1(t)\sin\frac{m\pi x}{1} + x_2(t)\sin\frac{n\pi x}{1},$$

by using the standard Bubnov–Galerkin method one obtains:

$$\ddot{x}_1 + \frac{EF}{4}\left[\left(\frac{m\pi}{1}\right)^4 x_1^3 + \left(\frac{m\pi}{1}\right)^2\left(\frac{n\pi}{1}\right)^2 x_1x_2^2\right]$$
$$= -EI\left(\frac{m\pi}{1}\right)^4 x_1 + [\sin pt + J(x_1, x_2)],$$
$$\ddot{x}_2 + \frac{EF}{4}\left[\left(\frac{n\pi}{1}\right)^4 x_2^3 + \left(\frac{m\pi}{1}\right)^2\left(\frac{n\pi}{1}\right)^2 x_2x_1^2\right]$$
$$= -EI\left(\frac{n\pi}{1}\right)^4 x_2 + \varepsilon[\sin pt + J(x_2, x_1)].$$

Here

$$J(u, v) = \int_{-\infty}^{t} R(t-s)\left(\frac{3}{4}u^3(s) + \frac{3}{2}uv^2\right)\mathrm{d}s, \qquad \lambda a_n = \lambda a_m = 1.$$

In the zeroth approximation, $\varepsilon = 0$, there are two nonlinear normal vibration modes with rectilinear trajectories (similar NNMs), $x_1 = 0$ and $x_2 = 0$. A numerical computation using the presented analytical approach was performed by Petrov for a rod of acrylic plastic ST-1 having $E = 6.3 \times 10^9\ \mathrm{N\,m^{-2}}$ and $l = 0.4$ m; the rod had a square cross-section with a side of 0.02 m; the representation $R(\tau) = A\,\mathrm{e}^{-\beta\tau}/\tau^{1-\alpha}$, with $A = 0.0286$, $\beta = 0.05$ and $\alpha = 0.075$ was assumed for the relaxation kernel; $n = 1$, $m = 2$.

Parameters of the resonance modes close to the mode $x_2 = 0$ were calculated in the zeroth and the first approximations with respect to ε. Some variation of the analytical procedure was introduced here, namely, the amplitude F_0 of the zeroth approximation was given and the period T was calculated.

When the amplitude F_0 of the zeroth approximation is equal to 5, the period $T = 1.916 \times 10^{-3}$ s and the phase shift between the zeroth and the first approximation solutions, x_{10} and x_{11} is: $\pi/2 < \phi < \pi$. When $F_0 = 8$, the calculations yield $T = 1.197 \times 10^{-3}$ s and the phase shift between x_{10} and x_{11} lies within the above range.

5. Conclusions and Discussion

At the present time, it is known that NNMs are typical periodical solutions in n-degrees-of-freedom nonlinear conservative systems. Moreover, normal or quasinormal vibrations exist in broad classes of nonlinear, near-conservative systems, as was indicated above, in nonautonomous systems (forced resonances), self-excited systems and systems containing viscoelastic interactions which are presented in the corresponding equations in a form of integral operators. It is found that approaches for the analysis of NNMs in conservative systems are applicable to near-conservative nonlinear finite-dimensional systems.

It is fruitful to extend the concept of NNMs to distributed nonlinear systems. One of possible extensions of the concept to continuous oscillators was performed by Vakakis [22]. In the spirit of Rosenberg [3, 4] and Shaw and Pierre [11], the continuous nonlinear normal modes were defined as motions during which all material points of the system vary equiperiodically, vanishing or reaching their extremal values at the same instant in time. This way leads to equations which are similar to the equations considered previously in the case of conservative or near-conservative systems.

Other generalizations of the NNMs to distributed systems are possible too:

1. It is possible to determine NNMs in distributed systems as solutions in separated variables with a uniform time function. The solutions correspond to similar normal modes in finite-dimensional systems with rectilinear trajectories in the configuration space.
2. During NNM, a finite-dimensional system behaves like a single-DOF conservative one. Therefore it is advisable to introduce nonlinear stationary traveling waves with a single phase (these are so-called *simple waves*) as a new generalization of NNMs to the distribution case.
3. There is a coupling of the existence of NNMs and symmetry properties of finite-dimensional systems [22]. It is possible to determine NNMs in a distributed system as solutions which are invariant with respect to the symmetry groups of the system.

Note, finally, that problems of NNMs extending to distributed nonlinear systems as a development of approaches for the analysis of the solutions, are interesting, but no details are available at present.

References

1. Lyapunov, A. M., *The General Problem of the Stability of Notion*, Princeton University Press, Princeton, 1947.
2. Mikhlin, Yu. V., 'Normal vibrations of a general class of conservative oscillators', *Nonlinear Dynamics* **11**, 1996, 1–16.
3. Rosenberg, R. M., 'The normal modes of nonlinear n-degrees-of-freedom systems', *Journal of Applied Mechanics* **30**, 1962, 7–14.
4. Rosenberg, R. M., 'Nonlinear vibrations of systems with many degrees of freedom', in *Advances of Applied Mechanics*, Vol. 9, Academic Press, New York, 1966, pp. 156–243.
5. Rosenberg, R. M. and Kuo, J. K., 'Nonsimilar normal mode vibrations of nonlinear systems having two degrees of freedom', *Journal of Applied Mechanics* **31**, 1964, 283–290.
6. Rand, R., 'A higher order approximation for nonlinear normal modes in two degrees of freedom systems', *International Journal of Non-Linear Mechanics* **6**, 1971, 545–547.
7. Manevich, L. I. and Mikhlin, Yu. V., 'Periodic solutions close to rectilinear normal vibration modes', *Prikladnaja Matematica i Mekhanica (PMM USSR)* **36**, 1972, 1051–1058.
8. Vedenova, E. G., Manevich, L. I., and Pilipchuk, V. N., 'Normal oscillations of a string with concentrated masses on non-linearly elastic support', *Prikladnaja Matematica i Mekhanica (PMM USSR)* **49**, 1985, 203–211.
9. Vakakis, A. F., 'Non-similar normal oscillations in a strongly non-linear discrete system', *Journal of Sound and Vibration* **158**, 1992, 341–361.
10. Vakakis, A. F. and Rand, R., 'Normal modes and global dynamics of a two-degree-of freedom non-linear system. I. Low energies. II. High energies', *International Journal of Non-Linear Mechanics* **27**, 1992, 861–874, 875–888.
11. Shaw, S. W. and Pierre, C., 'Nonlinear normal modes and invariant manifolds', *Journal of Sound and Vibration* **150**, 1991, 170–173.

12. Shaw, S. W. and Pierre, C., 'Normal modes for nonlinear vibratory systems', *Journal of Sound and Vibration* **164**, 1993, 85–124.
13. Nayfeh, A. H. and Nayfeh, S. A., 'On nonlinear modes of continuous systems', *Journal of Vibration and Acoustics* **116**, 1994, 129–136.
14. Mikhlin, Yu. V., 'Matching of local expansions in the theory of nonlinear vibrations', *Journal of Sound and Vibration* **182**, 1995, 577–588.
15. Malkin, I. G., *Certain Problems of the Theory of Nonlinear Vibrations*, Geotechteorizdat, Moscow, 1956.
16. Kinney, W. and Rosenberg, R. M., 'On the steady state vibrations of nonlinear systems with many degrees of freedom', *Journal of Applied Mechanics* **33**, 1966, 406–412.
17. Mikhlin, Yu. V., 'Resonance modes of near-conservative nonlinear systems', *Prikladnaja Matematica i Mekhanica (PMM USSR)* **38**, 1974, 425–429.
18. Yang, T.-L. and Rosenberg, R. M., 'On forced vibrations of a particle in the plane', *International Journal of Non-Linear Mechanics* **3**, 1968, 47–63.
19. Szemplinska-Stupnicka, W., 'The modified single mode method in the investigation of the resonant vibration of nonlinear system', *Journal of Sound and Vibration* **63**, 1979, 475–489.
20. Vakakis, A. F. and Caughey, T. K., 'A theorem on the exact nonsimilar steady state motions of a nonlinear oscillator', *Journal of Applied Mechanics* **59**, 1992, 418–424.
21. Vakakis, A. F., Manevich, L. I., Mikhlin, Yu. V., Pilipchuk, V. N., and Zevin, A. A., *Normal Modes and Localization in Nonlinear Systems*, Wiley, New York, 1996.
22. Manevich, L. I., Mikhlin, Yu. V., and Pilipchuk, V. N., *The Method of Normal Oscillation for Essentially Nonlinear Systems*, Nauka, Moscow, 1989.
23. Bellman, R., *Introduction to Matrix Analysts*, McGraw-Hill, New York, 1960.
24. Poincaré, H., *Les Méthodes Nouvelles de la Mécanique Celeste*, Gauthier-Villars, Paris, 1960.
25. Ilushin, A. A. and Pobedria, B. E., *Foundations of the Mathematical Theory of the Thermoviscouselasticity*, Nauka, Moscow, 1970.
26. Morgunov, B. I., 'Asymptotic methods for solution of some problems of nonlinear mechanics', in *Methods of Analysis of Nonlinear Systems*, Yu. Mitropolsky and A. Samoilenko (eds.), Mathematical Institute (Ukrainian Academy of Science) Press, Kiev, 1976, pp. 5–73.
27. Mikhlin, Yu. V. and Zhupiev, A. L., 'An application of the Ince algebraization to the stability of nonlinear normal vibration modes', *International Journal of Non-Linear Mechanics* **32**, 1997, 493–509.
28. Dmitrenko, E. M., Mikhlin, Yu. V., and Morgunov, B. I, 'Iterative calculation of self-induced oscillations of essentially non-linear viscoelastic systems, in *Dynamical Systems*, Vol. 5, Yu. Shevliakov (ed.), Naukova Dumka, Kiev, 1986, pp. 86–90.
29. Mikhlin, Yu. V., 'Normal vibrations of finite-dimensional non-linear systems', Dr.Sci. Thesis, Moscow Institute for Problems in Mechanics (Russian Academy of Science), Moscow, 1988.

Nonlinear Dynamics **25:** 49–64, 2001.

Nonlinear Normal Modes in a System with Nonholonomic Constraints

R. H. RAND and D. V. RAMANI
Department of Theoretical and Applied Mechanics, Cornell University, Ithaca, NY 14853, U.S.A.

(Received: 22 September 1998; accepted: 8 September 1999)

Abstract. We investigate the dynamics of a system involving the planar motion of a rigid body which is restrained by linear springs and which possesses a skate-like nonholonomic constraint known as Čaplygin's sleigh. It is shown that the system can be reduced to one with $2\frac{1}{2}$ degrees of freedom. The resulting phase flow is shown to involve a curve of nonisolated equilibria. Using second-order averaging, the system is shown to possess two families of nonlinear normal modes (NNMs). Each NNM involves two amplitude parameters. The structure of the NNMs is shown to depart from the generic form in the neighborhood of a 1:1 internal resonance.

Keywords: Nonlinear normal modes, nonholonomic, resonance, averaging.

1. Introduction

Nonlinear normal modes (NNMs) are special periodic motions of multi-degrees-of-freedom systems in which all coordinates achieve zero velocity simultaneously, i.e. vibrations in unison. NNMs have been investigated since the 1950s (see [8] for a thorough literature review), mainly in systems with holonomic constraints. The purpose of this work is to extend previous work by considering NNMs in a system which involves a nonholonomic constraint. (See [1] for a discussion of holonomic *versus* nonholonomic constraints.)

In order to investigate the special complications arising from the nonholonomic constraint, in contrast to complications arising from other sources, we have attempted to choose a problem in which all of the interesting dynamics results solely from the nonholonomic constraint. Thus we have chosen a problem which would consist of three uncoupled linear oscillators in the absence of the nonholonomic constraint. As we will see, the presence of the constraint produces a nonlinear system with $2\frac{1}{2}$ degrees of freedom. In choosing the constraint, we sought the simplest possible nonholonomic constraint. Following Niemark and Fufaev [4], we chose a constraint which corresponds to the presence of a skate on a moving body, that is a device which prevents the motion of the body in a direction perpendicular to that of the skate. Niemark and Fufaev refer to this constraint as *Čaplygin's sleigh.* This constraint has also been used to model the presence of fins on an underwater missile [7] and locked wheels on a skidding automobile [3, p. 232].

Niemark and Fufaev [4] studied the nature of small oscillations about equilibrium in an N-degrees-of-freedom system with M nonholonomic constraints. They showed that every such system possesses a manifold S of nonisolated equilibrium states having dimension $\geq M$. They also showed that when the equations of motion are linearized about an equilibrium point, the resulting system always possesses a zero eigenvalue of order $\geq M$. These results are related in that displacement on the manifold S corresponds to moving from one equilibrium to another,

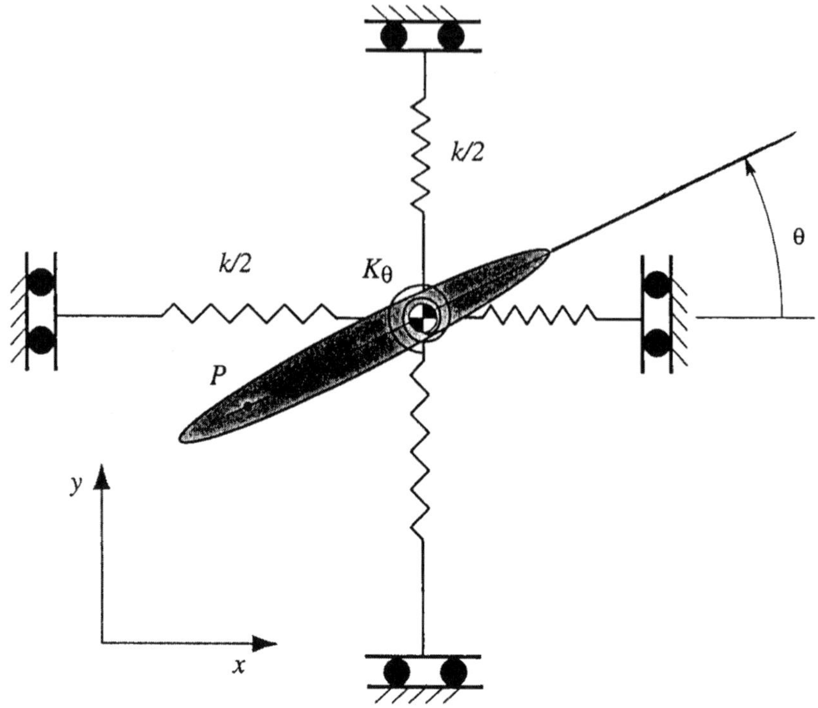

Figure 1. The system investigated in this work involves the planar motion of a body B which is restrained by linear springs and which possesses a skate-like nonholonomic constraint known as Caplygin's sleigh. The springs in the x and y directions are assumed to have moveable anchors which permit smooth motion in a direction perpendicular to the length of the spring. This assumption keeps the restoring forces linear, avoiding the nonlinearities which would accompany fixed anchors.

leading to a zero eigenvalue. To show this, Niemark and Fufaev changed coordinates to a system with variables parallel and perpendicular to the manifold of equilibrium states S. This allowed motions perpendicular to S to be uncoupled from motions in S. Niemark and Fufaev concluded that the occurrence of such zero eigenvalues does not represent a critical case.

In investigating the dynamics of the proposed system, we shall be especially interested in the NNMs and their stability.

2. Derivation of Equations of Motion

The problem we treat consists of a body B which moves in the xy plane, see Figure 1. The configuration of B is described by giving the x, y coordinates of the center of mass of B, as well as the angle θ which a skate attached to B at a point P makes with the x-axis. As B moves, it is required that the velocity components v_x and v_y of B at P satisfy the following nonholonomic constraint:

$$v_y = v_x \tan\theta. \tag{1}$$

This constraint (*Čaplygin's sleigh*, [4]) is due to the presence of the skate, which prevents motion at P in a direction perpendicular to the skate. It may be simplified by using the relations (Figure 1):

$$v_x = \dot{x} + d\,\dot{\theta}\sin\theta, \quad v_y = \dot{y} - d\,\dot{\theta}\cos\theta, \tag{2}$$

where d is the distance from the point P of attachment of the skate to the center of mass and dots represent differentiation with respect to time t. Then Equation (1) becomes:

$$\sin\theta\,\dot{x} - \cos\theta\,\dot{y} + d\,\dot{\theta} = 0. \tag{3}$$

In addition to the forces of constraint, the body B is assumed to be restrained by three sets of linear springs, attached to the center of mass of B, which respectively apply restoring forces in the x and y directions, and a restoring torque in the θ direction, see Figure 1. These springs have net spring constants k, k, and K_θ respectively. The identical x and y springs are assumed to have moveable anchors which permit smooth motion in a direction perpendicular to the length of the spring. This assumption keeps the restoring forces linear, avoiding the nonlinearities which would accompany fixed anchors.

This system has the simple Lagrangian

$$L = \frac{1}{2}m\dot{x}^2 + \frac{1}{2}m\dot{y}^2 + \frac{1}{2}I\dot{\theta}^2 - \frac{1}{2}kx^2 - \frac{1}{2}ky^2 - \frac{1}{2}K_\theta\theta^2, \tag{4}$$

where m is the mass of B and I is moment of inertia of B about its center of mass. In the absence of the nonholonomic constraint (3), this Lagrangian yields three simple harmonic oscillators. When the constraint (3) is included, we may obtain the equations of motion for the system by using Lagrange's equations [1]:

$$m\ddot{x} + kx = \lambda\sin\theta, \tag{5}$$

$$m\ddot{y} + ky = -\lambda\cos\theta, \tag{6}$$

$$I\ddot{\theta} + K_\theta\theta = \lambda d, \tag{7}$$

where λ is a Lagrange multiplier. Equations (3), (5), (6), (7) may be reduced to a fifth order system as follows: First differentiate the constraint (3), solve for $\ddot{\theta}$ and substitute into (7). Then solve (7) for λ and substitute into (5) and (6). Finally use the constraint (3) to eliminate $\dot{\theta}$. The resulting equations may be simplified by using the following nondimensionalization scheme:

$$X = \frac{x}{d}, \quad Y = \frac{y}{d}, \quad \tau = \sqrt{\frac{k}{m}}\,t, \tag{8}$$

$$\mu = \frac{I}{I + md^2}, \quad K = \frac{K_\theta\,m}{k(I + md^2)}. \tag{9}$$

In order to reduce the computational complexity while still maintaining the spirit of the problem, we shall take $\mu = \frac{1}{2}$, that is $I = md^2$, in what follows. The equations of motion become:

$$4X'' = -(3 + \cos 2\theta)X + \frac{1}{2}(\cos\theta - \cos 3\theta)(X'^2 - Y'^2) + (\sin\theta - \sin 3\theta)X'Y' - Y\sin 2\theta + 4K\theta\sin\theta, \tag{10}$$

$$4Y'' = -(3 - \cos 2\theta)Y + \frac{1}{2}(\sin\theta + \sin 3\theta)(Y'^2 - X'^2) + (\cos\theta + \cos 3\theta)X'Y' - X\sin 2\theta - 4K\theta\cos\theta, \tag{11}$$

$$\theta' = -X'\sin\theta + Y'\cos\theta, \tag{12}$$

where primes represent differentiation with respect to τ. Equations (5), (6), (7) are nondissipative and possess a first integral corresponding to conservation of energy:

$$T + V = \frac{1}{2}m\dot{x}^2 + \frac{1}{2}m\dot{y}^2 + \frac{1}{2}I\dot{\theta}^2 + \frac{1}{2}kx^2 + \frac{1}{2}ky^2 + \frac{1}{2}K_\theta\theta^2 = \text{constant}. \tag{13}$$

Transforming to dimensionless variables (8), (9), and using (12) to eliminate θ', Equation (13) can be written in the form:

$$\begin{aligned}&(3 - \cos 2\theta)X'^2 + (3 + \cos 2\theta)Y'^2 - 2X'Y' \sin 2\theta \\ &\quad + 2X^2 + 2Y^2 + 4K\theta^2 = \text{constant}.\end{aligned} \tag{14}$$

Equations (10), (11), (12) represent a flow on the five-dimensional phase space with coordinates X, Y, θ, X', Y'. The energy manifold (14) represents an invariant codimension one surface in this phase space. Thus the system we are investigating may be thought of as a $2\frac{1}{2}$-degrees-of-freedom nonlinear conservative dynamical system.

3. Equilibria

We begin the analysis of Equations (10), (11), (12) by looking for equilibria and considering their stability. Setting all derivatives to zero in Equations (10), (11), (12), we obtain the following conditions for an equilibrium point (X_0, Y_0, θ_0):

$$-(3 + \cos 2\theta_0)X_0 - Y_0 \sin 2\theta_0 + 4K\theta_0 \sin\theta_0 = 0, \tag{15}$$

$$-(3 - \cos 2\theta_0)Y_0 - X_0 \sin 2\theta_0 - 4K\theta_0 \cos\theta_0 = 0. \tag{16}$$

Equations (15), (16) are two algebraic equations in three unknowns X_0, Y_0, θ_0. We can solve for X_0 and Y_0 in terms of θ_0:

$$X_0 = 2K\theta_0 \sin\theta_0, \quad Y_0 = -2K\theta_0 \cos\theta_0. \tag{17}$$

For a given value of θ_0, Equations (17) give the coordinates of an equilibrium point and thus represent a curve of nonisolated equilibria. This is a major departure from the holonomic case, where equilibria are typically isolated [4]. In order to better understand the nature of these equilibria, we linearize Equations (17) for small θ_0, giving

$$Y_0 = -2K\theta_0, \quad X_0 = 0. \tag{18}$$

Now the occurrence of these equilibria may be explained on physical grounds by drawing a free-body diagram of the body B, see Figure 2. It is seen that the restoring force from the torsion spring is balanced by a couple produced by the spring force in the y-direction and the constraint force acting on the skate at point P.

In order to investigate the stability of these equilibria, we set

$$X = X_0 + \xi, \quad Y = Y_0 + \eta, \quad \theta = \theta_0 + \zeta, \tag{19}$$

and substitute into Equations (10), (11), (12). Linearizing the resulting equations on ξ, η and ζ, we obtain a system of five coupled linear constant coefficient ODEs. Seeking a solution in the form $e^{\Lambda t}$, we find the following five values for Λ:

$$\Lambda = 0, \quad \Lambda^2 = -1 + O(\theta_0^2), \quad \Lambda^2 = -K - \frac{1}{2} + O(\theta_0^2). \tag{20}$$

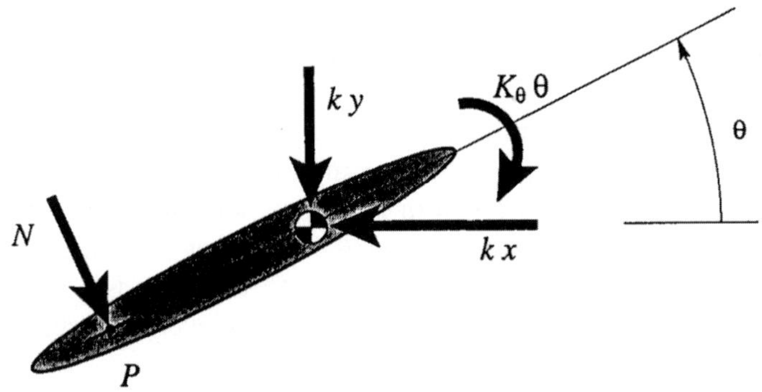

Figure 2. The appearance of nonisolated equilibria may be explained on physical grounds by reference to this free-body diagram, in which it has been assumed that θ_0 is a small angle. Equilibrium in the y-direction gives $ky_0 = -N\cos\theta_0 \approx -N$. Moment balance about the center of mass gives $-k_\theta\theta_0 = Nd \approx -ky_0 d$. Solving for θ_0, we obtain $\theta_0 \approx -(y_0/d)(kd^2/k_\theta) \approx -Y_0/(2K)$.

These results show that the equilibria are stable, at least for small values of θ_0. Note that the appearance of a zero eigenvalue Λ is due to the presence of nonisolated equilibria [4].

4. Nonlinear Normal Modes

In order to study the NNMs which occur in Equations (10), (11), (12), we begin by considering the system's linear normal modes. To this end, we linearize Equations (10), (11), (12) about the origin in phase space, that is, for small displacements and velocities.

$$X'' + X = 0, \tag{21}$$

$$Y'' + \frac{1}{2}Y + K\theta = 0, \tag{22}$$

$$\theta' = Y'. \tag{23}$$

These equations have the following general solution:

$$X = u_1 \cos\tau + u_2 \sin\tau, \tag{24}$$

$$Y = v_1 \cos\omega\tau + v_2 \sin\omega\tau - 2K v_3, \tag{25}$$

$$\theta = v_1 \cos\omega\tau + v_2 \sin\omega\tau + v_3, \tag{26}$$

where u_1, u_2, v_1, v_2, v_3 are arbitrary constants, and where

$$\omega^2 = K + \frac{1}{2}. \tag{27}$$

There are therefore two linear normal modes: Equations (24–26) with (i) $v_1 = v_2 = 0$, the X-mode, and (ii) $u_1 = u_2 = 0$, the $Y\theta$-mode. By comparing Equations (25), (26) to the linearized equilibria of Equations (18), we can see that these two linear normal modes are oscillations about a nonisolated equilibrium point. Here v_3 determines the equilibrium point, and u_1, u_2 or v_1, v_2 respectively determine the amplitude of the oscillation for the X-mode or $Y\theta$-mode. Each of these modes thus forms a continuum of periodic solutions, one branch of each mode passing through each nonisolated equilibrium position.

Note that the X-mode involves both Y and θ remaining constant, while X varies periodically. This would appear to violate the nonholonomic constraint (12), since the motion in the X-direction is not parallel to the orientation of the skate. However, this represents an approximate solution valid for small displacements, and is due to the linearization process. The linearized system will have linear normal modes which satisfy the linearized version of the constraint equation, which is given in Equation (23). Note that the torsion spring K_θ (Figure 1) is in unstretched equilibrium when $\theta = 0$, that is, when the axis of the body lies in the x-direction. This permits small motions to occur in the x-direction without an accompanying change in θ. Motions in the y-direction, however, are accompanied by changes in θ.

If nonlinear effects are included in the foregoing discussion, each of the linear normal modes will give rise to a corresponding NNM. We will now use the method of averaging to investigate the approximation of these NNMs and their stability, as well as the general behavior of the system, valid for small displacements.

5. Averaging

In order to use the method of averaging on Equations (10), (11), (12) we scale the variables X, Y and θ by introducing a small parameter $\varepsilon \ll 1$:

$$X = \varepsilon \tilde{X}, \quad Y = \varepsilon \tilde{Y}, \quad \theta = \varepsilon \tilde{\theta}. \tag{28}$$

We substitute (28) into (10), (11), (12) and Taylor expand for small ε. Dropping the tildes, we obtain the truncated system:

$$X'' + X = \varepsilon f_1 \left(X, Y, \theta, X', Y'; \varepsilon\right), \tag{29}$$

$$Y'' + \frac{1}{2} Y + K\theta = \varepsilon f_2 \left(X, Y, \theta, X', Y'; \varepsilon\right), \tag{30}$$

$$\theta' - Y' = \varepsilon f_3 \left(X, Y, \theta, X', Y'; \varepsilon\right), \tag{31}$$

where

$$2f_1 = 2K\theta^2 - Y\theta + \varepsilon\, \theta(X\theta - X'Y') + \cdots, \tag{32}$$

$$2f_2 = -X\theta + X'Y' + \varepsilon\theta \left[Y'^2 - X'^2 - \theta Y + K\theta^2\right] + \cdots, \tag{33}$$

$$2f_3 = -2X'\theta - \varepsilon\, \theta^2 Y' + \cdots. \tag{34}$$

We begin the averaging treatment of Equations (29–34) by using variation of parameters. We assume a solution of the form of Equations (24–26) where u_1, u_2, v_1, v_2, v_3 are now unknown functions of time τ. This gives the following (unaveraged) equations:

$$u_1{}' = -\varepsilon f_1 \left(X, Y, \theta, X', Y'; \varepsilon\right) \sin \tau, \tag{35}$$

$$u_2{}' = \varepsilon f_1 \left(X, Y, \theta, X', Y'; \varepsilon\right) \cos \tau, \tag{36}$$

$$v_1{}' = -\varepsilon \frac{f_2 \left(X, Y, \theta, X', Y'; \varepsilon\right)}{\omega} \sin \omega\tau + \varepsilon \frac{K f_3 \left(X, Y, \theta, X', Y'; \varepsilon\right)}{\omega^2} \cos \omega\tau, \tag{37}$$

$$v_2' = \varepsilon \frac{f_2(X, Y, \theta, X', Y'; \varepsilon)}{\omega} \cos \omega\tau + \varepsilon \frac{K f_3(X, Y, \theta, X', Y'; \varepsilon)}{\omega^2} \sin \omega\tau, \tag{38}$$

$$v_3' = \varepsilon \frac{f_3(X, Y, \theta, X', Y'; \varepsilon)}{2\omega^2}, \tag{39}$$

where $\omega = \sqrt{K + 1/2}$.

Following the method of averaging to $O(\varepsilon^2)$ (see [6] and [5]), we perform a near-identity transformation:

$$u_1 = \bar{u}_1 + \varepsilon W_1 + \varepsilon^2 V_1 + \cdots, \tag{40}$$

$$u_2 = \bar{u}_2 + \varepsilon W_2 + \varepsilon^2 V_2 + \cdots, \tag{41}$$

$$v_1 = \bar{v}_1 + \varepsilon W_3 + \varepsilon^2 V_3 + \cdots, \tag{42}$$

$$v_2 = \bar{v}_2 + \varepsilon W_4 + \varepsilon^2 V_4 + \cdots, \tag{43}$$

$$v_3 = \bar{v}_3 + \varepsilon W_5 + \varepsilon^2 V_5 + \cdots, \tag{44}$$

where $W_i = W_i(\bar{u}_1, \bar{u}_2, \bar{v}_1, \bar{v}_2, \bar{v}_3, \tau)$ and $V_i = V_i(\bar{u}_1, \bar{u}_2, \bar{v}_1, \bar{v}_2, \bar{v}_3, \tau)$ are generating functions which are to be chosen to simplify the resulting equations on the new (barred) variables. We obtain the following equations on the barred variables:

$$\bar{u}_1' = -\varepsilon^2 \bar{u}_2 \left[a_1 \left(\bar{v}_1^2 + \bar{v}_2^2\right)\right], \tag{45}$$

$$\bar{u}_2' = \varepsilon^2 \bar{u}_1 \left[a_1 \left(\bar{v}_1^2 + \bar{v}_2^2\right)\right], \tag{46}$$

$$\bar{v}_1' = -\varepsilon^2 \bar{v}_2 \left[b_1 \left(\bar{u}_1^2 + \bar{u}_2^2\right) + b_2 \left(\bar{v}_1^2 + \bar{v}_2^2\right)\right], \tag{47}$$

$$\bar{v}_2' = \varepsilon^2 \bar{v}_1 \left[b_1 \left(\bar{u}_1^2 + \bar{u}_2^2\right) + b_2 \left(\bar{v}_1^2 + \bar{v}_2^2\right)\right], \tag{48}$$

$$\bar{v}_3' = 0, \tag{49}$$

where

$$a_1 = \frac{(2K+1)(4K^2 - 14K + 3)}{8(1-2K)(4K+1)}, \tag{50}$$

$$b_1 = \frac{2K+5}{16\sqrt{2(2K+1)}(4K+1)}, \tag{51}$$

$$b_2 = \frac{3(2K-1)\sqrt{2K+1}}{16\sqrt{2}(4K+1)}. \tag{52}$$

Note that there are no contributions to $O(\varepsilon)$, thus requiring higher order averaging. Also note that the above treatment fails when $K = 1/2$ due to vanishing denominators. This case, which corresponds to a 1:1 resonance between the linear modes, will be treated separately later. Equations (45–49) simplify when written in polar coordinates:

$$\bar{u}_1 = r_1 \cos \theta_1, \quad \bar{u}_2 = r_1 \sin \theta_1, \tag{53}$$

$$\bar{v}_1 = r_2 \cos \theta_2, \quad \bar{v}_2 = r_2 \sin \theta_2, \tag{54}$$

becoming:

$$r_1{}' = 0, \quad r_2{}' = 0, \quad \bar{v}_3' = 0, \tag{55}$$

$$\theta_1{}' = \varepsilon^2 a_1 r_2^2, \tag{56}$$

$$\theta_2{}' = \varepsilon^2 \left(b_1 r_1^2 + b_2 r_2^2\right). \tag{57}$$

In order to obtain a periodic solution, it is necessary to take either $r_2 = 0$, in which case $v_1 = 0$, $v_2 = 0$, and we obtain the X-mode, or $r_1 = 0$, in which case $u_1 = 0$, $u_2 = 0$, and we obtain the $Y\theta$-mode. The resulting expressions for X, Y and θ are very complicated algebraically to $O(\varepsilon^2)$ because they involve the generating functions W_i of the near-identity transformation. In order to reduce the algebra, we give expressions for these modes for the case of $K = 1$:

X-mode:

$$X = \varepsilon r_1 \cos\tau + 2\varepsilon^2 v_3^2, \tag{58}$$

$$Y = -2\varepsilon v_3 + \varepsilon^2 r_1 v_3 \cos\tau, \tag{59}$$

$$\theta = \varepsilon v_3. \tag{60}$$

$Y\theta$-mode:

$$X = \varepsilon^2(-5r_2 v_3 \cos\Omega\tau - 0.05 r_2^2 \cos 2\Omega\tau + 0.25 r_2^2 + 2v_3^2), \tag{61}$$

$$Y = \varepsilon\left(r_2 \cos\Omega\tau - 2v_3\right), \tag{62}$$

$$\theta = \varepsilon\left(v_3 + r_2 \cos\Omega\tau\right), \tag{63}$$

where

$$\Omega = 0.04593\varepsilon^2 r_2^2 - 1.22474. \tag{64}$$

Note that each NNM has two parameters: r_i and $\bar{v}_3$. These may be thought of as amplitude parameters: (i) $\bar{v}_3$ is a static amplitude, representing the nonisolated equilibrium point about which the oscillation occurs, and (ii) r_1 (or r_2) is a dynamic amplitude, representing the size of the oscillation relative to the nonisolated equilibrium.

Both NNMs are stable to this order of approximation. E.g. in the case of the X-mode, which is defined by taking $r_2 = 0$, permitting a small value of r_2 produces a solution which lies close to the $r_2 = 0$ solution. Similarly for the $Y\theta$-mode.

In order to compare the predictions of the averaging solution with that of numerical integration, we set $\tau = 0$ in Equations (58–63) to obtain the initial conditions for a NNM (with $X' = Y' = 0$):

X-mode:

$$X = \varepsilon r_1 + 2\varepsilon^2 \bar{v}_3^2,$$

$$Y = -2\varepsilon\bar{v}_3 + \varepsilon^2 r_1 \bar{v}_3^2,$$

$$\theta = \varepsilon\bar{v}_3. \tag{65}$$

$Y\theta$-mode:

$$\begin{aligned} X &= \varepsilon^2 \left(2v_3^2 - 5r_2\bar{v}_3 + \frac{r_2^2}{5} \right), \\ Y &= \varepsilon \left(r_2 - 2\bar{v}_3 \right), \\ \theta &= \varepsilon(\bar{v}_3 + r_2). \end{aligned} \tag{66}$$

See Figures 3 and 4 where the initial condition surfaces (65), (66) ('grenz-surfaces' after [2]) are displayed along with the results of numerical integration of the equations of motion (10–12). The initial conditions were obtained using numerical integration by fixing two of the three initial conditions and varying the third until a NNM was achieved. Also shown are the initial condition planes which give rise to the associated linear normal modes.

6. 1:1 Resonance

Our derivation of Equations (45–49) is not valid for $K = 1/2$ due to the presence of vanishing denominators in Equations (50–52). This case corresponds to a 1:1 resonance between the two frequencies of the unperturbed system, $\omega = 1$. In order to investigate this case, we set

$$K = \frac{1}{2} + \varepsilon\Delta, \tag{67}$$

where Δ is a detuning parameter. In this case $O(\varepsilon)$ averaging gives the following slow-flow:

$$u_1' = -\frac{\varepsilon}{2} v_2 v_3, \tag{68}$$

$$u_2' = \frac{\varepsilon}{2} v_1 v_3, \tag{69}$$

$$v_1' = \frac{\Delta\varepsilon v_2}{2}, \tag{70}$$

$$v_2' = -\frac{\Delta\varepsilon v_1}{2}, \tag{71}$$

$$v_3' = \frac{\varepsilon}{4} \left(u_1 v_2 - u_2 v_1 \right), \tag{72}$$

where the bars have been dropped for convenience.

The fixed points of Equations (68–72) give the NNMs of the resonant system. One fixed point is obtained by taking $v_1 = v_2 = 0$, which represents the X-mode. No other fixed points exist in the system, and so the $Y\theta$-mode has been lost. However, the remnant of the $Y\theta$-mode is a quasiperiodic motion corresponding to $u_1 = u_2 = v_3 = 0$. In this case, v_1 and v_2 are oscillatory, while there is no X motion.

Equations (70) and (71) are uncoupled from the other equations in the system and give rise to oscillatory solutions. Furthermore, the system possesses the following first integrals

$$v_1^2 + v_2^2 = C_1, \tag{73}$$

$$u_1 v_1 + u_2 v_2 - 2\Delta v_3 = C_2, \tag{74}$$

where C_1 and C_2 are constants of integration.

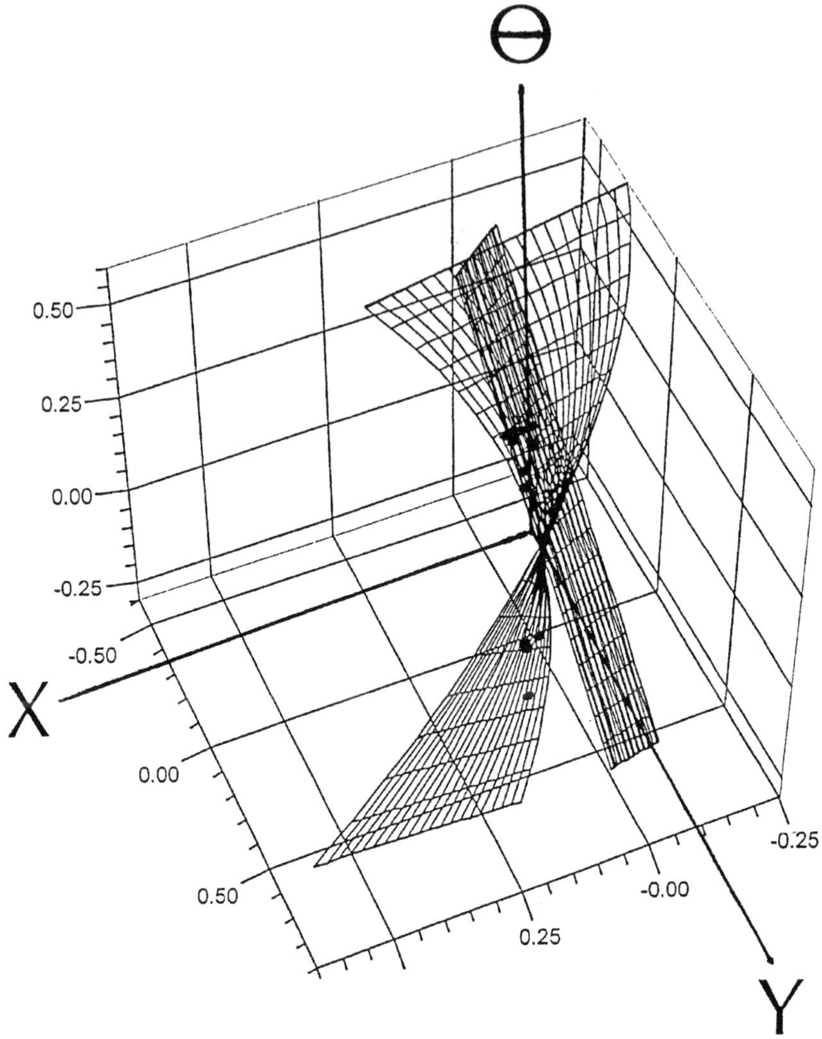

Figure 3. Initial condition surface (65) for the X-mode. The dots represent initial conditions obtained directly from the equations of motion (10–12). Also shown is the plane of initial conditions which produces the linear X-mode.

The system of Equations (68–72) can now be solved in closed form as follows. Because of the decoupling, oscillatory solutions are obtained for v_1 and v_2. Differentiating Equation (72) with respect to τ and then using the first integrals, Equations (73) and (74), the following equation is obtained on v_3

$$v_3'' + \frac{\varepsilon^2}{8}(C_1 + 2\Delta^2)v_3 = -\frac{\varepsilon^2}{8}C_2. \tag{75}$$

Equation (75) represents a driven harmonic oscillator with a constant driving force, and may be solved in closed form. Finally, since all the v_i are now known, u_1 and u_2 may be found by direct integration of Equations (68) and (69).

A comparison of solutions generated by numerical integrations and by the averaging is shown in Figure 5, for the case of $K = 0.50$ corresponding to $\Delta = 0$, with $\varepsilon = 0.1$. The

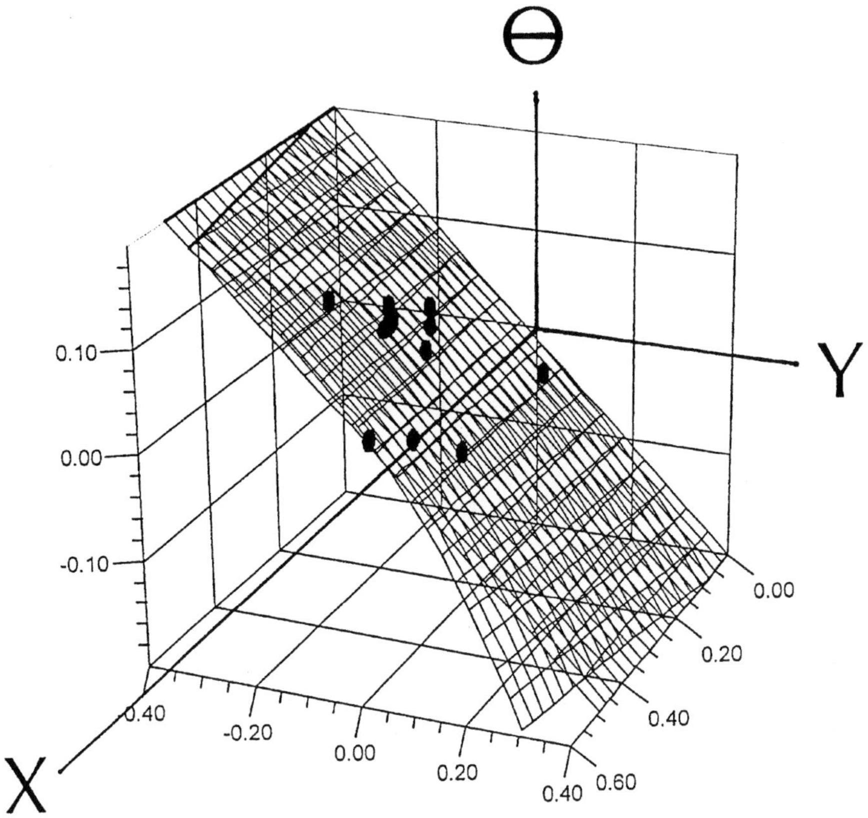

Figure 4. Initial condition surface (66) for the $Y\theta$-mode. The dots represent initial conditions obtained directly from the equations of motion (10–12). Also shown is the plane of initial conditions which produces the linear $Y\theta$-mode.

two curves lie on top of each other, demonstrating the accuracy of the averaged solution. As the value of Δ is moved away from 0, the level of agreement decreases. This can be seen in Figure 6, for the case of $K = 0.52$ corresponding to $\Delta = 0.2$ and the same value of ε, where the agreement in X is still quite good, but the agreement in Y and θ begins to break down around $\tau = 30$. The averaged solution has the correct amplitude of the oscillation but the wrong frequency.

It is clear from the nature of the solutions for the v_i that the original system is bounded in the neighborhood of the resonance. To further clarify the nature of the resonance, we present the results of numerical integration of the equations of motion (10–12) in Figures 7–10. Note that these figures show a projection of the orbits in the five-dimensional phase space down onto the X–Y plane. For general initial conditions, the resulting motion is quasiperiodic and will fill a region of the plane.

Figure 7 is for $K = 0.4$, below the resonance. The initial conditions were chosen to generate the X-mode of the system. Figure 8, also for $K = 0.4$, has initial conditions chosen to generate the $Y\theta$-mode. Figures 9 and 10 show the corresponding numerical integrations for the resonant case, $K = 0.50$. The blackened area of Figure 10 represents a quasiperiodic motion in the plane. As the parameter K is varied from 0.40 to 0.50, Figure 8 morphs into Figure 10. The instability due to the 1:1 resonance thus represents the destruction of the $Y\theta$-

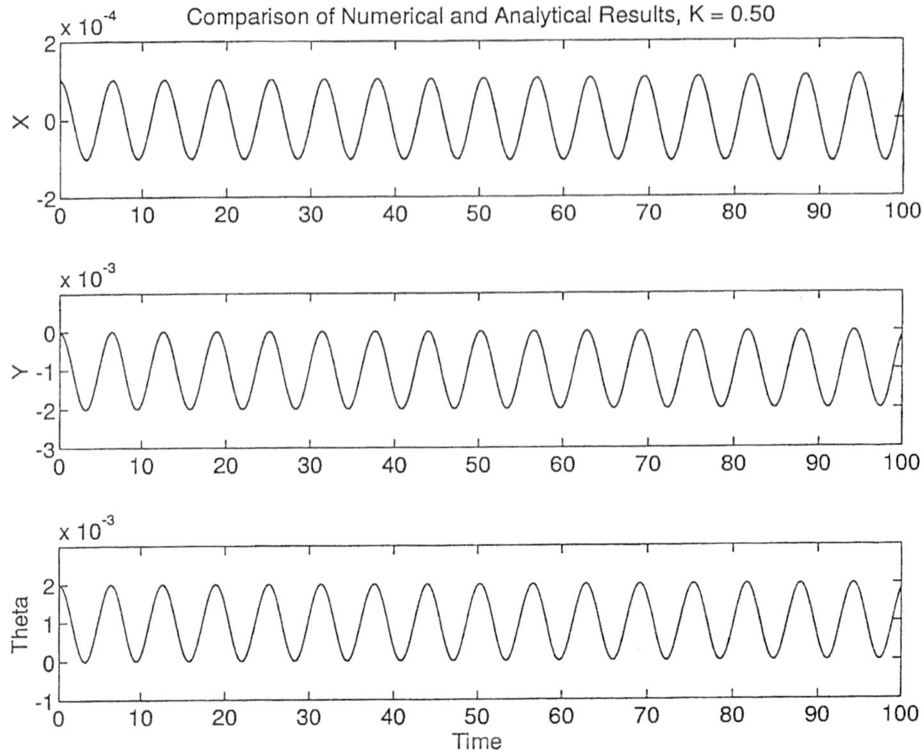

Figure 5. Comparison of time histories of solutions generated by numerical integration of Equations (10–12) and by averaging for $K = 0.50$. Solid lines represent numerical integrations, dashed lines represent averaging. Here the two curves lie atop each other.

mode and a loss of localization in phase space. Motions which for nonresonant parameter values remain in a region of phase space which includes a neighborhood of the $Y\theta$-mode, expand their orbits and visit a larger region for parameters close to the 1:1 resonance.

7. Conclusions

In contrast to a holonomic constraint, which reduces the size of a system by a full degree of freedom, a nonholonomic constraint has been seen to reduce the size of a system by half of a degree of freedom. Thus although the nonholonomic system conserves energy (in the absence of dissipation), it cannot be placed into the framework of a Hamiltonian system. The well-known result for holonomic systems, that NNM's depart from the energy manifold orthogonally in configuration space, no longer holds for nonholonomic systems. This is because the nature of the nonholonomic constraint is to restrict the possible velocity vectors at any point of configuration space, and such a restriction will not, in general, be consistent with a a direction normal to the energy manifold.

Niemark and Fufaev [4] have previously noted that a nonholonomic system has no isolated equilibria, possessing instead a manifold of equilibria. In contrast, the typical holonomic system has isolated equilibria. This difference of structure in the phase space has implications for the NNMs. In the holonomic case, the NNM's frequency is dependent on the mode's 'amplitude'. While such a dependence is seen in the nonholonomic system, there are two am-

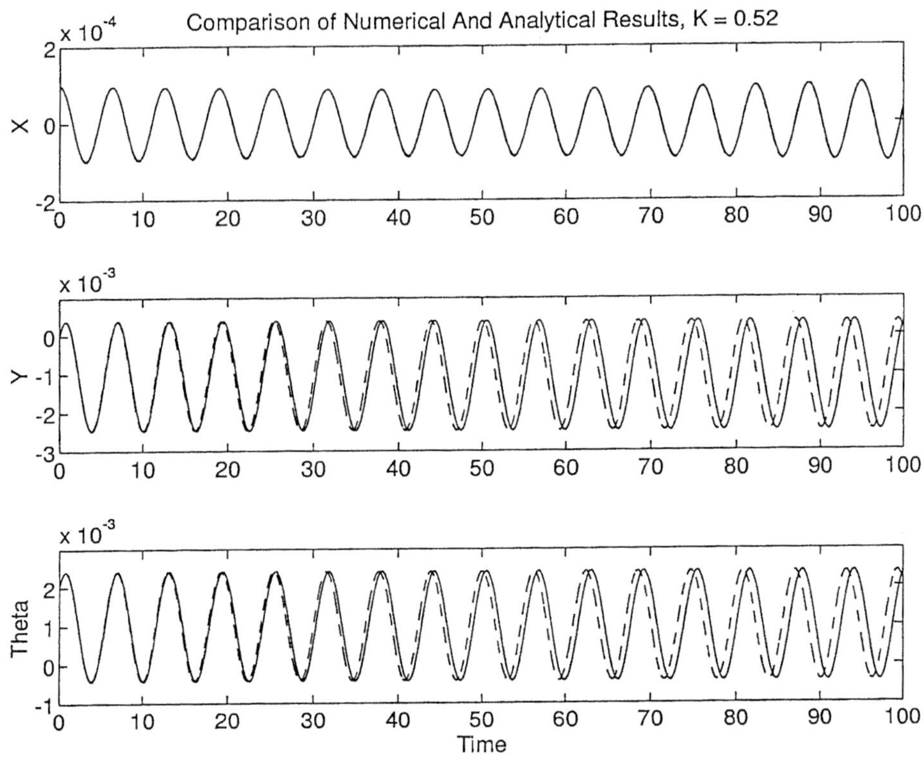

Figure 6. Comparison of time histories of solutions generated by numerical integration of Equations (10–12) and by averaging for $K = 0.52$. Solid lines represent numerical integrations, dashed lines represent averaging.

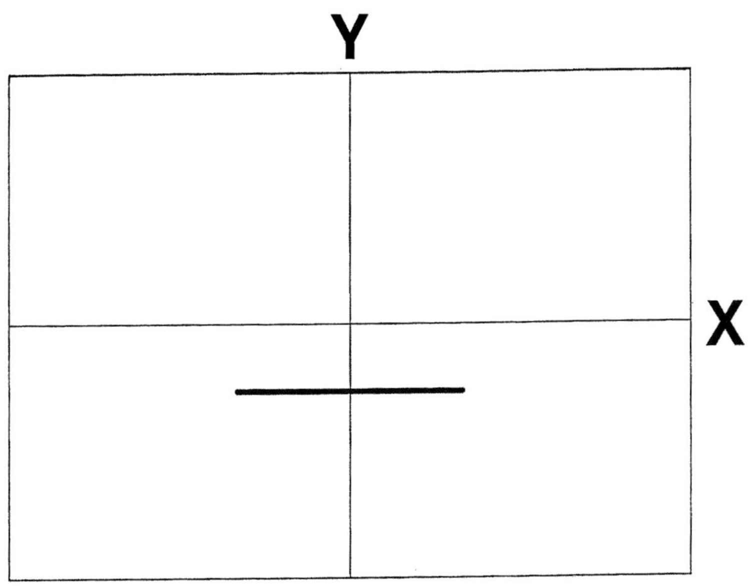

Figure 7. Numerical integration of Equations (10–12) for $K = 0.4$ displayed in the X–Y plane. This value of K lies below the resonance. Initial conditions have been chosen to obtain a motion which lies near the X-mode, namely $u_1 = v_3 = 0.001$, $u_2 = v_1 = v_2 = 0$ in Equations (24–26). Region displayed is $|X| < 0.003$, $|Y| < 0.003$.

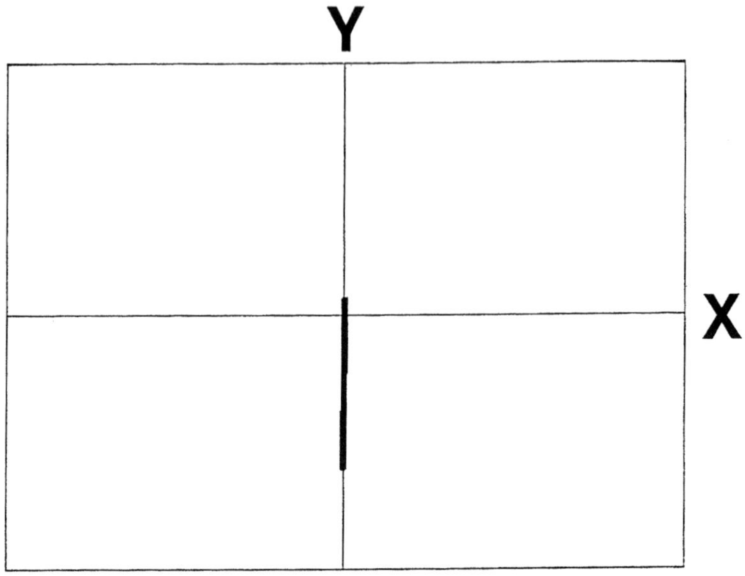

Figure 8. Numerical integration of Equations (10–12) for $K = 0.4$ displayed in the X–Y plane. This value of K lies below the resonance. Initial conditions have been chosen to obtain a motion which lies near the $Y\theta$-mode, namely $v_1 = v_3 = 0.001$, $u_1 = u_2 = v_2 = 0$ in Equations (24–26). Region displayed is $|X| < 0.003$, $|Y| < 0.003$.

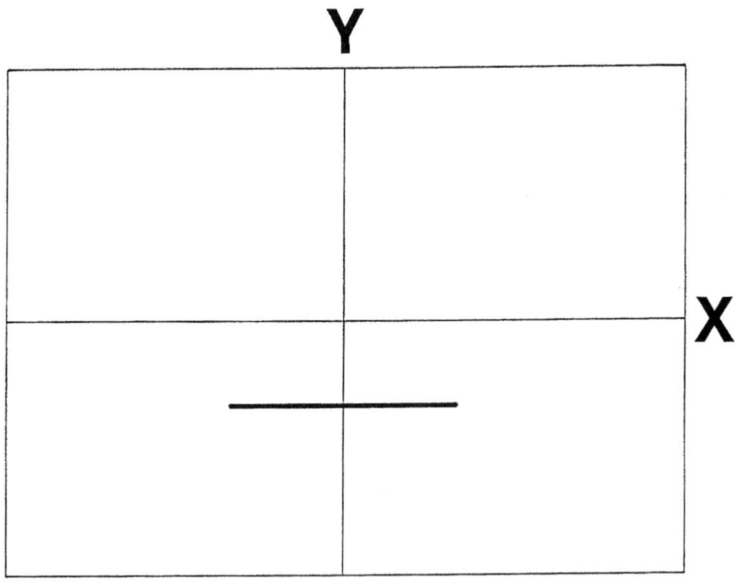

Figure 9. Numerical integration of Equations (10–12) for $K = 0.5$ displayed in the X–Y plane. This value of K corresponds to exact 1:1 resonance. Initial conditions were chosen which correspond to those in Figure 7, namely $u_1 = v_3 = 0.001$, $u_2 = v_1 = v_2 = 0$ in Equations (24–26). As shown in the text, the 1:1 resonance preserves the X-mode. Region displayed is $|X| < 0.003$, $|Y| < 0.003$.

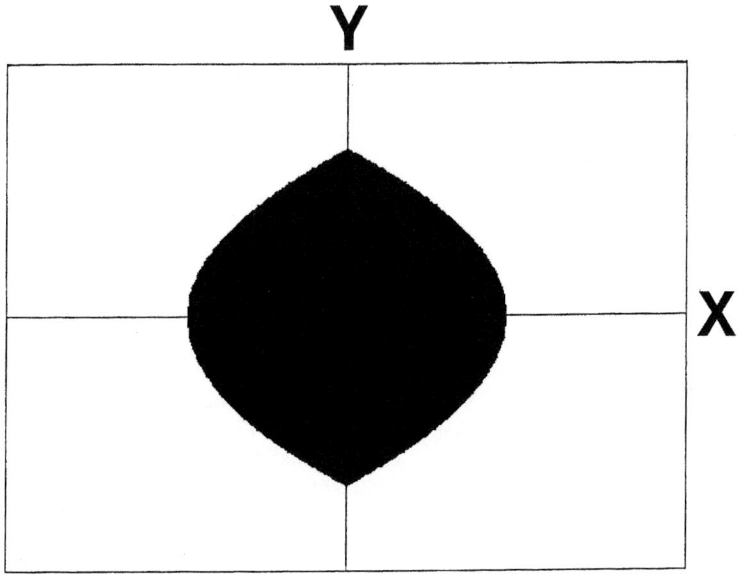

Figure 10. Numerical integration of Equations (10–12) for $K = 0.5$ displayed in the X–Y plane. This value of K corresponds to exact 1:1 resonance. Initial conditions were chosen which correspond to those in Figure 8, namely $v_1 = v_3 = 0.001$, $u_1 = u_2 = v_2 = 0$ in Equations (24–26). As shown in the text, the 1:1 resonance destroys the $Y\theta$-mode. The blackened region is filled with a quasiperiodic trajectory. Region displayed is $|X| < 0.003$, $|Y| < 0.003$.

plitudes: one is a static amplitude which locates the nonisolated equilibrium on the manifold of equilibria; the other is a dynamic amplitude measuring the size of the oscillation relative to the nonisolated equilibrium.

The NNMs of the system investigated in this work illustrate the foregoing properties. We may generalize this situation by considering a vibratory nonholonomic system in which there are N generalized coordinates and M nonholonomic constraints. In this case, there will typically exist an M-dimensional manifold of nonisolated equilibria. At each equilibrium point there is expected to be $N - M$ NNMs, corresponding to the independent directions normal to the manifold. Furthermore, each NNM will have a frequency-amplitude relationship, in which 'amplitude' is actually a collection of $M + 1$ amplitudes; M of these amplitudes locate the nonisolated equilibrium on the manifold, and the remaining amplitude is a measure of the size of the oscillation. This picture is generic, and will not hold for all systems. In fact, even for a generic system it is expected not to hold if parameters are varied such that the system lies in the neighborhood of an internal resonance. In such a case, one or more of the NNMs may not be present. This is seen in the example studied in this work at the 1:1 resonance, in which the $Y\theta$-mode is no longer present.

Acknowledgement

This material is based upon work supported under a National Science Foundation Graduate Fellowship for author Deepak V. Ramani.

References

1. Goldstein, H., *Classical Mechanics*, second edition, Addison-Wesley, Reading, MA, 1980.
2. Kauderer, H., *Nichtlineare Mechanik*, Springer-Verlag, Berlin, 1958.
3. Moon, F. C., *Applied Dynamics: With Applications to Multibody and Mechatronics Systems*, Wiley, New York, 1998.
4. Niemark, J. I. and Fufaev, N. A., *Dynamics of Nonholonomic Systems*, American Mathematical Society, Providence, RI, 1972.
5. Rand, R. H., *Topics in Nonlinear Dynamics with Computer Algebra*, Gordon and Breach, Langhorne, PA, 1994.
6. Rand, R. H. and Armbruster, D., *Perturbation Methods, Bifurcation Theory, and Computer Algebra*, Springer-Verlag, New York, 1987.
7. Rand, R. H. and Ramani, D. V., 'Relaxing nonholonomic constraints', in *Proceedings of the First International Symposium on Impact and Friction of Solids, Structures and Intelligent Machines*, Ottawa, Canada, June 27–30, A. Guran (ed.), World Scientific, Singapore, 2000, pp. 113–116..
8. Vakakis, A. F., Manevitch, L. I., Mikhlin, Y. V., Pilipchuk, V. N., and Zevin, A.A., *Normal Modes and Localization in Nonlinear Systems*, Wiley, New York, 1998.

Nonlinear Dynamics **25:** 65–77, 2001.

Nonlinear Normal Modes of a Parametrically Excited Cantilever Beam

HIROSHI YABUNO
Institute of Engineering Mechanics, University of Tsukuba, Tsukuba-City 305-8573, Japan

ALI H. NAYFEH
Department of Engineering Science and Mechanics, MC 0219, Virginia Polytechnic Institute and State University, Blacksburg, VA 24061, U.S.A.

(Received: 18 February 1999; accepted: 8 August 1999)

Abstract. We investigate theoretically the nonlinear normal modes of a vertical cantilever beam excited by a principal parametric resonance. We apply directly the method of multiple scales to the governing nonlinear nonautonomous integral-partial-differential equation and associated boundary conditions. In the absence of damping, it is shown that the system has nonlinear normal modes, as defined by Rosenberg, even in the presence of the parametric excitation. We calculate the spatial correction to the linear mode shape due to the effects of the inertia and curvature nonlinearities and the parametric excitation. We compare the result obtained with the direct approach with that obtained using a single-mode Galerkin discretization. The deviation between the two predictions increases as the oscillation amplitude increases.

Keywords: Nonlinear normal mode, parametric resonance, mode shape, direct approach, Galerkin method, discretization.

1. Introduction

The problem of determining the nonlinear normal modes of weakly nonlinear continuous systems has received considerable attention in recent years [1, 2]. The validity of several methods for constructing nonlinear normal modes has been discussed by Nayfeh [3]. Nayfeh and Nayfeh [4] constructed the nonlinear normal modes of one-dimensional continuous systems with cubic inertia and geometric nonlinearities by using various approaches. Nayfeh et al. [5] compared two approaches, namely discretization and direct approaches, for constructing the nonlinear normal modes of a cantilever beam with inertia and curvature nonlinearities. Also, Nayfeh and Nayfeh [6] constructed the nonlinear mode shapes and natural frequencies of continuous systems with quadratic and cubic nonlinearities by using the above two approaches. It was shown that these methods lead to the same results when the discretization is carried out by using a complete set of basis functions. Nayfeh and Lacarbonara [7, 8] showed analytically and experimentally that single-mode discretization may lead to erroneous results. The discretization and direct approaches yield different approximate nonlinear normal modes. For a review of the shortcomings of discretization, we refer the reader to Nayfeh and Lacarbonara [8].

In this paper, we construct the nonlinear normal modes of a parametrically excited beam. Anderson et al. [9] investigated experimentally the planer response of a parametrically excited cantilever beam. The experimental results demonstrate that the nonlinear curvature plays a dominant role in the steady-state response of the lowest two modes. For the same system,

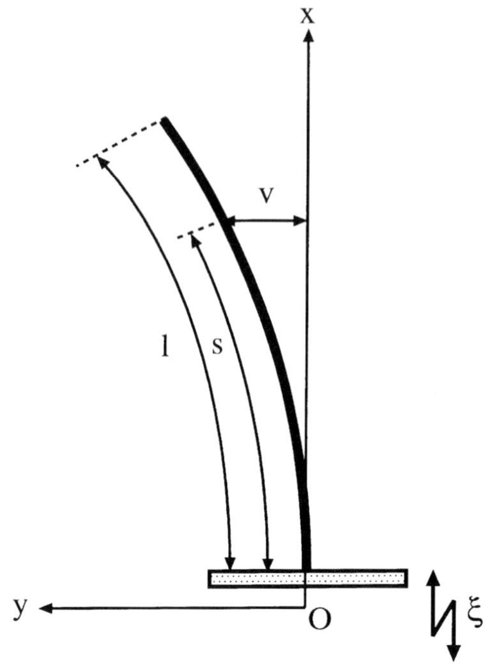

Figure 1. Parametrically excited cantilever beam.

Yabuno et al. [10] and Zavodney and Nayfeh [11] examined theoretically and experimentally the influence of a tip mass on the frequency-response curves. In the above analyses, the spatial variation is assumed to be given by the parametrically excited linear mode shape. In this work, without a priori assumptions, we analyze the effect of the nonlinearities and the parametric excitation on the spatial distribution of the deflection. The results are compared with those obtained with a single-mode Galerkin discretization. The nonlinear normal modes are constructed in the absence of linear viscous and quadratic damping. Although the parametrically excited system is nonautonomous, we obtain the nonlinear normal modes, as defined by Rosenberg [12], taking into account the effect of the parametric excitation. By comparing the predictions of the two approaches for some excitation frequencies, we show that the deviation between the predicted spatial corrections, due to the effects of the nonlinearities and the parametric excitation, is significant in the case of large oscillation amplitudes.

2. Governing Equation

The system is depicted schematically in Figure 1. Here ξ is the displacement of the supporting point expressed as follows:

$$\xi = A \cos \Omega t. \tag{1}$$

The deflection v of the beam in the y-direction, the time t, and the coordinate s are normalized by l/z_n, $(l^2/z_n^2)(m/EI)^{1/2}$, and l/z_n, respectively, where l is the constrained length of the beam, m is the mass per unit length, EI is the bending stiffness, z_n is the nth root of the characteristic equation $1 + \cos(z)\cosh(z) = 0$, and the nth mode is the mode of interest. We denote the resulting dimensionless quantities corresponding to v, t, and s by

v^*, t^*, and s^*, respectively. Furthermore, we introduce the dimensionless parameters $g^* = m(l/z_n)^2/(EI)g$, $\Omega^* = (l^2/z_n^2)(m/EI)^{1/2}\Omega$, and $a_b = A/(l/z_n)\Omega^{*2}$. Then the equation of motion and associated boundary conditions can be written in dimensionless form as [9, 10]

$$\ddot{v} + 2\varepsilon^2\mu\dot{v} + v^{(iv)} = -(v'(v'v'')')' - \left(\frac{1}{2}v'\int_{z_n}^{s}\frac{\partial^2}{\partial t^2}\left(\int_0^s v'^2\,\mathrm{d}s\right)\mathrm{d}s\right)'$$
$$- \varepsilon^2(v''(s - z_n) + v')(a_b\cos(\Omega t) - g) - \varepsilon c\dot{v}|\dot{v}|, \tag{2}$$

$$v = 0,\ v' = 0 \quad \text{at} \quad s = 0, \tag{3a}$$

$$v'' = 0,\ v''' = 0 \quad \text{at} \quad s = z_n. \tag{3b}$$

where $\dot{v} = \partial v^*/\partial t^*$ and $v' = \partial v^*/\partial s^*$. In Equations (2), (3a), and (3b) and hereafter, the asterisk is omitted to simplify the notation. The dimensionless parameter ε is a small positive number introduced as a bookkeeping device and will be set equal to unity in the final solution. The first and second terms on the right-hand side of Equation (2) express the effects of the curvature and inertia nonlinearities, respectively. The term involving $\cos\Omega t$ is due to the base excitation. The dimensionless coefficients 2μ $(= 2.10 \times 10^{-2})$, c $(= 4.06 \times 10^{-1})$, a_b $(= 5.50 \times 10^{-2})$, and $g (= 1.60 \times 10^{-1})$ represent the linear viscous damping, quadratic damping, base acceleration, and acceleration due to gravity, respectively. These values correspond to the parameter values of the beam tested by Anderson et al. [9].

3. Analysis by Direct Method

We use the method of multiple scales [13, 14] and seek a second-order uniform expansion of the solution of Equations (2) and (3) in the form

$$v(s, T_0, T_2; \varepsilon) = \varepsilon v_1(s, T_0, T_2) + \varepsilon^3 v_3(s, T_0, T_2) + \cdots, \tag{4}$$

where $T_0 = t$ is a fast scale and $T_2 = \varepsilon^2 t$ is a slow scale. Substituting Equation (4) into Equations (2) and (3) and equating coefficients of like powers of ε, we obtain

Order ε:

$$D_0^2 v_1 + v_1^{(iv)} = 0 \tag{5}$$

$$v_1 = 0,\ v_1' = 0 \quad \text{at} \quad s = 0, \tag{6a}$$

$$v_1'' = 0,\ v_1''' = 0 \quad \text{at} \quad s = z_n. \tag{6b}$$

Order ε^3:

$$D_0^2 v_3 + v_3^{(iv)} = -2D_0D_2v_1 - 2\mu D_0 v_1 - (v_1'(v_1'v_1'')')'$$
$$- \left(\frac{1}{2}v_1'\int_{z_n}^{s} D_0^2\left(\int_0^s v_1'^2\,\mathrm{d}s\right)\mathrm{d}s\right)'$$
$$- (v_1''(s - z_n) + v_1')(a_b\cos(\Omega T_0) - g) - \varepsilon c D_0 v_1|D_0 v_1|, \tag{7}$$

$$v_3 = 0, \; v_3' = 0 \quad \text{at} \quad s = 0, \tag{8a}$$

$$v_3'' = 0, \; v_3''' = 0 \quad \text{at} \quad s = z_n, \tag{8b}$$

where $D_k = \partial/\partial T_k$.

The solution of the first-order problem can be expressed as

$$v_1(s, T_0, T_2) = \Phi_n(s)\left\{A_n(T_2)\,\mathrm{e}^{iT_0} + \mathrm{cc}\right\}, \tag{9}$$

where Φ_n is the nth linear normal mode shape, which can be written as

$$\Phi_n(s) \;=\; \frac{1}{\sqrt{z_n}}\left\{\cosh s - \cos s + \frac{\cos(z_n) + \cosh(z_n)}{\sin(z_n) + \sinh(z_n)}(\sin s - \sinh s)\right\}. \tag{10}$$

Equations (7) and (8) yield the correction to the nth displacement of the linear mode due to the parametric excitation and the inertia and curvature nonlinearities. Substituting Equation (9) into Equation (7) yields

$$\begin{aligned} D_0^2 v_3 + v_3^{(iv)} \;=\; & -2i\Phi_n D_2 A_n\,\mathrm{e}^{iT_0} - 2i\mu\Phi_n A_n\,\mathrm{e}^{iT_0} + (\Phi_n''(s - z_n) + \Phi_n')gA_n\,\mathrm{e}^{iT_0} \\ & - (\Phi_n'(\Phi_n'\Phi_n'')')'(A_n^3\,\mathrm{e}^{3iT_0} + 3|A_n|^2 A_n\,\mathrm{e}^{iT_0}) \\ & + 2\left(\Phi_n' \int_{z_n}^{s}\int_{0}^{s} \Phi_n'^2\,\mathrm{d}s\,\mathrm{d}s\right)'(A_n^3\,\mathrm{e}^{3iT_0} + |A_n|^2 A_n\,\mathrm{e}^{iT_0}) \\ & - \frac{1}{2}a_b\left(\Phi_n''(s - z_n) + \Phi_n'\right)(A_n\,\mathrm{e}^{i(\Omega+1)T_0} + \bar{A}_n\,\mathrm{e}^{i(\Omega-1)T_0}) + \mathrm{cc} \\ & - ic\Phi_n^2(A_n\,\mathrm{e}^{iT_0} - \bar{A}_n\,\mathrm{e}^{-iT_0})|A_n\,\mathrm{e}^{iT_0} - \bar{A}_n\,\mathrm{e}^{-iT_0}|, \end{aligned} \tag{11}$$

where cc stands for the complex conjugate of the preceding terms. Here, we restrict our discussion to the case of principal parametric resonance of the nth mode. To express the nearness of this resonance, we introduce the detuning parameter σ defined by $\Omega = 2 + \varepsilon^2\sigma$. We seek the solution of Equations (11) and (8) in the form

$$v_3 = g_1(s, T_2)\,\mathrm{e}^{iT_0} + g_3(s, T_2)\,\mathrm{e}^{3iT_0} + \mathrm{cc}. \tag{12}$$

The spatial function $g_1(s, T_2)$ expresses the correction to the spatial distribution at half the excitation frequency Ω, which is assumed to be given by the linear mode shape $\Phi_n(s)$ at the first level of approximation. The second term is a correction to the spatial displacement at the frequency $3\Omega/2$. Substituting Equation (12) into Equation (11) yields

$$\begin{aligned} & (g_1^{(iv)} - g_1)\,\mathrm{e}^{iT_0} + (g_3^{(iv)} - 9g_3)\,\mathrm{e}^{3iT_0} + \mathrm{cc} \\ & \quad = -2i\Phi_n D_2 A_n\,\mathrm{e}^{iT_0} - 2i\mu\Phi_n A_n\,\mathrm{e}^{iT_0} + (\Phi_n''(s - z_n) + \Phi_n')gA_n\,\mathrm{e}^{iT_0} \\ & \qquad - (\Phi_n'(\Phi_n'\Phi_n'')')'(A_n^3\,\mathrm{e}^{3iT_0} + 3|A_n|^2 A_n\,\mathrm{e}^{iT_0}) \\ & \qquad + 2\left(\Phi_n' \int_{z_n}^{s}\int_{0}^{s} \Phi_n'^2\,\mathrm{d}s\,\mathrm{d}s\right)'(A_n^3\,\mathrm{e}^{3iT_0} + |A_n|^2 A_n\,\mathrm{e}^{iT_0}) \\ & \qquad - \frac{1}{2}a_b\left(\Phi_n''(s - z_n) + \Phi_n'\right)(A_n\,\mathrm{e}^{i(\Omega+1)T_0} + \bar{A}_n\,\mathrm{e}^{i(\Omega-1)T_0}) + \mathrm{cc} \\ & \qquad - ic\Phi_n^2(A_n\,\mathrm{e}^{iT_0} - \bar{A}_n\,\mathrm{e}^{-iT_0})|A_n\,\mathrm{e}^{iT_0} - \bar{A}_n\,\mathrm{e}^{-iT_0}|. \end{aligned} \tag{13}$$

Multiplying Equation (13) by e^{-iT_0} and integrating the result over the interval 0 to 2π, we obtain

$$g_1^{(iv)} - g_1 = -i(2D_2A_n + 2\mu A_n)\Phi_n - (3\beta_1 - 2\beta_2)\,|A_n|^2 A_n - \frac{1}{2}\beta_3 a_b \bar{A}_n\, e^{i\sigma T_2} + \beta_3 g A_n - c\beta_4\Gamma_{n1}, \tag{14}$$

where the β_i are functions of s and given by

$$\beta_1 = (\Phi_n'(\Phi_n'\Phi_n'')')', \quad \beta_2 = \left(\Phi_n' \int_{z_n}^{s}\int_{0}^{s} \Phi_n'^{\,2}\, ds\, ds\right)',$$

$$\beta_3 = \Phi_n''(s - z_n) + \Phi_n', \quad \beta_4 = \Phi_n^2 \tag{15}$$

and Γ_{n1} is given by

$$\Gamma_{n1} = \frac{i}{2\pi}\int_0^{2\pi} (A_n\, e^{iT_0} - \bar{A}_n\, e^{-iT_0})|A_n\, e^{iT_0} - \bar{A}_n\, e^{iT_0}|\, e^{-iT_0}\, dT_0. \tag{16}$$

The boundary conditions on g_1 are

$$g_1 = 0,\; g_1' = 0 \quad \text{at} \quad s = 0, \tag{17a}$$

$$g_1'' = 0,\; g_1''' = 0 \quad \text{at} \quad s = z_n. \tag{17b}$$

Because the homogeneous problem governed by Equations (14) and (17) has nontrivial solutions, the nonhomogeneous problem has a solution only if the right-hand side of Equation (14) is orthogonal to every solution of the adjoint homogeneous problem. In this case, the problem is self-adjoint, and hence the adjoints are given by the linear eigenfunction Φ_n. Therefore, to determine the solvability condition of Equations (14) and (17), we multiply the right-hand side of Equation (14) by $\Phi_n(s)$, integrate the result over the interval 0 to z_n, and obtain

$$2iD_2A_n + 2i\mu A_n + (3\alpha_1 - 2\alpha_2)|A_n|^2 A_n + \frac{1}{2}\alpha_3 a_b \bar{A}_n\, e^{i\sigma T_2} - \alpha_3 g A_n + c\alpha_4\Gamma_{n1} = 0, \tag{18}$$

where the α_j are given by

$$\alpha_j = \int_0^{z_n} \beta_j \Phi_n\, ds. \tag{19}$$

Expressing A_n in the polar form

$$A_n = \frac{1}{2}a_n\, e^{i(\sigma T_2 - \gamma_n)/2}, \tag{20}$$

where a_n and γ_n are real and separating Equation (18) into real and imaginary parts, we have

$$D_2 a_n = -\mu a_n - \frac{1}{4}a_b\alpha_3 a_n \sin\gamma_n - \frac{2}{3\pi}c\alpha_4 a_n^2, \tag{21a}$$

$$a_n D_2 \gamma_n = (\sigma + \alpha_3 g) a_n - \frac{1}{2} a_b \alpha_3 a_n \cos \gamma_n - \frac{1}{4}(3\alpha_1 - 2\alpha_2) a_n^3. \tag{21b}$$

Using Equation (20), we rewrite Equation (9) as

$$v_1(s, T_0, T_2) = \Phi_n(s) a_n \cos\left(\frac{1}{2}\Omega t - \frac{1}{2}\gamma_n\right), \tag{22}$$

where a_n and γ_n are governed by Equations (21).

Next we seek g_1. Using Equation (18), we rewrite Equation (14) as

$$g_1^{(iv)} - g_1 = f_{11}(s)|A_n|^2 A_n + f_{12}(s) a_b \bar{A}_n \,\mathrm{e}^{i\sigma T_2} + f_{13}(s) A_n + c f_{14}(s) \Gamma_{n1}, \tag{23}$$

where

$$f_{11}(s) = -3(\beta_1 - \alpha_1 \Phi_n) + 2(\beta_2 - \alpha_2 \Phi_n), \quad f_{12}(s) = -\frac{1}{2}(\beta_3 - \alpha_3 \Phi_n),$$

$$f_{13}(s) = (\beta_3 - \alpha_3 \Phi_n) g, \quad f_{14}(s) = -\beta_4 + \alpha_4 \Phi_n. \tag{24}$$

We can express the solution of Equations (23) and (17) as

$$g_1 = h_{11}|A_n|^2 A_n + h_{12} a_b \bar{A}_n \,\mathrm{e}^{i\sigma T_2} + h_{13} A_n + c h_{14} \Gamma_{n1}, \tag{25}$$

where the h_{1i} are solutions of the two-point boundary-value problems

$$h_{1i}^{(iv)} - h_{1i} = f_{1i}(s), \tag{26}$$

$$h_{1i} = h'_{1i} = 0 \quad \text{at} \quad s = 0, \tag{27a}$$

$$h''_{1i} = h'''_{1i} = 0 \quad \text{at} \quad s = z_n. \tag{27b}$$

In Equation (25), the terms including h_{1i} express the corrections of the spatial distribution of the deflection due to the curvature and inertia nonlinearities, the vertical excitation and the gravitational force, and the quadratic damping. Although the homogeneous parts of these four two-point boundary-value problems have nontrivial solutions, the nonhomogeneous problems are solvable because the nonhomogeneous parts are orthogonal to every solution of the adjoint homogeneous problems. However, the solutions are not unique. To determine uniquely their solutions, we require that they are orthogonal to the linear eigenfunction Φ_n. The results in the case of principal parametric resonance for the first mode ($n = 1$) are plotted in Figure 2; the solutions are not shown because they are long. It follows from Equations (24) and (27) that $h_{13} = -2gh_{12}$.

Next we seek g_3. Multiplying Equation (13) by e^{-3iT_0} and integrating the result over the interval 0 to 2π, we obtain

$$g_3^{(iv)} - 9g_3 = f_{31}(s) a_b A_n \,\mathrm{e}^{i\sigma T_2} + f_{32}(s) A_n^3 + c f_{33}(s) \Gamma_{n3}, \tag{28}$$

where

$$f_{31} = -\frac{1}{2}\beta_3, \quad f_{32} = -(\beta_1 - 2\beta_2), \quad f_{33} = -\beta_4, \tag{29a}$$

$$\Gamma_{n3} = \frac{i}{2\pi} \int_0^{2\pi} (A_n \,\mathrm{e}^{iT_0} - \bar{A}_n \,\mathrm{e}^{-iT_0}) |A_n \,\mathrm{e}^{iT_0} - \bar{A}_n \,\mathrm{e}^{iT_0}| \,\mathrm{e}^{-3iT_0} \,\mathrm{d}T_0. \tag{29b}$$

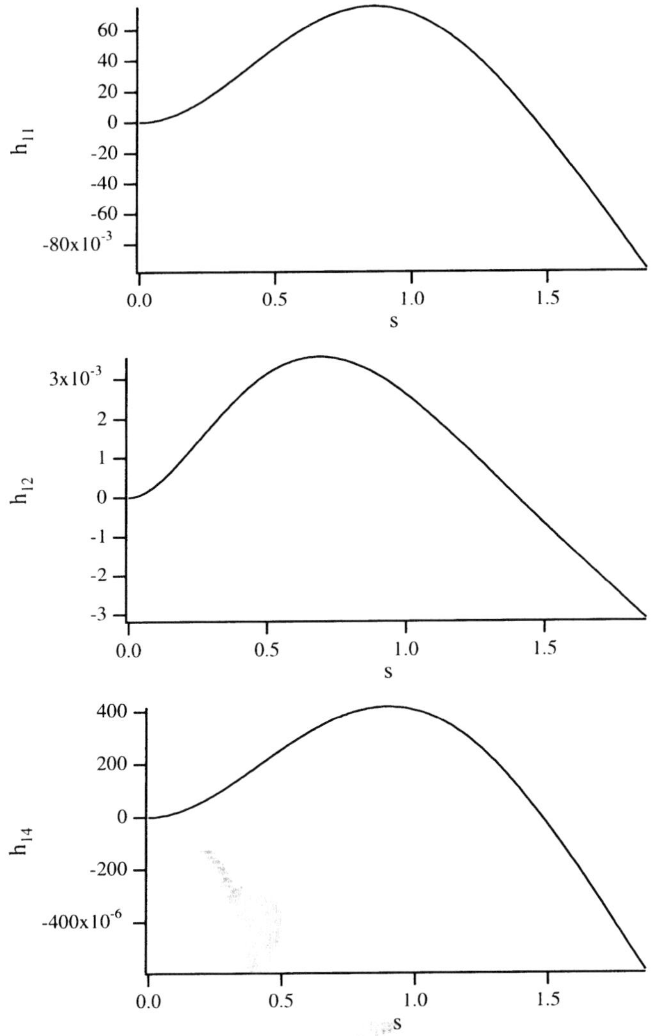

Figure 2. Variation of the functions h_i with s for the nonlinear mode at the frequency $\Omega/2$.

It follows from Equations (12) and (8) that the boundary condition on g_3 are

$$g_3 = 0, \; g_3' = 0 \quad \text{at} \quad s = 0, \tag{30a}$$

$$g_0'' = 0, \; g_0''' \quad \text{at} \quad s = z_n. \tag{30b}$$

Next, we express the solution of Equations (28) and (30) as

$$g_3 = h_{31} a_b A_n \, e^{i\sigma T_2} + h_{32} A_n^3 + c h_{33} \Gamma_{n3}, \tag{31}$$

where the h_{3i} are the solutions of the two-point boundary-value problems

$$h_{3i}^{(iv)} - h_{3i} = f_{3i}(s), \tag{32}$$

$$h_{3i} = h_{3i}' = 0 \quad \text{at} \quad s = 0, \tag{33a}$$

$$h_{3i}'' = h_{3i}''' = 0 \quad \text{at} \quad s = z_n. \tag{33b}$$

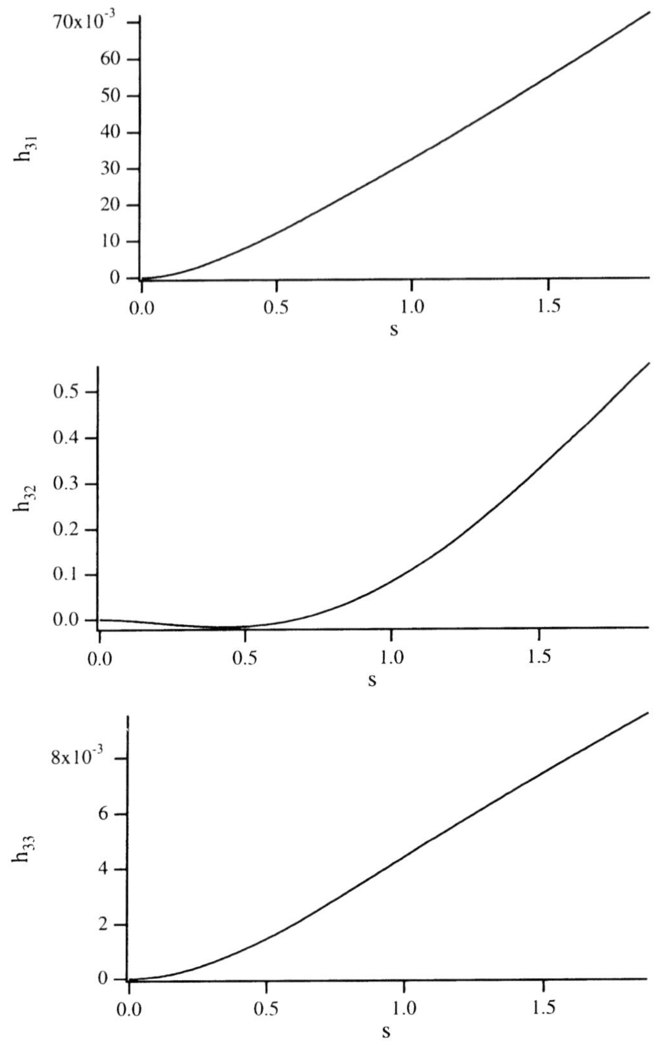

Figure 3. Variation of the h_{3i} with s for the first nonlinear mode; variation of the deflection at the frequency $3\Omega/2$.

The three two-point boundary-value problems, given by Equations (32) and (33), are solvable because the solvability conditions are automatically satisfied and the solutions are unique. The solutions are long and hence are not presented here. The results for the case of principal parametric resonance of the first mode ($n = 1$) are plotted in Figure 3.

Substituting Equations (9), (12), (25), and (31) into Equation (4) and using Equation (20), we obtain the following analytical solution:

$$\begin{aligned} v = {} & \varepsilon \Phi_n a_n \cos\left(\frac{1}{2}\Omega t - \frac{1}{2}\gamma_n\right) \\ & + \varepsilon^3 \left\{ \left(\frac{1}{4} h_{11} a_n^3 + h_{13} a_n\right) \cos\left(\frac{1}{2}\Omega t - \frac{1}{2}\gamma_n\right) - \frac{8}{3\pi} c h_{14} a_n^2 \sin\left(\frac{1}{2}\Omega t - \frac{1}{2}\gamma_n\right) \right. \end{aligned}$$

$$+ h_{12} a_b a_n \cos\left(\frac{1}{2}\Omega t + \frac{1}{2}\gamma_n\right) + h_{31} a_b a_n \cos\left(\frac{3}{2}\Omega t - \frac{1}{2}\gamma_n\right)$$

$$+ \frac{1}{4} h_{32} a_n^3 \cos\left(\frac{3}{2}\Omega t - \frac{3}{2}\gamma_n\right) + \frac{8}{15\pi} c h_{33} a_n^2 \sin\left(\frac{3}{2}\Omega t - \frac{3}{2}\gamma_n\right)\Bigg\} + \cdots. \tag{34}$$

Letting $\Omega t/2 - \gamma_n/2 \equiv \theta$, we rewrite Equation (34) as

$$\begin{aligned} v \;=\; & \varepsilon \Phi_n a_n \cos\theta \\ & + \varepsilon^3 \left\{ \left(\frac{1}{4} h_{11} a_n^3 + h_{13} a_n \right) \cos\theta - \frac{8}{3\pi} c h_{14} a_n^2 \sin\theta + h_{12} a_b a_n \cos(\theta + \gamma_n) \right. \\ & \left. + h_{31} a_b a_n \cos(3\theta + \gamma_n) + \frac{1}{4} h_{32} a_n^3 \cos 3\theta + \frac{8}{15\pi} c h_{33} a_n^2 \sin 3\theta \right\} + \cdots. \end{aligned} \tag{35}$$

Then, the velocity of the beam, $\partial v/\partial t$, can be expressed as

$$\begin{aligned} \frac{\partial v}{\partial t} \;=\; & -\frac{1}{2}\varepsilon \Omega a_n \Phi_n \sin\theta \\ & - \varepsilon^3 \left\{ \frac{1}{2}\Omega \left(\frac{1}{4} h_{11} a_n^3 + h_{13} a_n \right) \sin\theta + \frac{4}{3\pi} c \Omega h_{14} a_n^2 \cos\theta \right. \\ & + \frac{1}{2}\Omega h_{12} a_b a_n \sin(\theta + \gamma_n) + \frac{3}{2}\Omega h_{31} a_b a_n \sin(3\theta + \gamma_n) \\ & \left. + \frac{3}{8}\Omega h_{32} a_n^3 \sin 3\theta + \frac{4}{5\pi}\Omega h_{33} a_n^2 \cos 3\theta \right\} + \cdots. \end{aligned} \tag{36}$$

It follows from Equation (35) that the deflection of the beam at every position along it does not pass through zero at the same instant, due to the terms proportional to $\sin\theta$, $\cos(\theta + \gamma_n)$, $\cos(3\theta + \gamma_n)$, and $\sin 3\theta$, which are caused by the quadratic damping and the vertical(parametric) excitation. Similarly it follows from Equation (36) that the entire velocity field is not also zero at any instant because of the above effects. Consequently, the parametrically excited cantilever beam with linear viscous and quadratic damping does not have nonlinear normal modes, as defined by Rosenberg. However, in the absence of damping, it is shown in Section 5 that the parametrically excited beam has nonlinear normal modes, as defined by Rosenberg, even in the presence of the parametric excitation.

4. Single-Mode Discretization

In this section, we construct the nth nonlinear normal mode by a single-mode discretization of Equations (2) and (3). By comparing the result with that obtained in Section 3 with the direct approach, where no assumptions are made a priori on the form of the solution, we quantify the effect of the nonlinearities and the parametric excitation on the spatial and temporal distributions of the deflection. In discretizing the continuous problem, Equations (2) and (3), we approximate v as a linear combination in terms of the infinite linear eigenfunctions in the form

$$v = \sum_{m=1}^{\infty} \Phi_m(s) q_m(t). \tag{37}$$

For a single-mode discretization, we neglect all of the terms except the nth term. Then, substituting Equation (37) into Equation (2) and taking the inner product of the result with Φ_n, which is the parametrically excited mode, we have

$$\begin{aligned}\frac{d^2 q_n}{dt^2} + 2\varepsilon^2\mu\frac{dq_n}{dt} + q_n &= -\alpha_1 q_n^3 - \frac{1}{2}\alpha_2 q_n\left(\frac{d^2 q_n}{dt^2}\right)^2 \\ &\quad - \varepsilon^2\alpha_3(a_b\cos\Omega t - g)q_n - \varepsilon c\alpha_4\frac{dq_n}{dt}\left|\frac{dq_n}{dt}\right|. \qquad (38)\end{aligned}$$

Using the method of multiple scales, we obtain the following approximate solution of Equation (38)

$$q_n = \varepsilon q_{n1} + \varepsilon^3 q_{n3} + \cdots, \qquad (39)$$

where

$$\begin{aligned}q_{n1} &= a_n\cos\left(\frac{1}{2}\Omega t - \frac{1}{2}\gamma_n\right) \\ q_{n3} &= \frac{1}{4}(\alpha_1 - 2\alpha_2)a_n^3\cos\left(\frac{3}{2}\Omega t - \frac{3}{2}\gamma_n\right), \\ &\quad + \frac{\alpha_3}{16}a_b a_n\cos\left(\frac{3}{2}\Omega t - \frac{1}{2}\gamma_n\right) - \frac{c\alpha_4}{40\pi}a_n^2\sin\left(\frac{3}{2}\Omega t - \frac{3}{2}\gamma_n\right), \qquad (40)\end{aligned}$$

and the a_n and γ_n are governed by Equations (21). Consequently, the deflection of the parametrically excited cantilever beam, which is obtained using a single-mode discretization, can be expressed as

$$v_G = \Phi_n q_n. \qquad (41)$$

Subtracting Equation (41) from Equation (34), we obtain the following deviation between the deflections predicted with the discretization and direct approaches:

$$\begin{aligned}v - v_G &= \varepsilon^3\left\{\left(\frac{1}{4}h_{11}a_n^3 + h_{13}a_n\right)\cos\left(\frac{1}{2}\Omega t - \frac{1}{2}\gamma_n\right)\right. \\ &\quad - \frac{8}{3\pi}ch_{14}a_n^2\sin\left(\frac{1}{2}\Omega t - \frac{1}{2}\gamma_n\right) + h_{12}a_b a_n\cos\left(\frac{1}{2}\Omega t + \frac{1}{2}\gamma_n\right) \\ &\quad + \left(h_{31} - \frac{\alpha_3}{16}\Phi_n\right)a_b a_n\cos\left(\frac{3}{2}\Omega t - \frac{1}{2}\gamma_n\right) \\ &\quad + \left[\frac{1}{4}h_{32} - \frac{1}{4}(\alpha_1 - 2\alpha_2)\Phi_n\right]a_n^3\cos\left(\frac{3}{2}\Omega t - \frac{3}{2}\gamma_n\right) \\ &\quad \left. + \left(\frac{8}{15\pi}ch_{33} + \frac{c\alpha_4}{40\pi}\Phi_n\right)a_n^2\sin\left(\frac{3}{2}\Omega t - \frac{3}{2}\gamma_n\right)\right\}. \qquad (42)\end{aligned}$$

5. Nonlinear Normal Mode as Defined by Rosenberg

In this section, we consider the case without linear viscous and quadratic damping; that is, $\mu = c = 0$ in Equation (2). Then the frequency-response curves are obtained from Equations (21)

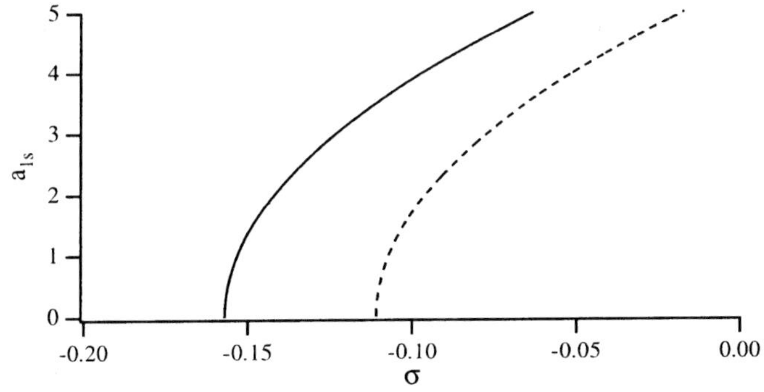

Figure 4. Frequency-response curve (no damping).

as shown in Figure 4, where the solid and dashed lines denote stable and unstable solutions, respectively. Because $\gamma_n = \pi$ for the equilibrium solutions of the modulation Equation (21) when $\mu = 0$ and $c = 0$, the terms proportional to $\cos(\theta+\gamma_n)$ and $\cos(3\theta+\gamma_n)$ in Equation (35) become $-\cos\theta$ and $-\cos 3\theta$, respectively. Also the terms proportional to $\sin\theta$ and $\sin 3\theta$ in Equation (35) disappear. Consequently, in the case of periodic motions, v and $\partial v/\partial t$ can be rewritten as

$$\begin{aligned} v &= \varepsilon\Phi_n a_{ns}\cos\theta + \varepsilon^3\left\{\left(\frac{1}{4}h_{11}a_{ns}^3 + h_{13}a_{ns} + h_{12}a_b a_{ns}\right)\cos\theta\right. \\ &\quad \left. - \left(h_{31}a_b a_{ns} - \frac{1}{4}h_{32}a_{ns}^3\right)\cos 3\theta\right\} + \cdots, \end{aligned} \tag{43}$$

$$\begin{aligned} \frac{\partial v}{\partial t} &= -\frac{1}{2}\varepsilon\Omega a_{ns}\Phi_n \sin\theta - \varepsilon^3\left\{\frac{1}{2}\Omega\left(\frac{1}{4}h_{11}a_{ns}^3 + h_{13}a_{ns} + h_{12}a_b a_{ns}\right)\sin\theta\right. \\ &\quad \left. + \frac{3}{2}\Omega\left(h_{31}a_b a_{ns} - \frac{1}{4}h_{32}a_{ns}^3\right)\sin 3\theta\right\} + \cdots. \end{aligned} \tag{44}$$

It follows from Equation (43) that the entire displacement field is zero at any instant when $\cos\theta = 0$; every point of the beam passes through the equilibrium state at the same instant. Moreover, it follows from Equation (44) that the entire velocity field is zero at any instant when $\sin\theta = 0$. Therefore, in the absence of damping, the parametrically excited cantilever beam has a nonlinear normal mode, as defined by Rosenberg, even in the presence of the principal parametric excitation. In this case, the nth nonlinear mode predicted by a single-mode discretization can be expressed as

$$v_G = \varepsilon\Phi_n a_{ns}\cos\theta + \varepsilon^3\Phi_n\left[\frac{1}{4}\left((\alpha_1 - 2\alpha_2)\right)a_{ns}^3 - \frac{1}{16}\alpha_3 a_b a_{ns}\right]\cos 3\theta + \cdots. \tag{45}$$

Putting $\theta = 0$ in Equations (43) and (45), we obtain approximations to the nth nonlinear mode shape predicted by the direct and discretization approaches. For the case of principal parametric resonance of the first mode, these shapes are shown in Figure 5. The solid and dashed lines refer to the results obtained with the direct and discretization approaches, respectively. As the excitation frequency Ω increases (i.e., as σ increases), the response amplitude a_{ns} of the parametrically excited cantilever beam becomes large, as shown in Figure 4. Then the effect of the nonlinearity becomes significant, as shown in Figure 5.

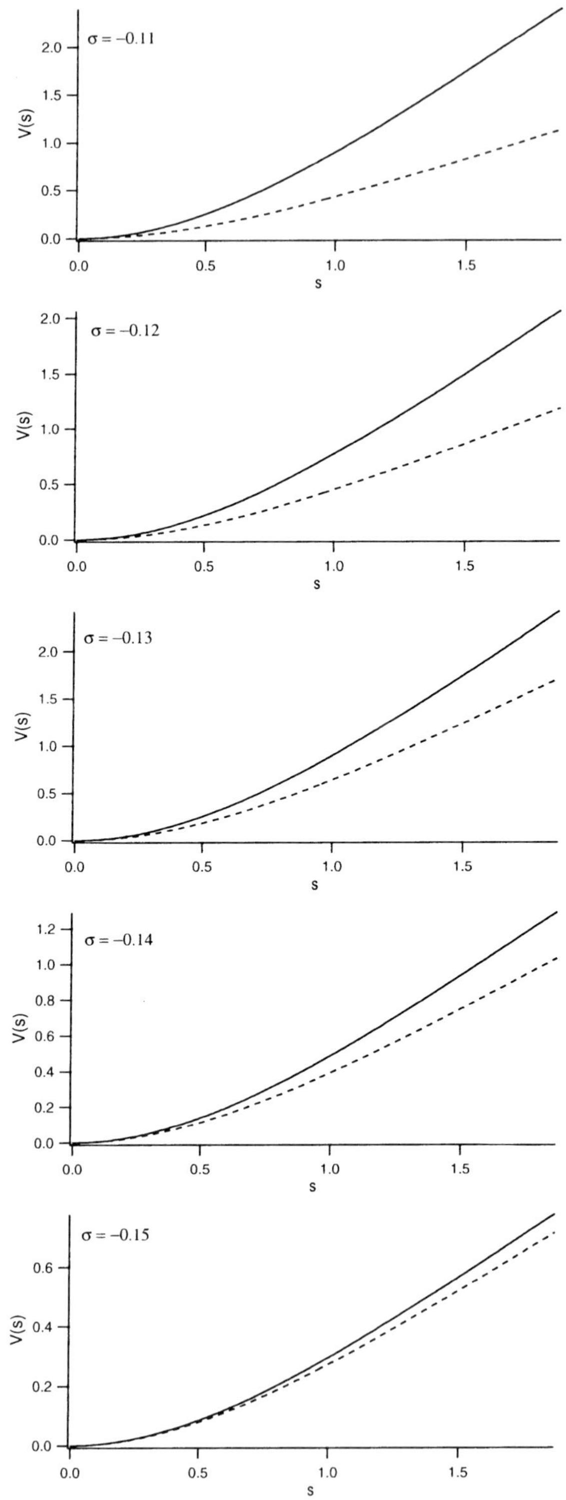

Figure 5. Comparison of the nonlinear normal mode obtained with the direct approach (—) with that obtained with the discretization approach (- - -).

6. Summary

The nonlinear normal modes of a parametrically excited cantilever beam are constructed by directly applying the method of multiple scales to the governing integral-partial differential equation and associated boundary conditions. The effect of the inertia and curvature nonlinearities and the parametric excitation on the spatial distribution of the deflection is examined. The results are compared with those obtained by using a single-mode discretization. In the absence of linear viscous and quadratic damping, it is shown that there are nonlinear normal modes, as defined by Rosenberg, even in the presence of a principal parametric excitation. Furthermore, the nonlinear mode shape obtained with the direct approach is compared with that obtained with the discretization approach for some values of the excitation frequency. In the single-mode discretization, the spatial distribution of the deflection is assumed a priori to be given by the linear mode shape Φ_n, which is parametrically excited, as Equation (41). Thus, the mode shape is not influenced by the nonlinear curvature and nonlinear damping. On the other hand, in the direct approach, the mode shape is not assumed a priori; the nonlinear effects modify the linear mode shape Φ_n. Therefore, in the case of large-amplitude oscillations, the single-mode discretization may yield inaccurate mode shapes.

References

1. Vakakis, A. F., Manevitch, L. I., Mikhlin, Y. V., Pilipchuk, V. N., and Zevin A. A., *Normal Modes and Localization in Nonlinear Systems*, Wiley, New York, 1996.
2. Nayfeh, A. H., *Nonlinear Interactions*, Wiley, New York, 2000.
3. Nayfeh, A. H., 'On direct methods for constructing nonlinear normal modes of continuous systems', *Journal of Vibration and Control* **1**, 1995, 389–430.
4. Nayfeh, A. H. and Nayfeh, S. A., 'On nonlinear modes of continuous systems', *Journal of Vibration and Acoustics* **116**, 1994, 129–136.
5. Nayfeh, A. H., Chin, C.-M., and Nayfeh, S. A., 'Nonlinear normal modes of a cantilever beam', *Journal of Vibration and Acoustics* **117**, 1995, 477–481.
6. Nayfeh, A. H. and Nayfeh, S. A., 'Nonlinear normal modes of a continuous system with quadratic nonlinearities', *Journal of Vibration and Acoustics* **117**, 1995, 199–205.
7. Lacarbonara, W. and Nayfeh, A. H., 'Nonlinear normal modes of buckled beams: Three-to-one and one-to-one internal resonances', *Nonlinear Dynamics* **18**, 1999, 253–273.
8. Nayfeh, A. H. and Lacarbonara, W., 'On the discretization of spatially continuous systems with quadratic and cubic nonlinearities', *JSME International Journal* **41**, 1998, 510–531.
9. Anderson, T. J., Nayfeh, A. H., and Balachandran, B., 'Experimental verification of the importance of the nonlinear curvature in the response of a cantilever beam', *Journal of Vibration and Acoustics* **118**, 1996, 21–27.
10. Yabuno, H., Ide, Y., and Aoshima, N., 'Nonlinear analysis of a parametrically excited cantilever beam', *JSME International Journal* **41**, 1998, 555–562.
11. Zavodney, L. D. and Nayfeh, A. H., 'The nonlinear response of a slender beam carrying a lumped mass to a principal parametric excitation: Theory and experiment', *International Journal of Non-Linear Mechanics* **24**, 1989, 105–125.
12. Rosenberg, R. M., 'On nonlinear vibrations of systems with many degrees of freedom', *Advances in Applied Mechanics* **9**, 1966, 155–242.
13. Nayfeh, A. H., *Perturbation Methods*, Wiley-Interscience, New York, 1973.
14. Nayfeh, A. H., *Introduction to Perturbation Techniques*, Wiley-Interscience, New York, 1981.

Nonlinear Dynamics 25: 79–93, 2001.

Normal Modes and Boundary Layers for a Slender Tensioned Beam on a Nonlinear Foundation

F. PELLICANO
Dipartimento di Scienze dell'Ingegneria, Università di Modena e Reggio Emilia, 41100 Modena, Italy

A. F. VAKAKIS
Department of Mechanical and Industrial Engineering, University of Illinois, Urbana, IL 61801, U.S.A.

(Received: 2 November 1998; accepted: 25 June 1999)

Abstract. In this paper, the nonlinear normal modes (NNMs) of a thin beam resting on a nonlinear spring bed subjected to an axial tension is studied. An energy-based method is used to obtain NNMs. In conjunction with a matched asymptotic expansion, we analyze, through simple formulas, the local effects that a small bending stiffness has on the dynamics, along with the secular effects caused by a symmetric nonlinearity. Nonlinear mode shapes are computed and compared with those of the unperturbed linear system. A double asymptotic expansion is employed to compute the boundary layers in the nonlinear mode shape due to the small bending stiffness. Satisfactory agreement between the theoretical and numerical backbone curves of the system in the frequency domain is observed.

Keywords: Nonlinear normal modes, boundary layers, nonlinear dynamics, perturbation methods.

1. Introduction

One of the first works where the concept of nonlinear normal modes (NNMs) appeared is due to Lyapunov [1]. In this fundamental work, mainly focusing on the stability of motion, the existence of periodic solutions for Hamiltonian systems was proven. More recently, Rosenberg [2] defined nonlinear normal modes as synchronous motions where all material points of a discrete system vibrate inunison, similarly to the linear normal modes of classical vibration theory. During the last thirty years, there was a lack of interest in NNMs, but recently activity in this field has grown noticeably. Indeed, many works have recently appeared in the literature about NNMs for weakly and strongly nonlinear systems [3, 4], matching techniques for general discrete systems [5], continuous systems [6, 7] and discrete systems with internal resonances [8]. Shaw and Pierre [9] presented a new definition of NNMs in terms of invariant manifolds and developed a sort of nonlinear modal analysis to study the general dynamics of weakly nonlinear systems. However, more recently Pellicano and Mastroddi [10] proved that the applicability of such nonlinear modal analysis is restricted to a very small class of dynamical systems. King and Vakakis [11] presented a new formulation for computing NNMs based on the conservation of energy, for undamped systems. A summary of recent works in this field is presented in [12].

Many structures are characterized by small thickness, e.g. aerospace skin-vibrating structures, cables, etc. When such structures vibrate with large amplitudes, NNM analyses can give interesting indications of their dynamic behavior. Moreover, some simplification in the governing equations can be used in the case of very small thickness. For example, thin tensioned beams can be considered as tensioned strings to a first approximation. A deeper investigation shows that such structures present local effects in the proximity of discontinuities and boundaries. Such local effects cannot be neglected since they often cause important stress concentrations that have to be accounted for in design considerations. Moreover, local effects are responsible for numerical troubles in the solution of the governing equations when they are not appropriately accounted for.

In this paper, the energetic method developed by King and Vakakis [11] is extended to the problem of very thin tensioned clamped-clamped beams. Local effects due to the presence of small thickness are considered by applying matched asymptotic expansions [13–16], that enable one to develop analytic approximations of the solution without the need of any numerical computations. Such solutions are more accurate for smaller thicknesses, i.e., when the singularity of the problem (local effect) becomes stronger.

2. Equation of Motion

Consider a clamped-clamped beam resting on a nonlinear spring bed having linear and cubic behavior, and subjected to an axial tension P. The equation of motion is

$$\rho A\tilde{w}_{,\tilde{t}\tilde{t}}(\tilde{x},\tilde{t}) + EI\tilde{w}_{,\tilde{x}\tilde{x}\tilde{x}\tilde{x}}(\tilde{x},\tilde{t}) - P\tilde{w}_{,\tilde{x}\tilde{x}}(\tilde{x},\tilde{t}) + \tilde{k}_1\tilde{w}(\tilde{x},\tilde{t}) + \tilde{k}_3\tilde{w}(\tilde{x},\tilde{t})^3 = 0, \tag{1}$$

where $\tilde{w}$ is the transverse displacement field, $\tilde{t} \in [0,\infty)$ the physical time, $\tilde{x} \in [0,\ell]$ the physical axial coordinate, ρ the mass density, A the cross-section area, E the Young modulus, I the inertia moment, $\tilde{k}_1, \tilde{k}_2$ the spring bed elastic constants, and $(\cdot)_{,x} \equiv \partial(\cdot)/\partial x$. Four boundary conditions and two initial conditions must be imposed for the clamped-clamped system:

$$\begin{aligned} \tilde{w}(0,\tilde{t}) &= 0, \quad \tilde{w}(\ell,\tilde{t}) = 0, \quad \tilde{w}_{,\tilde{x}}(0,\tilde{t}) = 0, \quad \tilde{w}_{,\tilde{x}}(\ell,\tilde{t}) = 0, \\ \tilde{w}(\tilde{x},0) &= \tilde{f}(\tilde{x}), \quad \tilde{w}_{,\tilde{t}}(\tilde{x},0) = \tilde{g}(\tilde{x}), \end{aligned} \tag{2}$$

where $\tilde{f}(\tilde{x})$, $\tilde{g}(\tilde{x})$ are sufficiently smooth functions.

Introducing the nondimensional quantities

$$\tilde{w} = \ell w, \quad \tilde{x} = \ell x, \quad \tilde{t} = \left(\ell\sqrt{\frac{\rho A}{P}}\right)t, \tag{3}$$

we obtain the nondimensional form of the governing equation as

$$w_{,tt} + \varepsilon_2 w_{,xxxx} - w_{,xx} + k_1 w + \varepsilon_1 w^3 = 0, \tag{4}$$

where $\varepsilon_1 = \ell^4\tilde{k}_1/P$, $\varepsilon_2 = EI/\ell^2 P$, $k_1 = \ell^2\tilde{k}_1/P$; ε_1 represents a sufficiently small ordering parameter. The following scaled boundary and initial conditions are considered:

$$w(0,t) = w(1,t) = 0, \quad w_{,x}(0,t) = w_{,x}(1,t) = 0, \tag{5}$$

$$w(x,0) = f(x), \quad w_{,t}(x,0) = g(x). \tag{6}$$

When the bending stiffness is small with respect to the beam length and the axial tension, the parameter ε_2 tends to zero ($\varepsilon_2 \to 0^+$). Moreover, note that when $\varepsilon_2 = 0$, the differential equation drops by two orders and two boundary conditions cannot be respected anymore. This implies that the limit causes a singular behavior of the solution of Equations (4–6). Indeed, as $\varepsilon_2 \to 0^+$ in the vicinity of the beam-ends boundary layers arise, i.e., sharp variations in the spatial behavior of the solution are present and are represented by exponential-like functions that do not admit Taylor expansions in ε_2. Furthermore, numerical codes can cause numerical trouble when boundary layers are not suitably accounted for. The classical way to study boundary layers is through matched asymptotic expansions; this methodology will be applied in the following analysis in conjunction with the energetic method developed by Vakakis and King [11] for computing NNMs.

Before proceeding with the analysis, however, we note that the governing equation contains two small parameters, namely, $\varepsilon_1, \varepsilon_2$, both of which will be treated separately in a double asymptotic expansion scheme. In the present problem, the two perturbation expansions can be considered to be independent from each other, if the linear part of the spring bed is different from zero.

3. The Energetic Method for NNMs

In this section, particular solutions of Equations (4, 6) called nonlinear normal modes (NNMs), are obtained. During an NNM's oscillation, each point of the beam reaches its maximum amplitude and passes through zero at the same instant of time [2]. Moreover, the motion of each point can be obtained by the knowledge of the motion of an 'arbitrary' point of the beam, by means of a functional relationship. This holds only when single mode oscillations occur, i.e., in the absence of internal resonances. For the study of multi-mode nonlinear normal mode oscillations, the reader is referred to [17]. We will take advantage of the conservation of energy in the problem under investigation to obtain an analytical description of NNMs. To this end, let us rewrite Equations (4–6) as

$$w_{,tt} = \mathcal{L}^{(0)}[w] + \varepsilon_1\, \mathcal{L}^{(1)}[w] + \varepsilon_2\, \mathcal{L}^{(2)}[w], \qquad \mathcal{B}[w] = 0, \tag{7}$$

where $\mathcal{L}[\cdot]$, $\mathcal{B}[\cdot]$ are differential operators representing the governing equation and the boundary conditions, respectively.

The total energy of the system can be expressed as

$$E_{\text{tot}} = \frac{1}{2}\int_0^1 w_{,t}^2\, \mathrm{d}x + L^{(0)}[w] + \varepsilon_1 L^{(1)}[w] + \varepsilon_2 L^{(2)}[w], \tag{8}$$

where

$$\begin{aligned} L^{(0)}[w] &= \frac{1}{2}\int_0^1 w_x^2\, \mathrm{d}x + \frac{1}{2}k_1 \int_0^1 w^2\, \mathrm{d}x, \\ L^{(1)}[w] &= \frac{1}{4}\int_0^1 w^4\, \mathrm{d}x, \\ L^{(2)}[w] &= \frac{1}{2}\int_0^1 w_{xx}^2\, \mathrm{d}x. \end{aligned} \tag{9}$$

Let us impose the following condition for in-unison (synchronous) motion of all material points of the system:

$$w(x,t) = W(x, w_0(t)), \tag{10}$$

where $w_0(t) = w(x_0, t)$ is a reference displacement and $W(\cdot)$ is a (yet unknown) functional relationship between $w_0(t)$ and $w(x,t)$. Equation (10) implies the following relationships between derivatives:

$$\frac{\partial w}{\partial t} = \frac{\partial W}{\partial w_0} w_{0,t}, \quad \frac{\partial^2 w}{\partial t^2} = \frac{\partial^2 W}{\partial w_0^2} w_{0,t}^2 + \frac{\partial W}{\partial w_0} w_{0,tt}. \tag{11}$$

Using these relationships, the energy (8) can be expressed as

$$E_{\text{tot}} = \frac{1}{2} w_{0,t}^2 \int_0^1 \left[\frac{\partial W}{\partial w_0}\right]^2 \mathrm{d}x + L^{(0)}[W] + \varepsilon_1 L^{(1)}[W, w_0] + \varepsilon_2 L^{(2)}[W], \tag{12}$$

leading to the following expression for the velocity of the reference point

$$w_{0,t}^2 = 2 \frac{E_{\text{tot}} - (L^{(0)}[W] + \varepsilon_1 L^{(1)}[W, w_0] + \varepsilon_2 L^{(2)}[W])}{\frac{1}{2}\int_0^1 \left[\frac{\partial W}{\partial w_0}\right]^2 \mathrm{d}x}. \tag{13}$$

On the other hand, the second time derivative of the reference displacement can be expressed as

$$w_{0,tt} = \{\mathcal{L}^{(0)}[W] + \varepsilon_1 \mathcal{L}^{(1)}[W, w_0] + \varepsilon_2 \mathcal{L}^{(2)}[W]\}\big|_{x=x_0}. \tag{14}$$

Combining the previous relationships, Equation (7) becomes

$$\begin{aligned} w_{,tt} &= \frac{\partial^2 W}{\partial w_0^2} 2 \frac{E_{\text{tot}} - (L^{(0)}[W] + \varepsilon_1 L^{(1)}[W, w_0] + \varepsilon_2 L^{(2)}[W])}{\frac{1}{2}\int_0^1 \left[\frac{\partial W}{\partial w_0}\right]^2 \mathrm{d}x} \\ &\quad + \frac{\partial W}{\partial w_0} \{\mathcal{L}^{(0)}[W] + \varepsilon_1 \mathcal{L}^{(1)}[W, w_0] + \varepsilon_2 \mathcal{L}^{(2)}[W]\}\big|_{x=x_0} \\ &= \mathcal{L}^{(0)}[W] + \varepsilon_1 \mathcal{L}^{(1)}[W, w_0] + \varepsilon_2 \mathcal{L}^{(2)}[W] \end{aligned} \tag{15a}$$

and the boundary and initial conditions (5, 6) are expressed as

$$\begin{aligned} W(0, w_0(t)) &= W(1, w_0(t)) = 0, \quad \frac{\partial W}{\partial x}(0, w_0(t)) = \frac{\partial W}{\partial x}(1, w_0(t)) = 0, \\ W(x, w_0(0)) &= f(x), \quad \frac{\partial W}{\partial w_0}(x, w_0(0)) w_{0,t}(0) = g(x). \end{aligned} \tag{15b}$$

Note that Equation (15a) becomes singular when the potential energy of the system reaches the maximum, in fact, in this situation the coefficient of $\partial^2 W/\partial w_0^2$ becomes zero, i.e., when $w_0 = w_0^*$, where w_0^* represents the maximum absolute amplitude attained by the reference displacement $w_0(t)$. In order to compute an asymptotic approximation of W and analytically continue it up to the maximum energy level, we require that the functional W respects Equation (15a) at the maximum energy level, obtaining the following compatibility condition

$$\left[\frac{\partial W}{\partial w_0}\{\mathcal{L}^{(0)}[W]+\varepsilon_1\mathcal{L}^{(1)}[W]+\varepsilon_2\mathcal{L}^{(2)}[W]\}_{x=x_0}\right.$$
$$\left.-\{\mathcal{L}^{(0)}[W]+\varepsilon_1\mathcal{L}^{(1)}[W]+\varepsilon_2\,\mathcal{L}^{(2)}[W]\}\right]_{w_0=w_0^*}=0. \tag{16}$$

4. The Double Asymptotic Expansion

The problem (15) with compatibility condition (16) is solved perturbatively by expanding the solution in a double series in terms of $\varepsilon_1, \varepsilon_2$. To this end, we expand the solution in ε_1 as follows

$$W(x, w_0(t))=\sum_{k=1}^{N} W^{(k)}(x, w_0(t)), \tag{17}$$

where $W(k)=O(\varepsilon_1^k)$ and $\partial^n W^{(k)}/\partial x^n = O(\varepsilon_1^k)$. Moreover, the approximations in (17) are expressed as

$$W^{(0)}(x, w_0)=a_1^{(0)}(x)w_0(t), \tag{18a}$$

$$W^{(k)}(x, w_0)=\sum_{m=1}^{M} a_m^{(k)}(x)w_0^k(t), \tag{18b}$$

where the separation of variables imposed in Equation (18a) indicates that we are going to perform a single mode nonlinear expansion where $a_1^{(0)}(x)$ is the generating function. Note that the obvious condition $W(x_0, w_0)=w_0$ implies that $a_1^{(0)}(x_0)=1$, and $a_m^{(k)}(x_0)=0$, $\forall m>0, k>1$.

4.1. First Order Approximation in ε_1

At order $O(\varepsilon_1^0)$ we obtain the relation

$$\frac{\partial W^{(0)}}{\partial w_0}\{\mathcal{L}^{(0)}[W^{(0)}]+\varepsilon_2\mathcal{L}^{(2)}[W^{(0)}]\}\Big|_{x=x_0}=\mathcal{L}^{(0)}[W^{(0)}]+\varepsilon_2\mathcal{L}^{(2)}[W^{(0)}] \tag{19}$$

with the clamped boundary conditions

$$W^{(0)}(0, w_0)=W^{(0)}(1, w_0)=\frac{\partial W^{(0)}}{\partial x}(0, w_0)=\frac{\partial W^{(0)}}{\partial x}(1, w_0)=0. \tag{20}$$

Let us rewrite (19) taking into account Equation (18a)

$$a_1^{(0)}(x)\{a_{1,xx}^{(0)}(x_0)-\varepsilon_2\, a_{1,xxxx}^{(0)}(x_0)\}=a_{1,xx}^{(0)}(x)-\varepsilon_2\, a_{1,xxxx}^{(0)}(x). \tag{21}$$

Note that we are concerned with the limit $\varepsilon_2\to 0^+$, therefore, the previous equation should be handled carefully. We use the method of matched asymptotic expansions which consists of subdividing the solution into two parts, an 'outer solution' which is valid far from the boundaries, and an 'inner solution' which is valid in a narrow region close to the boundaries.

4.1.1. *Outer Expansion*

Let us expand $a_1^{(0)}$ as series of ε_2 in the following form

$$a_1^{(0)}(x, \varepsilon_2) = a_1^{(0,0)}(x) + \sqrt{\varepsilon_2}\, a_1^{(0,1)}(x) + O(\varepsilon_2) \tag{22}$$

at order $O(\varepsilon_2^0)$, we obtain the equation

$$a_1^{(0,0)}(x)\{a_{1,xx}^{(0,0)}(x_0)\} = a_{1,xx}^{(0,0)}(x) \tag{23}$$

with boundary conditions

$$a_1^{(0,0)}(0) = a_1^{(0,0)}(1) = 0. \tag{24}$$

If we introduce the parameter $\lambda_{n,0}^2 = -a_{1,xx}^{(0,0)}(x_0)$ and take into account the compatibility condition $a_1^{(0,0)}(x_0) = 1$, the solution of (23) is

$$a_1^{(0,0)}(x) = \frac{\sin(n\pi x)}{\sin(n\pi x_0)} \tag{25}$$

and $\lambda_{n,0} = n\pi, n = 1, 2, \ldots$. The $O(1)$ solution for the functional relation then becomes

$$W^{(0,0)}(x, w_0) = \frac{\sin(n\pi x)}{\sin(n\pi x_0)} w_0(t). \tag{26}$$

Note that the $O(1)$ solution has nonzero first derivatives at the ends and, hence, it does not satisfy the original boundary conditions of the problem. In order to respect the boundary conditions of $O(\varepsilon_1^0)$, we now apply the matched asymptotic expansion method by computing an inner solution to the problem.

4.1.2. *Inner Expansion*

In order to study the behavior of the solution in the vicinity of the left clamped boundary condition, let us introduce the stretched coordinate

$$\eta_1 = \frac{x}{\sqrt{\varepsilon_2}} \tag{27}$$

and rewrite Equation (21) as

$$a_1^{(0)}(\eta_1)\left\{\frac{1}{\varepsilon_2}a_{1,\eta_1\eta_1}^{(0)}(\eta_1) - \frac{1}{\varepsilon_2}a_{1,\eta_1\eta_1\eta_1\eta_1}^{(0)}(\eta_1)\right\}_{x=x_0} = \frac{1}{\varepsilon_2}a_{1,\eta_1\eta_1}^{(0)}(\eta_1) - \frac{1}{\varepsilon_2}a_{1,\eta_1\eta_1\eta_1\eta_1}^{(0)}(\eta_1). \tag{28}$$

Note that the inner solution has to be matched with the outer. In particular, the limit of the inner solution as $\varepsilon_2 \to 0^+$, i.e., $\eta_1 \to \infty$ when x does not approaches zero, is exponentially zero. This implies that the term

$$\gamma_n = \left\{\frac{1}{\varepsilon_2}a_{1,\eta_1\eta_1}^{(0)}(\eta_1) - \frac{1}{\varepsilon_2}a_{1,\eta_1\eta_1\eta_1\eta_1}^{(0)}(\eta_1)\right\}_{x=x_0}$$

in formula (28) can be considered to be finite as $\varepsilon_2 \to 0^+$. Then, let us study the equation

$$a_{1,\eta_1\eta_1}^{(0)}(\eta_1) - a_{1,\eta_1\eta_1\eta_1\eta_1}^{(0)}(\eta_1) = \varepsilon_2\gamma_n a_1^{(0)}(\eta_1). \tag{29}$$

To this end we expand the inner solution

$$a_1^{(0)} = a_{1,0}^{(0)} + \sqrt{\varepsilon_2} a_{1,1}^{(0)} + \cdots . \tag{30}$$

At the $O(\varepsilon_2^0)$ we have

$$a_{1,0,\eta_1\eta_1}^{(0)}(\eta_1) - a_{1,0,\eta_1\eta_1\eta_1\eta_1}^{(0)}(\eta_1) = 0, \tag{31}$$

with the boundary and matching conditions

$$a_{1,0,\eta_1}^{(0)}(0) = 0, \qquad \lim_{\eta_1 \to \infty} a_{1,0}^{(0)}(\eta_1) = 0. \tag{32}$$

Clearly the solution is $a_{1,0}^{(0)}(\eta_1) \equiv 0$.

At $O(\sqrt{\varepsilon_2})$ we obtain,

$$a_{1,1,\eta_1\eta_1}^{(0)}(\eta_1) - a_{1,1,\eta_1\eta_1\eta_1\eta_1}^{(0)}(\eta_1) = -\gamma_n a_{1,0}^{(0)}(\eta_1), \tag{33}$$

with the boundary and matching conditions:

$$\lim_{\eta_1 \to 0} a_{1,1,\eta_1}^{(0)}(\eta_1) = - \lim_{x \to 0} a_{1,x}^{(0,0)}(x), \tag{34a}$$

$$\lim_{\eta_1 \to \infty} a_{1,1}^{(0)}(\eta_1) = 0. \tag{34b}$$

The boundary condition (34a) is suitably chosen in order to allow the composite expansion to respect exactly the boundary condition (5) on the first derivative. Equation (34b) is the matching condition of the inner solution with the outer solution, which approaches zero when $x \to 0$.

Let us replace the symbol $a_{1,1}^{(0)}$ with $a_{1,\ell}^{(0)}$, in order to indicate the left inner solution. For the left boundary, we have

$$a_{1,\ell}^{(0)} = \frac{n\pi}{\sin n\pi x_0} \exp(-\eta_1). \tag{35}$$

A similar approach can be used for the right end where the stretched coordinate takes the form

$$\eta_2 = \frac{1-x}{\sqrt{\varepsilon_2}}. \tag{36}$$

For this boundary, the following solution must be solved:

$$a_{1,\eta_2\eta_2}^{(0)}(\eta_2) - a_{1,\eta_2\eta_2\eta_2\eta_2}^{(0)}(\eta_2) = \varepsilon_2 \gamma_n a_1^{(0)}(\eta_2). \tag{37}$$

We expand the inner solution similarly to the case for the left boundary and at the order $O(\varepsilon_2^0)$, we obtain

$$a_{1,0,\eta_2\eta_2}^{(0)}(\eta_2) - a_{1,0,\eta_2\eta_2\eta_2\eta_2}^{(0)}(\eta_2) = 0 \tag{38}$$

with the boundary and matching conditions

$$a_{1,0,\eta_2}^{(0)}(0) = 0, \qquad \lim_{\eta_2 \to \infty} a_{1,0}^{(0)}(\eta_2) = 0. \tag{39}$$

The solution of this system is $a_{1,0}^{(0)}(\eta_2) \equiv 0$.

At $O(\sqrt{\varepsilon_2})$ we find the equation

$$a_{1,1,\eta_2\eta_2}^{(0)}(\eta_2) - a_{1,1,\eta_2\eta_2\eta_2\eta_2}^{(0)}(\eta_2) = -\gamma_n\, a_{1,0}^{(0)}(\eta_2) \tag{40}$$

with boundary and matching conditions

$$\lim_{\eta_2\to 0} a_{1,0,\eta_2}^{(0)}(\eta_2) = \lim_{x\to 1} a_{1,x}^{(0,0)}(x), \qquad \lim_{\eta_2\to\infty} a_{1,1}^{(0)}(\eta_2) = 0. \tag{41}$$

Let us replace the symbol $a_{1,1}^{(0)}$ with $a_{1,r}^{(0)}$ and write

$$a_{1,r}^{(0)} = \frac{n\pi(-1)^n}{\sin n\pi x_0}\exp(-\eta_2). \tag{42}$$

Combining the previous results, the solution for the leading order coefficient $a_1^{(0)}(x, \varepsilon_1, \varepsilon_2)$ is given by

$$\begin{aligned} a^{(0)}(x,\varepsilon_2) &= a_1^{(0,0)}(x) + \sqrt{\varepsilon_2}(a_{1,l}^{(0)}(x) + a_{1,r}^{(0)}(x)) + \cdots \\ &= \frac{\sin(n\pi x)}{\sin(n\pi x_0)} + \sqrt{\varepsilon_2}\frac{n\pi}{\sin n\pi x_0}\left(\exp\left(-\frac{x}{\sqrt{\varepsilon_2}}\right) - (-1)^n\exp\left(-\frac{1-x}{\sqrt{\varepsilon_2}}\right)\right) + \cdots, \end{aligned} \tag{43}$$

where the $O(\sqrt{\varepsilon_2})$ contributions are local, i.e., negligible in the field and finite at the boundaries, exactly satisfying the boundary conditions for the first derivative in x and asymptotically those for w. Note that the common part of the outer and inner solutions is zero.

4.2. Second-Order Approximation in ε_1

Once the order $O(\varepsilon_1^0)$ solution is obtained, we use this solution to solve the second-order problem obtained from Equations (15a, b) and (16), we have

$$\begin{aligned} &\frac{\partial^2 W^{(1)}}{\partial w_0^2} \times \\ &\frac{\left\{\int_0^1[(W_{,x}^{(0)})^2 + k_1(W^{(0)})^2]\,\mathrm{d}x + \varepsilon_2\int_0^1(W_{,xx}^{(0)})^2\,\mathrm{d}x\right\}_{w_0=w_0^*} - \int_0^1[(W_{,x}^{(0)})^2 + k_1(W^{(0)})^2]\,\mathrm{d}x - \varepsilon_2\int_0^1(W_{,xx}^{(0)})^2\,\mathrm{d}x}{\int_0^1(\partial W^{(0)}\big/\partial w_0)^2\,\mathrm{d}x} \\ &+ \frac{\partial W^{(0)}}{\partial w_0}\{-\varepsilon_1 W^{(0)^3} - k_1 W^{(1)} + W_{,xx}^{(1)} - \varepsilon_2 W_{,xxxx}^{(1)}\}_{x=x_0} + \frac{\partial W^{(1)}}{\partial w_0}\{-k_1 W^{(0)} + W_{,xx}^{(0)} - \varepsilon_2 W_{,xxxx}^{(0)}\}_{x=x_0} \\ &= -k_1 W^{(1)} + W_{,xx}^{(1)} - \varepsilon_2 W_{,xxxx}^{(0)} - \varepsilon_1 W^{(0)^3} \end{aligned} \tag{44}$$

with the compatibility condition

$$\begin{aligned} &\left[\frac{\partial W^{(0)}}{\partial w_0}\{-\varepsilon_1 W^{(0)^3} - k_1 W^{(1)} + W_{,xx}^{(1)} - \varepsilon_2 W_{,xxxx}^{(1)}\}_{x=x_0}\right. \\ &\quad + \frac{\partial W^{(1)}}{\partial w_0}\{-k_1 W^{(0)} + W_{,xx}^{(0)} - \varepsilon_2 W_{,xxxx}^{(0)}\}_{x=x_0} \\ &\quad \left. + k_1 W^{(1)} - W_{,xx}^{(1)} + \varepsilon_2 W_{,xxxx}^{(0)} + \varepsilon_1 W^{(0)^3}\right]_{w_0=w_0^*} = 0. \end{aligned} \tag{45}$$

4.2.1. *Outer Expansion*

Let us expand $W^{(1)}$ in powers of ε_2

$$W^{(1)} = W^{(1,0)} + \sqrt{\varepsilon_2} W^{(1,1)} + \cdots. \tag{46}$$

This expansion enables us to solve the outer problem governed by the following equation:

$$\frac{\partial^2 W^{(1,0)}}{\partial w_0^2} \frac{\left\{ \int_0^1 W_{,x}^{(0)^2}\, dx \right\}_{w_0=w_0^*} - \int_0^1 W_{,x}^{(0)^2}\, dx}{\int_0^1 \left[\frac{\partial W^{(0)}}{\partial w_0} \right]^2 dx} + \frac{\partial W^{(0)}}{\partial w_0} \{-\varepsilon_1 W^{(0)^3} - k_1 W^{(1,0)} + W_{,xx}^{(1,0)}\}_{x=x_0}$$
$$+ \frac{\partial W^{(1,0)}}{\partial w_0} \{-k_1 W^{(0)} + W_{,xx}^{(0)}\}_{x=x_0} = -\varepsilon_1 W^{(0)^3} - k_1 W^{(1,0)} + W_{,xx}^{(1,0)}, \tag{47}$$

with the compatibility condition

$$\left[\frac{\partial W^{(0)}}{\partial w_0} \{-\varepsilon_1 W^{(0)^3} - k_1 W^{(1,0)} + W_{,xx}^{(1,0)}\}_{x=x_0} \right.$$
$$\left. + \frac{\partial W^{(1,0)}}{\partial w_0} \{-k_1 W^{(0)} + W_{,xx}^{(0)}\}_{x=x_0} - (-\varepsilon_1 W^{(0)^3} - k_1 W^{(1,0)} + W_{,xx}^{(1,0)}) \right]_{w_0=w_0^*} = 0. \tag{48}$$

Note that in the solution of the outer Equations (47) and (48), the boundary layers present in the functional $W^{(}0)$ can be neglected and the outer expansion at order $O(\varepsilon_1^0)$ used. Let us now expand $W^{(1,0)}$ in the following form:

$$W^{(1,0)} = a_1^{(1,0)}(x) w_0(t) + a_1^{(1,0)}(x) w_0^3(t) + \cdots. \tag{49}$$

Substituting Equation (49) into (47), neglecting terms higher than w_0, and recalling that $w^{(0)} = a_1^{(0)}(x) w_0(t)$, we obtain

$$6(n^2\pi^2 + k_1) w_0(t) w_0^{*2} a_3^{(1,0)}(x) + a_1^{(0)}(x) a_{1,xx}^{(1,0)}(x_0) w_0(t)$$
$$- n^2\pi^2 a_{1,xx}^{(1,0)}(x) w_0(t) = a_{1,xx}^{(1,0)}(x) w_0(t). \tag{50}$$

Equation (50) determines $a_3^{(1,0)}$ as a function of $a_1^{(1,0)}$. A similar procedure can be performed for the compatibility Equation (51) leading to

$$(a_{1,xx}^{(1,0)}(x_0) + a_{3,xx}^{(1,0)}(x_0) w_0^{*2} - \varepsilon_1 w_0^{*2}) \frac{\sin n\pi x}{\sin n\pi x_0} - (n^2\pi^2 + k_1)(a_1^{(1,0)}(x) + 3a_3^{(1,0)}(x) w_0^{*2})$$
$$= a_{1,xx}^{(1,0)}(x) + a_{3,xx}^{(1,0)}(x) w_0^{*2} - k_1 a_1^{(1,0)}(x) - k_1 a_3^{(1,0)}(x) w_0^{*2} - \varepsilon_1 \left(\frac{\sin n\pi x}{\sin n\pi x_0} \right)^3 w_0^{*2}. \tag{51}$$

Evaluate now $a_3^{(1,0)}$ from Equation (50) and substitute it into Equation (51). Neglecting terms higher than w_0, we solve the equation

$$a_{1,xxxx}^{(1,0)}(x) + A a_{1,xx}^{(1,0)}(x) + B a_1^{(1,0)}(x) = h(x) \tag{52}$$

where coefficients are listed in Table 1. The solution of Equation (52) is the following:

$$a_1^{(1,0)}(x) = \alpha_1 \sin n\pi x + \alpha_2 \sin 3n\pi x + O(\sqrt{\varepsilon_2}), \tag{53}$$

Table 1. Coefficients of Equation 52).

$A = (10n^2\pi^2 + 8k_1)$	$B = n^2\pi^2(9n^2\pi^2 + 8k_1)$
$h(x) = [a^{(1,0)}_{1,xxxx}(x_0) + (10n^2\pi^2 + 8k_1)a^{(1,0)}_{1,xx}(x_0) - 6(n^2\pi^2 + k_1)\varepsilon_1 w_0^{*2}] \times \frac{\sin n\pi x}{\sin n\pi x_0} - \varepsilon_1\left(\frac{\sin n\pi x}{\sin n\pi x_0}\right)^3 w_0^{*2}\, 6(n^2\pi^2 + k_1)$	

where

$$\alpha_1 = -\alpha_2 \frac{\sin 3n\pi x}{\sin n\pi x}, \qquad \alpha_2 = \frac{\varepsilon_1 w_0^{*2}(n^2\pi^2 + k_1)}{42k_1 n^2\pi^2}. \tag{54}$$

The solution $a_1^{(1,0)}(x)$ can be rewritten as

$$a_1^{(1,0)}(x) = \varepsilon_1 w_0^{*2} \frac{2(n^2\pi^2 + k_1)}{21k_1 n^2\pi^2}\left(\frac{\sin n\pi x}{\sin n\pi x_0} - \frac{\sin^3 n\pi x}{\sin^3 n\pi x_0}\right). \tag{55}$$

Substituting (55) into Equation (50) and solving for $a_3^{(1,0)}$, we obtain

$$a_3^{(1,0)}(x) = -\varepsilon_1 \frac{8}{63k_1}\left(\frac{\sin n\pi x}{\sin n\pi x_0} - \frac{\sin^3 n\pi x}{\sin^3 n\pi x_0}\right). \tag{56}$$

Finally, substituting expressions (53) and (55) into (49), we obtain the outer expansion of $O(\varepsilon_1, \varepsilon_2^0)$ as follows

$$\begin{aligned} W^{(1,0)} &= w_0 \varepsilon_1 w_0^{*2} \frac{2(n^2\pi^2 + k_1)}{21k_1 n^2\pi^2}\left(\frac{\sin n\pi x}{\sin n\pi x_0} - \frac{\sin^3 n\pi x}{\sin^3 n\pi x_0}\right) \\ &\quad - \varepsilon_1 \frac{8}{63k_1}\left(\frac{\sin n\pi x}{\sin n\pi x_0} - \frac{\sin^3 n\pi x}{\sin^3 n\pi x_0}\right) w_0^3. \end{aligned} \tag{57}$$

4.2.2. *Inner Expansion*

Again, we consider the stretched coordinate $\eta_1 = x/\sqrt{\varepsilon_2}$, close to the right boundary and substitute it into Equation (44) to study the limit $\varepsilon_2 \to 0^+$. Moreover, we expand the inner displacement field in the form $W^{(1)} = W_0^{(1,0)} + \sqrt{\varepsilon_2} W_1^{(1,1)} + \cdots$. Then, we obtain the following subproblems at different orders in ε_2:

$O(\varepsilon_2^0)$ subproblem:

$$W^{(1,0)}_{0,\eta_1\eta_1} - W^{(1,0)}_{0,\eta_1\eta_1\eta_1\eta_1} = 0. \tag{58}$$

$O(\sqrt{\varepsilon_2})$ subproblem:

$$W^{(1,1)}_{1,\eta_1\eta_1} - W^{(1,1)}_{1,\eta_1\eta_1\eta_1\eta_1} = 0. \tag{59}$$

The following matching and boundary conditions are applied:

$$\begin{aligned} &\lim_{\eta_1\to\infty} W_0^{(1)} = 0, \quad \lim_{\eta_1\to\infty} W_1^{(1)} = 0, \\ &W^{(1)}_{0,\eta_1}(0, w_0) = 0, \quad W^{(1)}_{1,\eta_1}(0, w_0) = -W^{(1,0)}_{,x}(0, w_0). \end{aligned} \tag{60}$$

The solution of the previous system is

$$W_0^{(1)} = 0, \tag{61a}$$

$$W_{\ell,1}^{(1)} = W_{,x}^{(1,0)}(0, w_0)\,\mathrm{e}^{-\eta_1}, \tag{61b}$$

where, as previously, the symbol ℓ means that we obtain a solution for the left boundary layer.

In order to study the right boundary layer, we introduce the stretched coordinate $\eta_2 = (1-x)/\sqrt{\varepsilon_2}$, and follow a similar procedure. The final result is

$$W_0^{(1)} = 0, \quad W_{r,1}^{(1)} = -W_{,x}^{(1)}(1, w_0)\,\mathrm{e}^{-\eta_2}, \tag{62}$$

where the symbol r means that we obtain a solution for the right boundary layer.

Combining the inner and outer solutions, the $O(\varepsilon_1^1)$ approximation to the functional relation is given by

$$W^{(1)} = W^{(1,0)} + \left[W_{,x}^{(1,0)}(0, w_0)\exp\left(-\frac{x}{\sqrt{\varepsilon_2}}\right) - W_{,x}^{(1,0)}(1, w_0)\exp\left(\frac{x-1}{\sqrt{\varepsilon_2}}\right)\right] + \cdots . \tag{63}$$

5. NNM Dynamics

Once the functional relationship W of Equation (10) is obtained, the dynamics of the reference point w_0 is easily evaluated by considering Equations (13) or (14). Equation (13) enables us to develop an exact solution in terms of the inverse of the elliptic integrals. In fact, Equation (13) can be rewritten as

$$\begin{aligned}\left(\frac{\partial w_0}{\partial t^2}\right)^2 &= E_{\text{tot}} - (n^2\pi^2 + k_1)w_0^2 - \left[\frac{3(n^2\pi^2 + k_1)(1 + 2\cos 2n\pi x_0)}{16k_1 \sin^2 n\pi x_0}\right] w_0^{*2} w_0^2 \\ &\quad - 2\left[\frac{4k_1 + n^2\pi^2 - 3n^2\pi^2\cot^2 n\pi x_0}{4k_1}\right] w_0^4\end{aligned} \tag{64}$$

where

$$\begin{aligned}E_{\text{tot}} &= (n^2\pi^2 + k_1)w_0^{*2} + \left[\frac{3(n^2\pi^2 + k_1)(1 + 2\cos 2n\pi x_0)}{16k_1 \sin^2 n\pi x_0}\right] w_0^{*4} \\ &\quad + 2\left[\frac{4k_1 + n^2\pi^2 - 3n^2\pi^2\cot^2 n\pi x_0}{4k_1}\right] w_0^{*4},\end{aligned}$$

which can be integrated, leading to

$$t = \int_0^{\bar{w}_0} \times \tag{65}$$

$$\frac{\mathrm{d}w_0}{\sqrt{E_{\text{tot}} - (n^2\pi^2 + k_1)w_0^2 - \left[\frac{3(n^2\pi^2+k_1)(1+2\cos 2n\pi x_0)}{16k_1\sin^2 n\pi x_0}\right] w_0^{*2} w_0^2 - 2\left[\frac{4k_1+n^2\pi^2-3n^2\pi^2\cot^2 n\pi x_0}{4k_1}\right] w_0^4}}.$$

This expression can be inverted in terms of elliptic functions.

The perturbation approach followed in obtaining W suggests an alternative perturbation technique to solve Equation (14). Let us rewrite Equation (14) to this end as

$$\ddot{w}_0 + (n^2\pi^2 + k_1)w_0 + \left[\frac{3(n^2\pi^2 + k_1)(1 + 2\cos 2n\pi x_0)}{16k_1 \sin^2 n\pi x_0}\right] w_0^{*2} w_0 + \left[\frac{4k_1 + n^2\pi^2 - 3n^2\pi^2 \cot^2 n\pi x_0}{4k_1}\right] w_0^3 = 0. \tag{66}$$

Using, for example, the Lindstedt–Poincaré technique, we obtain the following nonlinear oscillation frequency:

$$\hat{\omega} = \sqrt{n^2\pi^2 + k_1} + \frac{\varepsilon_1 w_0^{*2}\, 9}{32(n^2\pi^2 + k_1)\sin^2 n\pi x} + O(\varepsilon_1^2). \tag{67}$$

It is interesting to compare the previous formula with that obtained after attacking the initial equations with a single-degree-of-freedom Galerkin discretization and the Lindstedt–Poincaré method:

$$\hat{\omega}_{1\mathrm{DOF}} = \sqrt{n^2\pi^2 + k_1} + \frac{\varepsilon_1 w_0^{*2}\, 9}{16(n^2\pi^2 + k_1)\sin^2 n\pi x} + O(\varepsilon_1^2). \tag{68}$$

We note that an overestimation of the nonlinear correction is present, when the assumed mode shape is different than that of an NNM.

6. Numerical Results

In this section, a physical example of a clamped-clamped beam with a constant rectangular cross-section and with the following physical parameters is considered: $E = 2.1 \times 10^{11}$ Pa, $\rho = 7800$ kg/m^3, $b = 2.5 \times 10^{-3}$ m, $h = 8 \times 10^{-4}$ m, $\ell = 1$ m, $P = 50$ N, $\tilde{k}_1 = 50$ N/m^2, $\tilde{k}_3 = 500$ N/m^4. The corresponding nondimensional parameters are $\varepsilon_1 = 10$, $\varepsilon_2 = 4.2 \times 10^{-6}$, $k_1 = 1$. As will become clear in the following, the value of the nonlinear parameter ε_1 corresponds to a small perturbation of the linear system.

Analytical results will be compared with those obtained from numerical simulations. These, in turn, are obtained by discretizing the spatial operators through standard finite difference schemes [18], then the resulting finite-dimensional dynamical system is numerically integrated using an adaptive step size Runge–Kutta routine [19]. A 100-degrees-of-freedom discretization is used for the present problem.

In Figure 1 the backbone curve, representing the ratio between the nonlinear free oscillation frequency and the linear natural frequency of the unperturbed system *versus* the amplitude of oscillation, is shown for the second mode ($n = 2$). The continuous line represents the present perturbation results (Equation (67)), dots indicate the measured free oscillation frequency obtained from the numerical integration, and the dashed line indicates the analytical backbone obtained from a one-degree-of-freedom Galerkin anlysis (Equation (68)). A good agreement between analytical and numerical results, despite that the frequency variation, is larger than 5% for the highest ampitudes of oscillation. This enlightens the accuracy of the energetic perturbative NNM procedure and its effectiveness in comparison with the simpler one-degree-of-freedom Galerkin approach.

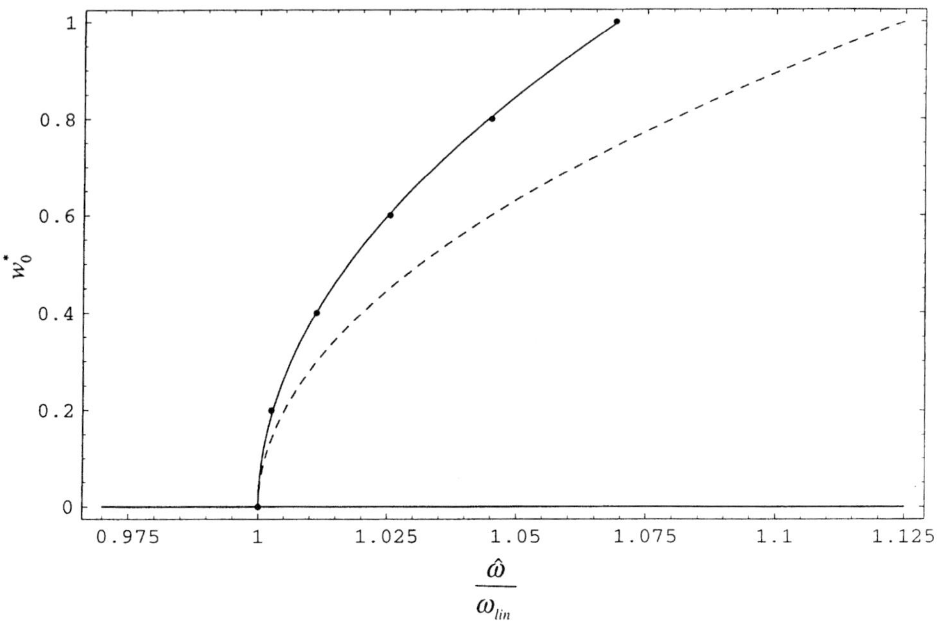

Figure 1. Backbone curve of the second NNM ($n = 2$): present NNM method, solid line; Galerkin method, dashed line; direct numerical simulations, dots.

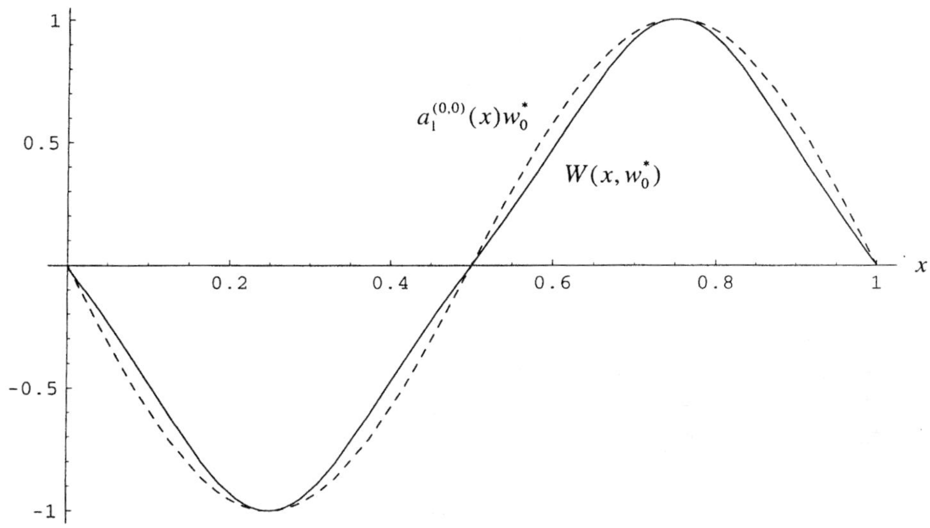

Figure 2. Spatial shape of the first spatial derivative of the NNM at the maximum amplitude of oscillation, $w_0^* = 1$: NNM shape, solid line; unperturbed linear string, dashed line.

In Figure 2, the shape of the NNM (solid line) and the mode shape of the unperturbed mode (dashed line) are depicted for the maximum amplitude of oscillation. We note that the effect of nonlinearity on the mode shape is not negligible. In particular, we note that the mode shape changes from the fundamental sinusoidal shape tending to a less smooth shape. Boundary layers are not visible in this plot because they only affect the first derivative.

In Figure 3, the first spatial derivative of the NNM shape (solid line) and of the mode shape of the unperturbed linear string mode (dashed line) are depicted for the maximum amplitude

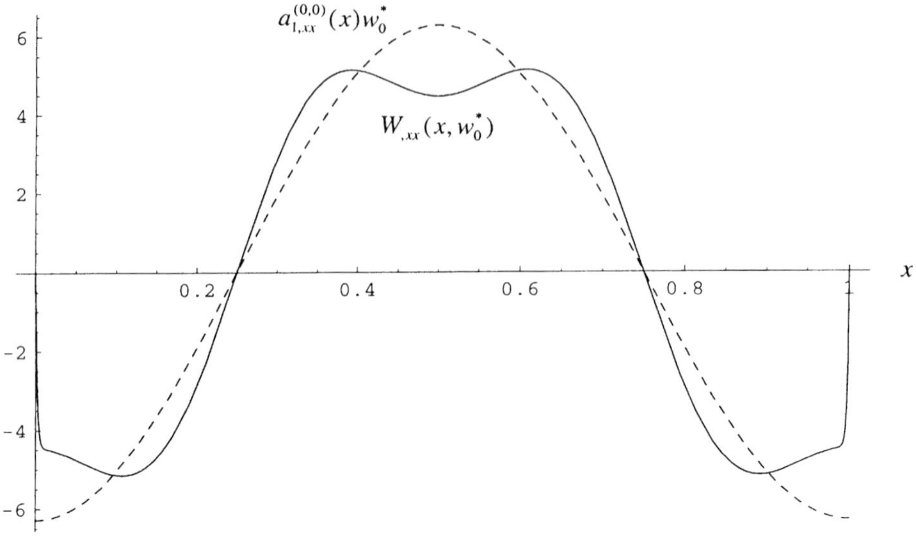

Figure 3. Spatial shape of the first spatial derivative of the NNM at the maximum amplitude of oscillation, $w_0^* = 1$: NNM shape, solid line; unperturbed linear string, dashed line.

of oscillation. We note that the effect of nonlinearity on the mode shape is stronger than in the primitive function. Strong ribble effects are present where the unperturbed shape presents a maximum; moreover, the boundary layers at the ends are clearly visible. The first derivatives of the unperturbed system approach the ends of the beam with nonzero amplitude. Conversely, the corrected shapes sharply fall to zero in a very small region in order to respect the boundary conditions.

7. Concluding Remarks

In this paper, we have employed an energy-based method to compute the NNM of a clamped-clamped tensioned beam with small bending stiffness, resting on a nonlinear spring bed. Local effects close to the clamped boundaries due to the small bending stiffness were analytically studied using matched asymptotic techniques. The analytical backbone curve derived from the present NNM calculation is in satisfactory agreement with the one derived by direct, finite-difference computations of the response of the original nonlinear partial differential equation of motion. We note at this point that the present study is of significance to thin elastic structures, whose dynamic response cannot be easily computed (even using sophisticated numerical techniques) due to the presence of boundary layers that lead to stiff problems. Hence, the present analysis can find applications in other elastic structures that possess very small bending stiffness.

References

1. Lyapunov, A. M., *The General Problem of the Stability of Motion*, Taylor & Francis, London, 1992.
2. Rosenberg, R. M., 'On non-linear vibrations of systems with many degrees-of-freedom', *Advances in Applied Mechanics* **9**, 1966, 155–242.
3. Vakakis, A. F. and Rand, R. H., 'Normal modes and global dynamics of a two-degree of freedom nonlinear system – I. Low energies', *International Journal of Non-Linear Mechanics* **27**, 1992, 861–874.

4. Vakakis, A. F. and Rand, R. H., 'Normal modes and global dynamics of a two-degree of freedom nonlinear system – II. High energies', *International Journal of Non-Linear Mechanics* **27**, 1992, 875–888.
5. Mikhlin, Yu. V., 'Matching of local expansions in the theory of nonlinear vibrations', *Journal of Sound and Vibration* **182**, 1995, 577–588.
6. Nayfeh, A. H. and Nayfeh, S. A., 'On nonlinear modes of continuous systems', *Journal of Vibration and Acoustics* **116**, 1994, 129–136.
7. Shaw, S. W. and Pierre, C., 'Normal modes of vibration for non-linear continuous systems', *Journal of Sound and Vibration* **169**, 1994, 319–347.
8. Nayfeh, A. H., Chin, C., and Nayfeh, S. A., 'On nonlinear normal modes of systems with internal resonance', *Journal of Vibration and Acoustics* **118**, 1996, 340–345.
9. Shaw, S. W. and Pierre, C., 'Normal modes for non-linear vibratory systems', *Journal of Sound and Vibration* **164**, 1993, 85–124.
10. Pellicano, F. and Mastroddi, F., 'Applicability conditions of a non-linear superposition technique', *Journal of Sound and Vibration* **200**(1), 1997, 3–14.
11. King, M. E. and Vakakis, A. F., 'An energy-based formulation for computing nonlinear normal modes in undumped continuous systems', *Journal of Vibration and Acoustics* **116**, 1994, 332–340.
12. Vakakis, A. F., Manevitch, L. I., Mikhlin, Y. V., Pilipchuk, V. N., and Zevin, A. A., *Normal Modes and Localization in Nonlinear Systems*, Wiley, New York, 1996.
13. Nayfeh, A. H., *Perturbation Methods*, Wiley, New York, 1973.
14. Van Dyke, M., *Perturbation Methods in Fluid Mechanics*, Academic Press, New York, 1964.
15. O'Malley, R. E., *Introduction to Singular Perturbations*, Academic Press, New York, 1974.
16. Pellicano, F. and Zirilli, F., 'Boundary layers and nonlinear vibrations of a moving beam', *International Journal of Non-Linear Mechanics* **33**(4), 1998, 691–711.
17. King, M. E. and Vakakis, A. F., 'An energy-based approach to computing resonant nonlinear normal modes', *Journal of Applied Mechanics* **63**(3), 1996, 810–819.
18. Hamming, R. W., *Numerical Methods for Scientists and Engineers*, Dover, New York, 1986.
19. Vetterling, W. T., *Numerical Recipes*, Cambridge University Press, Cambridge, 1987.

Nonlinear Dynamics **25:** 95–109, 2001.

The Description of Localized Normal Modes in a Chain of Nonlinear Coupled Oscillators Using Complex Variables

L. I. MANEVITCH
Institute of Chemical Physics, RAS, Kosygin str.,4, 117977 Moscow, Russia

(Received: 13 February 1999; accepted: 31 August 1999)

Abstract. We present an asymptotic approach to the analysis of coupled nonlinear oscillators with asymmetric nonlinearity based on the complex representation of the dynamic equations. The ideas of the approach are first developed for the case of the system with two degrees of freedom. Special attention is paid to the study of localized normal modes in a chain of weakly coupled nonlinear oscillators. We discuss also certain peculiarities of the localization of excitations in the case of strong coupling between the oscillators.

Keywords: Coupled oscillators, normal modes, localization.

1. Introduction

The complex representation of classical equations of motion for a system of linear oscillators was first used in quantum mechanics and for the analysis of so-called coupled oscillations and waves in mechanics, electronics, and solid state physics [1–6]. In this representation, the complex conjugate linear combinations $v_j + iu_j$ and $v_j - iu_j$ of displacements u_j and velocities v_j of oscillators can be visually presented as vectors of equal length rotating in opposite directions. Actually, it is enough to find only one complex function for each oscillator thus completely determining both displacement and velocity. Such a choice of variables in particular leads to a very simple and natural procedure of quantization, complex conjugate functions become operators of creation and annihilation, and their squared moduli – number of elementary excitations [1, 5, 8].

In the case of coupling between oscillators, both complex conjugate functions are included in each equation of motion. However, in the theory of coupled oscillations, it is supposed that the unidirectional rotations of oscillators have a stronger connection than vectors with opposite directions [2–4]. This simplification reduces the order of equations of motion by the factor of two and ensures the right expressions for the first two terms in the expansion of an exact solution of the linearized equations in powers of the coupling parameter [1, 2]. For a coupled system of nonlinear oscillators, the possibility of comparison with exact solutions is absent. In this connection, the equations of motion for unidirectional rotations are usually treated phenomenologically as the simplest mathematical model of the nonlinear oscillatory system [5]. Its validity is confirmed by comparison with the averaged equations for complex amplitudes [3, 5, 7], qualitative reasons [4] and asymptotic estimations [9]. However, the scheme of perturbation theory is not only expected to justify the domination of coupling between unidirectional rotations but also to provide a possibility of constructing higher approximations.

Such a scheme has been proposed in [10] in application to the systems with symmetric potential. We considered the complex representation of the equations of motion for the weakly coupled nonlinear oscillators as a natural form for efficient application of the asymptotic method after the introduction of 'slow' times [11, 12]. As this takes place, the coupled nonlinear equations for unidirectional rotations correspond to a principal approximation in 'slow time' and the possibility of corrections is ensured.

In this paper, we consider systems with asymmetric potentials. The paper is constructed as follows. In Section 2, the basic ideas of the approach are illustrated by the example of a system of two coupled oscillators. In Section 3, the nonlinear waves in finite and infinite systems of coupled nonlinear oscillators are considered.

Numerous interesting physical effects in nonlinear dynamics are connected with the space localization of vibrations and waves. This problem has become the subject of growing interest during the last decades. To a great extent, it essentially relates to vibrations and waves called as breathers (see, e.g., [31]). It is beyond our scope to discuss all interesting results in this field, which are successfully reviewed in the above reference. The main goal of this paper is to demonstrate the possibility of crucial simplification of nonlinear problems by means of the complex representation. Therefore, we are dealing only with those references where this representation was used.

2. System of Two Weakly Coupled Nonlinear Oscillators

The efficiency of complex representation of the equations of motion becomes more and more obvious with the complication of the system considered. It was convincingly shown in [5] that this representation allows us to treat with maximum simplicity the complicated problems concerning the bifurcation of solutions in a system of two weakly coupled anharmonic oscillators. However, the complex representation is also considered there phenomenologically as a model system (in the case of symmetric potential only) which on a qualitative level is only connected with known realistic models.

Let us consider a system of two identical weakly coupled nonlinear oscillators with asymmetric cubic potentials. Their dynamics is described by the equations

$$\begin{aligned} m\frac{\mathrm{d}^2U_1}{\mathrm{d}t^2} + 2n\frac{\mathrm{d}U_1}{\mathrm{d}t} + c_1U_1 + c_2U_1^2 + c_3U_1^3 + c_{12}(U_1 - U_2) &= 0, \\ m\frac{\mathrm{d}^2U_2}{\mathrm{d}t^2} + 2n\frac{\mathrm{d}U_2}{\mathrm{d}t} + c_1U_2 + c_2U_2^2 + c_3U_2^3 + c_{12}(U_2 - U_1) &= 0, \end{aligned} \tag{1}$$

where U_j $(j = 1, 2)$ are displacements of the oscillators.

We rewrite these equations using the complex representation

$$\begin{aligned} &\frac{\mathrm{d}\psi_j}{\mathrm{d}\tau} - i\psi_j + \varepsilon^2\gamma(\psi_j + \psi_j^*) - \alpha_1\varepsilon(\psi_j - \psi_j^*)^2 + i\alpha_2\varepsilon^2(\psi_j - \psi_j^*)^3 \\ &\qquad - i\beta\varepsilon^2[(\psi_j - \psi_j^*) - (\psi_k - \psi_k^*)] = 0, \quad j = 1, 2, k = 3 - j, \end{aligned} \tag{2}$$

where

$$\psi_j = v_j + iu_j, \quad \psi_j^* = v_j - iu_j (j = 1, 2); \quad v_j = \frac{\mathrm{d}u_j}{\mathrm{d}\tau}; \quad i = \sqrt{-1};$$

$\tau = \omega_0 t$; $\quad u_i = \dfrac{U_i}{r_0}$; $\quad r_0$ is the distance between particles in the undisturbed state;

$$\varepsilon^2\gamma = \frac{n}{\sqrt{c_1 m}}; \quad 4\alpha_1\varepsilon = \frac{c_2 r_0}{c_1}; \quad 8\alpha_2\varepsilon^2 = \frac{c_3 r_0^2}{c_1};$$

$$2\varepsilon^2\beta = \frac{c_{12}}{c_1}; \quad \varepsilon \ll 1; \quad \omega_0 = \sqrt{c_1/m}.$$

Introducing the new variables $\varphi_j = \psi_j \,\mathrm{e}^{-i\tau}$ and 'slow' times

$$\tau_1 = \varepsilon\tau_0; \quad \tau_2 = \varepsilon^2\tau_0, \ldots, \quad \text{where} \quad \tau_0 = \tau, \tag{3}$$

we consider φ_j as functions of many times $\tau_0, \tau_1, \tau_2, \ldots$. Let us present a solution in a form of power expansion as

$$\varphi_j(\tau_0, \tau_1, \tau_2, \ldots) = \varphi_{j,0} + \varepsilon\varphi_{j,1} + \varepsilon^2\varphi_{j,2} + \cdots. \tag{4}$$

The substitution of the above expressions into Equation (2) leads to the equation

$$\begin{aligned}
&\frac{\partial}{\partial\tau_0}(\varphi_{j,0} + \varepsilon\varphi_{j,1} + \varepsilon^2\varphi_{j,2} + \cdots) + \varepsilon\frac{\partial}{\partial\tau_1}(\varphi_{j,0} + \varepsilon\varphi_{j,1} + \varepsilon^2\varphi_{j,2} + \cdots)\\
&\quad + \varepsilon^2\frac{\partial}{\partial\tau_2}(\varphi_{j,0} + \varepsilon\varphi_{j,1} + \varepsilon^2\varphi_{j,2} + \cdots) + \cdots + \varepsilon M(\varphi_{j,0}, \varphi_{j,0}^*, \varphi_{j,1}, \varphi_{j,1}^*, \ldots, \varepsilon)\\
&\quad - i\varepsilon^2\beta[(\varphi_{j,0} - \varphi_{k,0}) - \mathrm{e}^{-2i\tau}(\varphi_{j,0}^* - \varphi_{k,0}^*) + \cdots] = 0,
\end{aligned} \tag{5}$$

where

$$\begin{aligned}
&M(\varphi_{j,0}, \varphi_{j,0}^*, \varphi_{j,1}, \varphi_{j,1}^*, \ldots, \varepsilon)\\
&\quad = \gamma\varepsilon[(\varphi_{j,0} + \varepsilon\varphi_{j,1} + \cdots) + \mathrm{e}^{-2i\tau}(\varphi_{j,0}^* + \varepsilon\varphi_{j,1}^* + \cdots)]\\
&\qquad - \alpha_1\big[\mathrm{e}^{i\tau_0}(\varphi_{j,0} + \varepsilon\varphi_{j,1} + \cdots)^2 - 2\,\mathrm{e}^{-i\tau_0}|\varphi_{j,0}^* + \varepsilon\varphi_{j,1}^* + \cdots|^2\\
&\qquad + \mathrm{e}^{-3i\tau}(\varphi_{j,0}^* + \varepsilon\varphi_{j,1}^* + \cdots)^2\big]\\
&\qquad + i\alpha_2\varepsilon\big[\mathrm{e}^{2i\tau}(\varphi_{j,0} + \varepsilon\varphi_{j,1} + \cdots)^3 - 3|\varphi_{j,0} + \varepsilon\varphi_{j,1} + \cdots|^2(\varphi_{j,0} + \varepsilon\varphi_{j,1} + \cdots)\\
&\qquad + 3\,\mathrm{e}^{-2i\tau}|\varphi_{j,0} + \varepsilon\varphi_{j,1} + \cdots|^2(\varphi_{j,0}^* + \varepsilon\varphi_{j,1}^* + \cdots) - \mathrm{e}^{-4i\tau}(\varphi_{j,0}^* + \varepsilon\varphi_{j,1}^* + \cdots)^3\big],\\
&\qquad\quad j, k = 1, 2, k = 3 - j.
\end{aligned} \tag{6}$$

Now we equate with zero the coefficients at each growing power of the parameter ε:

(1) ε^0: $\partial\varphi_{j,0}/\partial\tau_0 = 0$. Therefore, $\varphi_{j,0} = \varphi_{j,0}(\tau_1, \tau_2, \ldots)$, i.e. in the principal approximation, the functions $\varphi_{j,0}$ depend on 'slow' times only.

(2) ε^1:

$$\frac{\partial\varphi_{j,1}}{\partial\tau_0} = -\frac{\partial\varphi_{j,0}}{\partial\tau_1} + \alpha_1(\mathrm{e}^{i\tau_0}\varphi_{j,0}^2 - 2\,\mathrm{e}^{-i\tau_0}|\varphi_{j,0}|^2 + \mathrm{e}^{-3i\tau_0}\varphi_{j,0}^{*2}).$$

The condition of the absence of secular terms leads to the equation

$$\frac{\partial\varphi_{j,0}}{\partial\tau_1} = 0, \quad \text{so } \varphi_{j,0} = \varphi_{j,0}(\tau_2, \tau_3, \ldots). \tag{7}$$

Integrating equations for $\varphi_{j,1}$ with respect to the variable τ_0 and taking into account Equation (7), we obtain

$$\varphi_{j,1} = -\alpha_1 i \left(e^{i\tau_0}\varphi_{j,0}^2 + 2\,e^{-i\tau_0}|\varphi_{j,0}|^2 - \frac{1}{3}\,e^{-3i\tau_0}\varphi_{j,0}^{*2} \right). \tag{8}$$

Now we consider the terms of the second order:

(3) ε^2:

$$\begin{aligned} \frac{\partial\varphi_{j,2}}{\partial\tau_0} &= -\frac{\partial\varphi_{j,0}}{\partial\tau_2} - \frac{\partial\varphi_{j,1}}{\partial\tau_1} + \gamma(\varphi_{j,0} + e^{-2i\tau_0}\varphi_{j,0}^*) \\ &\quad + 2\alpha_1[e^{i\tau_0}\varphi_{j,0}\varphi_{j,1} - e^{-i\tau_0}(\varphi_{j,0}\varphi_{j,1}^* + \varphi_{j,1}\varphi_{j,0}^*) + e^{-3i\tau_0}\varphi_{j,0}^*\varphi_{j,1}^*] \\ &\quad - i\alpha_2(e^{2i\tau_0}\varphi_{j,0}^3 - 3|\varphi_{j,0}|^2\varphi_{j,0} + 3\,e^{-2i\tau}|\varphi_{j,0}|^2\varphi_{j,0}^* - e^{-4i\tau_0}\varphi_{j,0}^{*3}) \\ &\quad + i\beta[(\varphi_{j,0} - \varphi_{k,0}) - e^{-2i\tau}(\varphi_{j,0}^* - \varphi_{k,0}^*)] = 0, \end{aligned} \tag{9}$$

from which we obtain the following conditions for the absence of secular terms:

$$\frac{\partial\varphi_{j,0}}{\partial\tau_2} + \gamma\varphi_{j,0} - \alpha i|\varphi_{j,0}|^2\varphi_{j,0} - i\beta(\varphi_{j,0} - \varphi_{k,0}) = 0, \tag{10}$$

where $j, k = 1, 2$, $k = 3 - j$, $\alpha = 3\alpha_2 - (20/3)\alpha_1^2$, which is the system of equations of motion in the principal approximation.

The system which is close to (10) (at $\gamma = 0$) was considered in [5] as an example of a phenomenological model ('a discrete model with self-localization'). It was shown that this system is integrable. Besides, at some energy of oscillations, the localized normal modes arise in addition to in-phase and out-of-phase normal modes. Their appearance is a consequence of the instability of in-phase or out-of phase normal modes (dependent on the sign of parameter α).

Let us discuss the influence of damping forces in more detail. Introducing into (10) the new variables

$$\varphi_{j,0} = e^{(i\beta-\gamma)\tau_2}\Phi_{j,0}, \quad j = 1, 2,$$

we obtain

$$\begin{aligned} \frac{\partial\Phi_{1,0}}{\partial\tau_2} + i\beta\Phi_{2,0} - i\alpha\,e^{-2\gamma\tau_2}|\Phi_{1,0}|^2\Phi_{1,0} &= 0, \\ \frac{\partial\Phi_{2,0}}{\partial\tau_2} + i\beta\Phi_{1,0} - i\alpha\,e^{-2\gamma\tau_2}|\Phi_{2,0}|^2\Phi_{2,0} &= 0. \end{aligned} \tag{11}$$

We note that the Lagrangian corresponding to the equations of motion (9) at $\gamma = 0$ has the form

$$L = \frac{1}{2}i\sum_{j=1}^{2}\left(\Phi_{j,0}\frac{\partial\Phi_{j,0}^*}{\partial\tau_2} - \Phi_{j,0}^*\frac{\partial\Phi_{j,0}}{\partial\tau_2}\right) - \beta(\Phi_{1,0}\Phi_{2,0}^* + \Phi_{2,0}\Phi_{1,0}^*) + \frac{1}{2}\alpha\sum_{j=1}^{2}\Phi_{j,0}^4,$$

and the integral of energy is written as

$$\beta(\Phi_1\Phi_2^* + \Phi_2\Phi_1^*) - \frac{1}{2}\alpha\sum_{j=1}^{2}|\Phi_{j,0}|^4 = H(\tau_3, \tau_4, \ldots). \tag{12}$$

Exact integrability of a system (11) at $\gamma = 0$ follows from the existence of energy integral (12) and the integral

$$\sum_{j=1}^{2} |\Phi_{j,0}|^2 = N(\tau_3, \tau_4, \ldots). \tag{13}$$

Returning to the full system (11) ($\gamma \neq 0$) we conclude that the relation (13) remains valid in this case also. Indeed, multiplying the first and second equations of (11) respectively by $\Phi^*_{1,0}$ and $\Phi^*_{2,0}$, making the operation of conjugation, and combining all four equations, we obtain

$$\frac{\partial N}{\partial \tau_2} = 0 \quad \text{or} \quad N = (\tau_3, \tau_4, \ldots). \tag{14}$$

Therefore, as well as in the case of zero damping [5], the unknown functions $\Phi_{1,0}$ and $\Phi_{2,0}$ can be presented as

$$\Phi_{1,0} = \sqrt{N} \cos\frac{\theta}{2}\, \mathrm{e}^{i\delta_1(\tau_2)}; \qquad \Phi_{2,0} = \sqrt{N} \sin\frac{\theta}{2}\, \mathrm{e}^{i\delta_2(\tau_2)}.$$

Substituting these expressions into (6), we obtain

$$\begin{aligned}
&-\frac{1}{2} i \,\mathrm{tg}\frac{\theta}{2}\frac{\partial\theta}{\partial\tau_2} - \frac{\partial\delta_1}{\partial\tau_2} - \beta\,\mathrm{tg}\frac{\theta}{2}\, \mathrm{e}^{i(\delta_2-\delta_1)} + \alpha N \,\mathrm{e}^{-2\gamma\tau_2} \cos^2\frac{\theta}{2} = 0,\\
&\frac{1}{2} i \,\mathrm{ctg}\frac{\theta}{2}\frac{\partial\theta}{\partial\tau_2} - \frac{\partial\delta_2}{\partial\tau_2} - \beta\,\mathrm{ctg}\frac{\theta}{2}\, \mathrm{e}^{i(\delta_1-\delta_2)} + \alpha N \,\mathrm{e}^{-2\gamma\tau_2} \sin^2\frac{\theta}{2} = 0.
\end{aligned} \tag{15}$$

Equating the real and imaginary parts of relations (15) to zero, we come to the equations for real functions $\theta(\tau)$ and $\Delta(\tau) = \delta_1(\tau) - \delta_2(\tau)$

$$\begin{aligned}
\frac{\partial\theta}{\partial\tau_2} &= 2\beta \sin\Delta,\\
\frac{\partial\Delta}{\partial\tau_2} &= 2\beta\, \mathrm{ctg}\theta \cos\Delta + \alpha N \,\mathrm{e}^{-2\gamma\tau_2} \cos\theta.
\end{aligned} \tag{16}$$

The stationary points in which the right-hand sides of Equations (16) are equal to zero are

$$(1) \quad \Delta = 0, \quad \theta = \frac{\pi}{2};$$

$$(2) \quad \Delta = \pi, \quad \theta = \frac{\pi}{2}.$$

They correspond to the in-phase and out-of-phase cooperative normal modes which also survive in the presence of energy dissipation. In the case $\gamma = 0$, the additional out-of-phase ($\alpha > 0$) or in-phase ($\alpha < 0$) normal modes appear [3–5]. They correspond to the values

$$\Delta = \pi \quad (\alpha > 0); \qquad \Delta = 0 \quad (\alpha < 0),$$

$$\sin\theta = \frac{2\beta}{\alpha N} = \frac{N_0}{N}, \quad \text{where} \quad N_0 = \frac{2\beta}{\alpha},$$

and describe the nonlinear oscillations which are localized more and more (with increasing N) in the vicinity of a single particle. Indeed, the ratio of complex functions $\Phi_{1,0}$ and $\Phi_{2,0}$ corresponding to additional modes, is written as

$$\frac{\Phi_{1,0}}{\Phi_{2,0}} = -\frac{\sqrt{1+\sqrt{1-\rho^2}}}{\sqrt{1-\sqrt{1-\rho^2}}}, \quad \text{where} \quad \rho = \frac{N_0}{N}.$$

Obviously, such modes arise only at $\rho < 1$ and at $\rho \to 0$ (that, naturally, falls outside the limits of applicability of the principal approximation) the full localization of excitation upon a single particle would be reached. The consideration of a dissipation once more reveals important aspects of the problem. In this case, at $N \cdot \mathrm{e}^{-2\gamma\tau_2} \geq N_0$, the energy of the system can also be originally localized on a particle. But because of the presence of an exponential coefficient in the second equation of motion, the localization of excitation becomes impossible in the instant $\tau_2 = 1/2\gamma(\ln N/N_0)$ and the energy should be redistributed between both particles. This important effect can be observed in computer experiments (in some instances, the domination of one mass is replaced by a redistribution of the energy between masses [13]).

Let us note that the determination of higher approximations can also be realized without difficulty, if the principal approximations for localized normal modes are known. For example, the functions $\varphi_{j,2}$ can be determined from the equations of first approximation after substitution of the functions $\varphi_{j,0}$, $\varphi_{j,1}$.

Direct computation of localized normal modes becomes difficult if the number of degrees of freedom is more than two. Therefore, generalization for more complicated cases requires a more universal approach. The existence of localized modes can be associated with bifurcations and instability of cooperative modes. For the system treated, the bifurcation analysis becomes more simple with the use of Equations (15), i.e. in terms of real functions $\theta(\tau_2)$, $\delta(\tau_2)$ [5]. However, the appropriate extension on the systems with many degrees of freedom also becomes impossible. In this connection, let us perform a bifurcation analysis using the complex equations (6) at $\gamma = 0$.

$$\begin{aligned} \frac{\partial \Phi_{1,0}}{\partial \tau_2} + i\beta\Phi_{2,0} - i\alpha|\Phi_{1,0}|^2\Phi_{1,0} &= 0, \\ \frac{\partial \Phi_{2,0}}{\partial \tau_1} + i\beta\Phi_{1,0} - i\alpha|\Phi_{2,0}|^2\Phi_{2,0} &= 0. \end{aligned} \tag{17}$$

Let us focus on the analysis of the stability of the out-of-phase collective mode

$$\Phi_{2,0} = -\Phi_{1,0}.$$

We consider the perturbations of this mode w_1 and w_2,

$$\Phi_{1,0} + w_1 \quad \text{and} \quad -\Phi_{1,0} + w_2. \tag{18}$$

Substituting expressions (18) in the equations of motion and taking into account relations (17), we obtain

$$\begin{aligned} \frac{\partial w_1}{\partial \tau_2} + i\beta w_2 - i\alpha(\Phi_{1,0}^2 w_1^* + 2|\Phi_{1,0}|^2 w_1 + 2\Phi_{1,0}|w_1|^2 + w_1^2\Phi_{1,0}^* + |w_1|^2 w_1) &= 0, \\ \frac{\partial w_2}{\partial \tau_2} + i\beta w_1 - i\alpha(\Phi_{1,0}^2 w_2^* + 2|\Phi_{1,0}|^2 w_2 - 2\Phi_{1,0}|w_2|^2 - w_2^2\Phi_{1,0}^* + |w_2|^2 w_2) &= 0, \end{aligned} \tag{19}$$

where

$$\Phi_{1,0} = \sqrt{\frac{N}{2}}\,\mathrm{e}^{i(\beta+\alpha(N/2))\tau_2}.$$

By combining these equations and by denoting through W_1 and W_2 the sum and the difference of w_1 and w_2 accordingly, we get the system of two nonlinear equations with respect to new variables.

Linearization of this system leads to two independent equations of parametric oscillations for functions W_1 and W_2, describing the perturbation of an initial mode with respect to in-phase and out-of-phase modes accordingly,

$$\begin{aligned}
\frac{\partial W_1}{\partial \tau_2} + i(\beta - \alpha N)W_1 - i\alpha\left[\frac{N}{2}\,\mathrm{e}^{2i(\beta+\alpha(N/2))\tau_2}\right] W_1^* &= 0,\\
\frac{\partial W_2}{\partial \tau_2} - i(\beta + \alpha N)W_2 - i\alpha\left[\frac{N}{2}\,\mathrm{e}^{2i(\beta+\alpha(N/2))\tau_2}\right] W_2^* &= 0,
\end{aligned} \tag{20}$$

At the value $N = 2\beta/\alpha$, the conditions of the main parametric resonance are satisfied and the periodic solution $W_1 = A\,\mathrm{e}^{i\omega\tau_2}$ exists. Here A is a real constant and $\omega = \beta + \alpha N$. This solution corresponds to the boundary between the stability and instability regions for the first equation (20). On this boundary and for $N \geq 2\beta/\alpha$ the intensive transfer of energy to the in-phase mode becomes possible and localized normal modes exist. In the second equation (20), as it is easy to see, the instability occurs at infinitesimal amplitudes, but it is realized as a phase shift of out-of-phase oscillations.

The exact solution of Equations (17) gives the same value of critical amplitude. But as is shown below, the applied approach allows us to find the conditions of the energy localization not only in finite-dimensional systems of high dimensionality, but also in infinite-dimensional models.

3. Nonlinear Dynamics of an Infinite Chain of Coupled Oscillators

The equations of motion for an infinite system of coupled nonlinear oscillators with damping have the form

$$\begin{aligned}
&m\frac{\mathrm{d}^2U_j}{\mathrm{d}t^2} + 2\tilde{\gamma}\frac{\mathrm{d}U_j}{\mathrm{d}t} + c_1U_j + c_2U_j^2 + c_3U_j^3 + \tilde{c}(2U_j - U_{j-1} - U_{j+1}) = 0,\\
&\qquad -\infty < j < \infty.
\end{aligned} \tag{21}$$

We introduce a change of variables

$$\tau = \sqrt{\frac{c_1}{m}}\,t; \quad u_j = \frac{U_j}{r_0}, \tag{22}$$

where r_0 is the distance between particles.

Then we have

$$\frac{\mathrm{d}^2u_j}{\mathrm{d}\tau^2} + 2\varepsilon^2\gamma\frac{\mathrm{d}u_j}{\mathrm{d}\tau} + u_j + 4\alpha_1\varepsilon u_j^2 + 8\alpha_2\varepsilon^2u_j^3 + 2\varepsilon^2\beta(2u_j - u_{j-1} - u_{j+1}) = 0, \tag{23}$$

where

$$4\alpha_1\varepsilon = \frac{c_2}{c_1} r_0, \quad 8\alpha\varepsilon^2 = \frac{c_3}{c_1} r_0^2, \quad \gamma\varepsilon^2 = \frac{\tilde{\gamma}}{\sqrt{c_1 m}}, \quad 2\varepsilon^2\beta = \frac{\tilde{c}}{c_1}, \quad \varepsilon \ll 1.$$

Let us now use the complex representation of the equations of motion,

$$\begin{aligned}\frac{d\psi_j}{d\tau} &- i\psi_j + \varepsilon^2\gamma(\psi_j + \psi_j^*) - i\alpha_1\varepsilon(\psi_j - \psi_j^*)^2 + i\alpha_2\varepsilon^2(\psi_j - \psi_j^*)^3 \\ &- \varepsilon^2 i\beta[2(\psi_j - \psi_j^*) - (\psi_{j-1} - \psi_{j-1}^*) - (\psi_{j+1} - \psi_{j+1}^*)] = 0,\end{aligned}$$

where $\psi_j = v_j + iu_j$; $\psi_j^* = v_j - iu_j$.

For new variables

$$\psi_j = e^{i\tau}\varphi_j,$$

we come to a set of equations

$$\begin{aligned}\frac{d\varphi_j}{d\tau} &+ \varepsilon^2\gamma(\varphi_j + e^{-2i\tau}\varphi_j^*) - i\alpha_1\varepsilon(e^{i\tau_0}\varphi_j^2 - 2\,e^{-i\tau_0}|\varphi_j|^2 + e^{-3i\tau_0}\varphi_j^{*2}) \\ &+ i\alpha_2\varepsilon^2(e^{2i\tau}\varphi_j^3 - 3|\varphi_j|^2\varphi_j + 3\,e^{-2i\tau}|\varphi_j|^2\varphi_j^* - e^{-4i\tau}\varphi_j^{*3}) \\ &- i\varepsilon^2\beta[(2\varphi_j - \varphi_{j-1} - \varphi_{j+1}) - e^{-2i\tau}(2\varphi_j^* - \varphi_{j-1}^* - \varphi_{j+1}^*)] = 0. \end{aligned} \tag{24}$$

We now introduce the 'slow' times

$$\tau_1 = \varepsilon\tau_0, \quad \tau_2 = \varepsilon^2\tau_0, \quad \tau_3 = \varepsilon^3\tau_0, \ldots, \quad \text{where} \quad \tau_0 = \tau, \tag{25}$$

and consider φ_j as the functions of the 'fast' time τ_0 and the 'slow' times $\tau_1, \tau_2, \ldots$. By presenting the unknown variables as expansions on the parameter ε,

$$\varphi_j = \varphi_{j,0} + \varepsilon\varphi_{j,1} + \varepsilon^2\varphi_{j,2} + \cdots \tag{26}$$

and substituting (26) in the equations of motion we obtain

$$\begin{aligned}&\frac{\partial}{\partial\tau_0}(\varphi_{j,0} + \varepsilon\varphi_{j,1} + \varepsilon^2\varphi_{j,2} + \cdots) + \varepsilon\frac{\partial}{\partial\tau_1}(\varphi_{j,0} + \varepsilon\varphi_{j,1} + \varepsilon^2\varphi_{j,2} + \cdots) \\ &\quad + \varepsilon^2\frac{\partial}{\partial\tau_2}(\varphi_{j,0} + \varepsilon\varphi_{j,1} + \varepsilon^2\varphi_{j,2} + \cdots) + \varepsilon M(\varphi_{j,0}, \varphi_{j,0}^*, \varphi_{j,1}, \varphi_{j,1}^*, \ldots, \varepsilon) \\ &\quad - i\varepsilon^2\beta\big\{\big[2(\varphi_{j,0} + \varepsilon\varphi_{j,1} + \cdots) - (\varphi_{j-1,0} + \varepsilon\varphi_{j-1,1} + \cdots) - (\varphi_{j+1,0} + \varepsilon\varphi_{j+1,1} + \cdots)\big] \\ &\quad - \big[2(\varphi_{j,0}^* + \varepsilon\varphi_{j,1}^* + \cdots) - (\varphi_{j-1,0}^* + \varepsilon\varphi_{j-1,1}^* + \cdots) \\ &\quad - (\varphi_{j+1,0}^* + \varepsilon\varphi_{j+1,1}^* + \cdots)\big]\,e^{-2i\tau_0}\big\} = 0, \end{aligned} \tag{27}$$

where the expression for M is as presented above.

Equating to zero the coefficients at various degrees of the small parameter ε we have:

(1) ε^0:

$$\frac{\partial\varphi_{j,0}}{\partial\tau_0} = 0, \quad \text{therefore } \varphi_{j,0} = \varphi_{j,0}(\tau_1, \tau_2, \ldots),$$

(2) ε^1:

$$\frac{\partial \varphi_{j,1}}{\partial \tau_0} = -\frac{\partial \varphi_{j,0}}{\partial \tau_1} + \alpha_1(\mathrm{e}^{i\tau_0}\varphi_{j,0}^2 - 2\,\mathrm{e}^{-i\tau_0}|\varphi_{j,0}|^2 + \mathrm{e}^{-3i\tau_0}\varphi_{j,0}^{*2}). \tag{28}$$

The condition of the absence of secular terms leads to the equation

$$\frac{\partial \varphi_{j,0}}{\partial \tau_1} = 0, \quad \text{so } \varphi_{j,0} = \varphi_{j,0}(\tau_2, \tau_3, \ldots). \tag{29}$$

Taking into account (29) and integrating (28) with respect to variable τ_0, we obtain

$$\varphi_{j,1} = -i\alpha_1(\mathrm{e}^{i\tau_0}\varphi_{j,0}^2 + 2\,\mathrm{e}^{-i\tau_0}|\varphi_{j,0}|^2 - \frac{1}{3}\mathrm{e}^{-3i\tau_0}\varphi_{j,0}^{*2}). \tag{30}$$

Let us now select the terms corresponding to the second order of the small parameter:

(3) ε^2:

$$\begin{aligned}\frac{\partial \varphi_{j,2}}{\partial \tau_0} &= -\frac{\partial \varphi_{j,0}}{\partial \tau_2} - \frac{\partial \varphi_{j,1}}{\partial \tau_1} + \gamma\big[(\varphi_{j,0} + \varepsilon\varphi_{j,1} + \varepsilon^2\varphi_{j,2}^* + \cdots) \\ &\quad + \mathrm{e}^{-2i\tau_0}(\varphi_{j,0}^* + \varepsilon\varphi_{j,1}^* + \varepsilon^2\varphi_{j,2}^* + \cdots)\big] \\ &\quad + 2\alpha_1 i\big[\mathrm{e}^{i\tau_0}\varphi_{j,0}\varphi_{j,1} - \mathrm{e}^{i\tau_0}(\varphi_{j,0}\varphi_{j,1}^* + \varphi_{j,1}\varphi_{j,0}^*) + \mathrm{e}^{-3i\tau_0}\varphi_{j,0}^*\varphi_{j,1}^*\big] \\ &\quad - i\alpha_2(\mathrm{e}^{2i\tau_0}\varphi_{j,0}^3 - 3|\varphi_{j,0}|^2\varphi_{j,0} + 3\,\mathrm{e}^{-2i\tau}|\varphi_{j,0}|^2\varphi_{j,0}^* - \mathrm{e}^{-4i\tau_0}\varphi_{j,0}^{*3}) \\ &\quad + i\beta\big[(2\varphi_{j,0} - \varphi_{j+1,0} - \varphi_{j-1,0}) - \mathrm{e}^{-2i\tau_0}(2\varphi_{j,0}^* - \varphi_{j+1,0}^* - \varphi_{j-1,0}^*)\big].\end{aligned} \tag{31}$$

The condition of the absence of secular terms lead to the system of equations of the motion in principal approximation

$$\frac{\partial \phi_{j,0}}{\partial \tau_2} + \gamma\phi_{j,0} - i\alpha|\phi_{j,0}|^2\phi_{j,0} - i\beta(2\phi_{j,0} - \phi_{j+1,0} - \phi_{j-1,0}) = 0, \quad -\infty < j < \infty. \tag{32}$$

We now introduce the new change of the variable

$$\varphi_{j,0} = \mathrm{e}^{-\gamma\tau_2}\Phi_{j,0}. \tag{33}$$

Then the system of equations in the principal approximation is presented as

$$\frac{\partial \Phi_{j,0}}{\partial \tau_2} - \alpha i\,\mathrm{e}^{-2\gamma\tau_2}|\Phi_{j,0}|^2\Phi_{j,0} - i\beta(2\Phi_{j,0} - \Phi_{j+1,0} - \Phi_{j-1,0}) = 0. \tag{34}$$

Let us seek the functions $\Phi_{j,0}$ in the form

$$\Phi_{j,0} = f_0(\tau_2)\,\mathrm{e}^{i(kr_0 j + \theta_0)}, \tag{35}$$

where r_0 is the distance between particles, θ_0 is the phase corresponding to $j = 0$. Substituting (35) in (34), we obtain

$$\frac{\partial f_0}{\partial \tau_2} - i\alpha\,\mathrm{e}^{-2\gamma\tau_2}|f_0|^2 f_0 - 4i\beta\sin^2\frac{kr_0}{2}\cdot f_0 = 0. \tag{36}$$

After a change of variable,

$$f_0 = \mathrm{e}^{4i\beta \sin(kr_0/2)\tau_2} w_0(\tau_2),$$

we come to the equation

$$\frac{\partial w_0}{\partial \tau_2} - i\alpha\, \mathrm{e}^{-2\gamma\tau_2} |w_0|^2 w_0 = 0, \tag{37}$$

which coincides with the known equation for a system with one degree of freedom. Its solution has the form

$$w_0 = \sqrt{N} \exp\left[-\frac{i\alpha}{2\gamma} N (\mathrm{e}^{-2\gamma\tau_2} - 1) \right],$$

so that

$$\Phi_{j,0} = \sqrt{N} \exp i \left[4\beta \sin^2 \frac{kr_0}{2} \tau_2 - \alpha \frac{N}{2\gamma} (\mathrm{e}^{-2\gamma\tau_2} - 1) + kr_0 j + \theta_0 \right], \tag{38}$$

where $\sqrt{N}$ is the modulus of the complex amplitude.

Returning to the initial complex variables we obtain

$$\psi_{j,0} = \sqrt{N}\, \mathrm{e}^{-\gamma\varepsilon^2\tau} \exp i \left[\left(\tau + 4\varepsilon^2 \beta\tau \sin \frac{kr_0}{2} \right) - \alpha\varepsilon^2 \frac{N}{2\gamma} (\mathrm{e}^{-2\gamma\varepsilon^2\tau} - 1) + kr_0 j + \theta_0 \right]. \tag{39}$$

Thus, the analysis of the principal approximation in slow time leads to the conclusion that in an infinite chain of weakly coupled damped nonlinear oscillators the quasi-harmonic waves with exponentially decreasing amplitude and variable frequency can spread. At $\gamma = 0$ and $\alpha = 0$, they turn out to be the usual harmonic waves

$$\psi_{j,0} = \sqrt{N}\, \mathrm{e}^{i(\omega\tau + kr_0 j + \theta_0)}, \tag{40}$$

where $\omega = 1 + 4\varepsilon^2\beta \sin^2(kr_0/2)$, with a spectrum of wavenumbers $0 \le kr_0 \le \pi$, spectrum of frequencies $1 \le \omega \le 1 + 4\varepsilon^2\beta$, and a constant amplitude. The spectrum of simple harmonic waves determined by expression (40) is characterized by double degeneration of all modes, except those corresponding to wavenumbers $k = 0$ and $k = \pi/r_0$. This means that the modes with different phases θ_0 having identical frequencies and wavenumbers can be obtained by a superposition of two such modes. Introduction of nonlinearity removes the degeneration [5]. Actually, the dependence of normal oscillation frequencies on distribution of the energy between the particles arises. And, generally speaking, such a distribution depends on the value θ_0.

Let us note that the solution obtained turns out to be asymptotically valid if the lengths of the considered waves are not too large when compared with interparticle distances. Further, we accept r_0 as a unit of length, then this condition will be noted as $k \sim 1$.

In the case $k \ll 1$ there is one more small parameter in the problem. Its appearance requires a revision of the overall procedure of asymptotic expansion.

In the case of strong coupling ($\tilde{c}/c_1 = O(1)$), the complex representation becomes justified (i.e. the rotations in opposite directions are separated in the principal approximation) if the coupling between oscillators is 'effectively weak' because of the relative smallness of the second differences in the equations of motion. It means that the field of applicability of complex representation in that case coincides with the sphere of applicability of the continuum approximation. The equations of motion (5), in this case, take the form

$$\frac{\mathrm{d}^2 u_j}{\mathrm{d}\tau^2} + 2\varepsilon^2\gamma \frac{\mathrm{d}u_j}{\mathrm{d}\tau} + u_j + 4\varepsilon u_j^2 + 8\alpha_2\varepsilon^2 u_j^3 + 2\beta(2u_j - u_{j-1} - u_{j+1}) = 0, \tag{41}$$

where $\beta = \bar{c}/c_1 = O(1)$. Now it is expedient to measure the distances between particles in the units of $\varepsilon^{-1}r_0$, so we have for the second differences in (41), after the introduction of the continuum approximation, the expansion

$$-(2u_j - u_{j-1} - u_{j+1}) = \varepsilon^2 \frac{\partial^2 u}{\partial \zeta^2} + \frac{\varepsilon^4}{12}\frac{\partial^4 u}{\partial \zeta^4} - \cdots,$$

where ζ is the corresponding space coordinate.

Thus, in spite of the fact that $\beta \approx O(1)$, the derivatives terms now contain a small parameter and the complex representation turns out to be justified when using the continuum approximation. The similar situation arises in the case of short waves. As a result, the equations in the principal approximation take the form (if $\gamma \neq 0$)

(1) $k \ll 1$:

$$i\frac{\partial \Phi_0}{\partial \tau_2} - \beta \frac{\partial^2 \Phi_0}{\partial \zeta^2} + \alpha\, \mathrm{e}^{-2\gamma\tau_2}|\Phi_0|^2\Phi_0 = 0. \tag{42}$$

(2) $\pi - k \ll 1$:

$$i\frac{\partial \tilde{\Phi}_0}{\partial \tau_2} + \beta \frac{\partial^2 \tilde{\Phi}_0}{\partial \zeta^2} + \alpha\, \mathrm{e}^{-2\gamma\tau_2}|\tilde{\Phi}_0|^2\tilde{\Phi}_0 = 0. \tag{43}$$

Relations (43) and (42) at $\gamma = 0$ become the nonlinear Schrödinger equation (NSE), which is the integrable system [14]. Its solution in a wide class of initial conditions can be obtained by the inverse scattering method [14]. It is known [14, 15] that, if $\alpha > 0$, at the relation of coefficients signs which has a place in the case of Equation (43), all periodic wave packets are unstable. As a result, the localized soliton-like waves ('envelope solitons') exist:

$$\tilde{\Phi}(\zeta, \tau_2) = \left(\frac{2S}{\alpha/\beta}\right)^{1/2} \mathrm{e}^{i(k\zeta - \omega\tau_2)}\, \mathrm{sech}[S^{1/2}(\zeta - v\tau_2)], \tag{44}$$

where

$$k = \frac{v}{2\beta}, \qquad \omega = \frac{v^2}{4\beta^2} - S.$$

Here amplitude and velocity of the soliton are independent parameters.

At $\alpha < 0$ ('soft' nonlinearity) the periodic wave packets described by Equation (43) are stable and solitonic solutions do not exist. At signs of coefficients in the NSE corresponding to Equation (42), on the contrary, the wave packets are unstable which leads to the formation of solitons. If $\alpha > 0$, the wave packets are stable and solitonic solutions are absent.

Thus, if the coupling between oscillators is not weak, the conditions of localized excitations formation are well known and are formulated in terms of the continuum approximation.

What happens in the case of the weak coupling when it is necessary to use a discrete description?

As well as in the system with two degrees of freedom, in the infinite chain of nonlinear oscillators, apart from cooperative modes, waves with some space localization can be realized. They result from the instability of the cooperative modes. Considering small perturbations of cooperative nonlinear normal modes, we substitute into Equations (34), instead of functions $\Phi_{j,0}$, the expressions $\Phi_{j,0} + w_{j,0}$. Then, taking into account that the functions $\Phi_{j,0}$ satisfy Equations (34) and $|w_{j,0}| \ll |\Phi_{j,0}|$ supposing $\gamma = 0$, we obtain

$$\frac{\partial w_{j,0}}{\partial \tau_2} - i\beta(2w_{j,0} - w_{j+1,0} - w_{j-1,0}) - i\alpha(\Phi_{j,0}^2 w_{j,0}^* + 2|\Phi_{j,0}|^2 w_{j,0}) = 0, \tag{45}$$

where the cooperative mode is determined by expression (38).

Let us now assume that instead of the infinite system of coupled oscillators, we consider a finite chain with the number of atoms $\tilde{N} = 2n$ and the conditions of periodicity are satisfied:

$$\begin{aligned} \Phi_{0,0} &= \Phi_{\tilde{N},0}, \quad \Phi_{-1,0} = \Phi_{\tilde{N}-1,0}, \quad \ldots, \\ w_{0,0} &= w_{\tilde{N},0}, \quad w_{-1,0} = w_{\tilde{N}-1,0}, \quad \ldots, \end{aligned} \tag{46}$$

and $j = 0, 1, 2, \ldots, 2n - 1$.

Then the wavenumber can have the magnitudes $k = (2\pi/\tilde{N})m$ where $m = 0, 1, 2, \ldots, n$. Let us consider solutions of the system (45) satisfying the relations

$$\begin{aligned} & w_{j+1,0} = e^{ik_1} w_{j,0}, \quad w_{j-1,0} = e^{-ik_1} w_{j,0}, \\ & k_1 = \frac{2\pi}{\tilde{N}} m_1, \quad (m_1 = 0, 1, 2, \ldots, \tilde{N}/2). \end{aligned} \tag{47}$$

The system (45) after substitution into (47) becomes uncoupled:

$$\begin{aligned} \frac{\partial w_{j,0}}{\partial \tau_2} &= 2i \left(2\beta \sin^2 \frac{k_1}{2} + \alpha N \right) w_{j,0} \\ &\quad + i\alpha N\, e^{2ikj} \exp\left[2i \left(\alpha N + 4\beta \sin^2 \frac{k}{2} \right) \tau_2 \right] w_{j,0}^*. \end{aligned} \tag{48}$$

Thus, stability analysis of the cooperative mode characterized by a wavenumber m, with respect to the mode with wavenumber m_1 is reduced to the equation of parametrical oscillations considered above. The criterion of instability of this cooperative mode (i.e. of reaching the boundary between regions of stability and instability) is determined by the condition

$$4\beta \sin^2 \frac{k}{2} + \alpha N = 3\alpha N + 4\beta \sin^2 \frac{k_1}{2}. \tag{49}$$

From here for the magnitude of squared complex amplitude corresponding to the instability of the collective mode with a wavenumber m, with respect to the mode with a wavenumber m_1, we obtain the expression

$$N = 2\frac{\beta}{\alpha} \left(\sin^2 \frac{k}{2} - \sin^2 \frac{k_1}{2} \right), \tag{50}$$

where

$$k = \frac{2\pi}{\tilde{N}} m, \quad k_1 = \frac{2\pi}{\tilde{N}} m_1.$$

As follows from (50), at $\alpha > 0$ each mode corresponding to a certain value m with increasing intensity of excitations sequentially becomes unstable with respect to the modes with $m_1 = m - 1, m - 2, \ldots$, smaller than the wavenumber of the considered nonlinear normal mode. The threshold excitation energy increases with decreasing m_1 at fixed m. It is necessary to focus especially on the case $m_1 = m$ when the critical value of the excitation energy is equal to zero. In this case, instability, no matter how small the amplitudes, is nothing but Lyapunov instability, and actually leads to a phase shift of an analyzable mode (a similar situation arose in the system with two degrees of freedom, Section 2). The most important is the first nontrivial instability with respect to the mode with $m_1 = m - 1$, corresponding to the minimal (if to eliminate the case $m_1 = m$) excitation energy.

The matter is that at superposition of modes with close wavenumbers m and m_1 (space beatings!), there is a tendency towards localization of excitation, which becomes more and more noticeable with an increase of energy. With a decrease of m_1, when the critical energy increases, the localization becomes weaker. Let us consider, for example, the stability of an out-of-phase mode of minimum length ($m = \tilde{N}/2$). Then, at the excitation energy corresponding to squared modulus of complex amplitude,

$$N = 2\frac{\beta}{\alpha}\left[1 - \sin^2 \frac{2\pi\left(\frac{\tilde{N}}{2} - 1\right)}{\tilde{N}/2}\right]$$

the instability is realized. This instability leads to the localization of oscillations generally on one particle, just as happens in a system with two degrees of freedom. It is clear that the mode formed is multiply degenerated, as the localization of excitation can be realized on any of the particles. A comparison with models consisting of three and four particles [5], allows us to conclude that all other critical values of energy correspond to the formation of unstable modes.

Now, returning to the initial model, we will consider the limiting case $N \to 0$. With increased N, the arguments of trigonometric functions for the nearest m and m_1 in (50) differ less and less so that, in the limit, the energetic barrier corresponding to the instability, tends to zero. But it means that all cooperative modes in an infinite chain are unstable (except for a homogenous mode for which there are no values m_1 satisfying the condition $m > m_1$) and spatially localized oscillations turn out to be unique elementary excitations.

If anharmonicity is negative ($\alpha < 0$), the situation is quite similar. In this case, all cooperative modes, except the out-of-phase mode with a minimal wavelength, are unstable in an infinite limit with respect to the modes with wavenumbers $m_1 > m$. As a result of this instability, the localized modes arise again.

Let us note that numerous applications of localized modes to the theory of oscillatory spectra in polyatomic molecules and molecular crystals are presented in [3–5, 16–23]. A common concept of localized nonlinear oscillations with applications to the problems of mechanics and polymer physics are presented in [24, 25] and [26–30], respectively.

4. Conclusion

The complex representation of the dynamics of coupled nonlinear oscillators turns out to be efficient in the asymptotic analysis of systems with weak as well as strong couplings. In the first case, the equations of the principal approximation remain discrete and, in the second case, the asymptotic approach leads to a continual description in terms of a nonlinear Schrödinger equation and its generalizations.

Acknowledgements

The author wishes to thank to E. B. Gusarova for help in the preparation of the paper. This work is supported by RBRF (grant No. 98-03-33366).

References

1. Rabinovich, M. I. and Trubetskov, D. I., *Introduction to the Theory of Vibrations and Waves*, Moscow, Nauka, 1984 [in Russian].
2. Louisell, W. H., *Coupled Mode and Parametric Electronics*, Wiley, New York, 1960.
3. Ovchinnikov, A. A., 'Localized long-living oscillatory states in molecular crystals', *Soviet Physics (Journal of Experimental and Theoretical Physics)* **30**, 1970, 147–154.
4. Skott, A. C. and Lomdahl, P. S., 'Between local and normal modes', *Chemical Physics Letters* **311**, 1985, 29–42.
5. Kosevich, A. M. and Kovalyov, A. S., *Introduction to Nonlinear Physical Mechanics*, Naukova Dumka, Kiev, 1989 [in Russian].
6. Pierce, J. K., 'Coupling of modes of propagation', *Journal of Applied Physics* **25**, 1954, 179–183.
7. Kosevich, A. M. and Kovalyov, A. S., 'Self-localization of oscillations in one-dimensional anharmonic lattice', *Soviet Physics (Journal of Experimental and Theoretical Physics)* **40**(5), 1975, 891–896.
8. Louisell, W. H., 'Correspondence between Pierce's coupled mode amplitudes and quantum operators', *Journal of Applied Physics* **33**, 1962, 2435–2436.
9. Manevitch, L. I., 'Complex representation in nonlinear dynamics', in *Dynamical Systems, Theory and Applications*, J. Awrejcewicz (ed.), ABA-GRAF, Lodz, Poland, 1997, pp. 221–224.
10. Manevitch, L. I., 'Complex representation of dynamics of coupled nonlinear oscillators', in *Mathematical Models of Non-Linear Excitations, Transfer, Dynamics, and Control in Condensed Systems and Other Media*, L. A. Uvarova and A. E. Arinstein (eds.), Kluwer, Dordrecht, 1999, pp. 269–300.
11. Cole, J. D., *Perturbation Methods in Applied Mathematics*, Blaisdell, Toronto, 1968.
12. Nayfeh, A. H., *Introduction to Perturbation Techniques*, Wiley, New York, 1981.
13. Gendelman, O. V. and Vakakis, A. F., Private communication.
14. Zaharov, V. E., Novicov, S. P., Shabat, B. V., and Pitaevsky, L. P., *Theory of Solitons*, Nauka, Moscow, 1980 [in Russian].
15. Whitham, G. B., *Linear and Nonlinear Waves*, Wiley, New York, 1974.
16. Henry, B. R. and Siebrand, W., 'Anharmonicity in polyatomic molecules', *Journal of Chemical Physics* **49**, 1968, 5369–5376.
17. Saje, M. I. and Jornter, J., 'Bond modes', *Advances of Chemical Physics* **47**, 1981, 293–321.
18. Jaffe, C. and Brumer, P., 'Local and normal modes: A classical perspective', *Journal of Chemical Physics* **73**, 1980, 5646–5658.
19. Sibert, E. L., Reinhardt, W. P., and Hynes, J. T., 'Classical dynamics of energy transfer between bonds in triatomics', *Journal of Chemical Physics* **77**, 1982, 3583–3584.
20. Benjiamin, I. and Levine, R. D., 'Oberton spectrum in terms of normal or of equivalent modes', *Chemical Physics Letters* **101**, 1983, 518–523.
21. van Rosmalen, O. S., Lachello, F., Levine, R. D., and Dieperim, A. E. L., 'Algebraic approach to molecular rotation-vibration spectra', *Journal of Chemical Physics* **79**, 1983, 2515–2536.
22. Thiele, E. and Wilson, D. J., 'Anharmonicity in unimolecular reactions', *Journal of Chemical Physics* **35**, 1961, 1256–1263.
23. Collins, A., 'Solitons in chemical physics', *Advances of Chemical Physics* **532**, 1983, 225–280.
24. Manevitch, L. I., Mikhlin, Yu. V., and Pilipchuk, V. N., *Normal Vibration Technique for Strongly Nonlinear Systems*, Nauka, Moscow, 1989 [in Russian].
25. Vakakis, A. F., Manevitch, L. I., Mikhlin, Yu. V., Pilipchuk, V. N., and Zevin, A. A., *Normal Modes and Localization in Nonlinear Systems*, Wiley, New York, 1996.

26. Manevitch, L. I. and Smirnov, V. V., 'Propagation of exothermic reaction in condensed matter', *Physics Letters A* **165**, 1992, 427–432.
27. Manevitch, L. I., Savin, A. V., Smirnov, V. V., and Volkov, S. N., 'Solitons in degenerate bistable systems', *Physics Uspekhi* **37**, 1994, 859–878.
28. Manevitch, L. I. and Savin, A. V., 'Soliton mechanism of exothermic structural transitions propagation in bistable systems', *Journal of Experimental and Theoretical Physics* **80**(4), 1995, 706–718.
29. Gendelman, O. V. and Manevitch, L. I., 'Exact soliton-like solutions in generalized dynamic models of quasi-one-dimensional crystal', *Journal of Experimental and Theoretical Physics* **85**(4), 1997, 824–829.
30. Manevitch, L. I. and Savin, A. V., 'Solitons in polyethylene', *Physical Review E* **55**, 1997, 4713–4727.
31. Aubry, S., 'Breathers in nonlinear lattices: Existence, linear stability and quantization', *Physica D* **103**, 1997, 201–250.

Nonlinear Dynamics **25:** 111–132, 2001.

Spatially Localized Models of Extended Systems

RALF W. WITTENBERG*
Institute for Mathematics and its Applications, University of Minnesota, 400 Lind Hall, 207 Church Street SE, Minneapolis, MN 55455, U.S.A.

PHILIP HOLMES
Program in Applied and Computational Mathematics, Princeton University, Fine Hall, Washington Road, Princeton, NJ 08544, U.S.A.

(Received: 18 February 1999; accepted: 31 August 1999)

Abstract. We investigate the construction of low-dimensional spatially localized models of extended systems. Specifically, the Kuramoto–Sivashinsky (KS) equation on large one-dimensional domains displays spatiotemporally complex dynamics that are remarkably well-localized in both real and Fourier space, as demonstrated by a (spline) wavelet representation. We show how wavelet projections may be used to construct various localized, relatively low-dimensional models of KS spatiotemporal chaos. There is persuasive evidence that short, periodized systems, internally forced at their largest scales, form minimal models for chaotic dynamics in arbitrarily large domains. Such models assist in the understanding of extended systems.

Keywords: Spatiotemporal chaos, low-dimensional models, wavelets, Kuramoto–Sivashinsky equation.

1. Introduction

For the prediction, simulation, control and understanding of high-dimensional systems with complex spatiotemporal dynamics, one frequently seeks low-dimensional models capturing the essential features. One approach toward obtaining such models is to propose, on the basis of considerations such as relevant symmetries and physical principles, relatively 'simple' *a priori* governing equations to model the system. Alternatively, one can project known governing equations onto the 'relevant' modes, obtained for instance through the proper orthogonal decomposition (POD), via Galerkin projection. Crucial aspects of this procedure are that the most essential modes are isolated and retained, and that the effects of neglected modes are modeled appropriately. The goal is to obtain a, hopefully tractable, system of the form

$$\dot{\mathbf{a}} = \mathbf{f}(\mathbf{a}, \mathbf{b}) \tag{1}$$

for the retained 'internal' modes of the model $\mathbf{a}$, where in the present procedure, the excluded, or neglected modes $\mathbf{b}$ are externally specified, rather than being 'slaved' to the internal modes. That is, while the internal modes $\mathbf{a}$ are analogous to the span of a 'critical' eigenspace, the modes $\mathbf{b}$ will not be represented in the form $\mathbf{b}(\mathbf{a})$, as in a center, center-unstable or inertial manifold reduction, since they derive from dynamics that are largely supported outside the model domain, or that involve length scales exceeding those of the model.

The latter approach has been successfully implemented and extensively studied in the last decade or so, particularly in the context of fluids [1, 10]. One identifies coherent structures

* Current address: Department of Mathematics, University of Michigan, 2072 East Hall, 525 East University Avenue, Ann Arbor, MI 48109, U.S.A. (E-mail: ralf@math.lsa.umich.edu).

via the POD (or Karhunen–Loève decomposition), performs a Galerkin projection of the governing Navier–Stokes equations onto the most energetic modes, and models the neglected modes to obtain a closed model capturing the essential dynamics.

In this paper, we investigate the construction of low-dimensional spatially localized models to understand spatiotemporally complex dynamics in a particular system. Our approach, while motivated by and related to the above-mentioned fluid studies, is novel and involves the use of wavelets for the study of PDEs and extended dynamical systems, exploiting them to extract localized models in space and scale. Our model problem is the one-dimensional Kuramoto–Sivashinsky (KS) equation,

$$u_t + u_{xxxx} + u_{xx} + uu_x = 0, \quad x \in [0, L], \tag{2}$$

where $u_t = \partial u/\partial t$, $u_x = \partial u/\partial x$, and after rescaling the only control parameter is the system length L; we assume periodic boundary conditions and set the (conserved) spatial mean to zero. For future reference, we rewrite the KS equation in the symbolic form

$$u_t = \mathcal{L}u + u\mathcal{D}u, \tag{3}$$

where $\mathcal{L} = -\partial_x^4 - \partial_x^2$, and $\mathcal{D} = -\partial_x$. For remarks on the derivation and applications of the KS equation, see, for instance, [22].

In (2), the u_{xx} term is responsible for energy production, destabilizing large scales, while the u_{xxxx} term strongly damps small scales, and the uu_x term stabilizes by nonlinear coupling, transferring energy between large and small scales. From the evolution equation for the Fourier modes $\hat{u}_q$ (where the Fourier representation is defined by $u(x, t) = \sum_q \hat{u}_q(t) \exp(iqx)$, $q = 2\pi n/L, n \in \mathbb{Z}$), it is readily apparent that the number of linearly unstable Fourier modes is proportional to the system length L. The effect of nonlinear coupling between excited and damped modes is to yield complicated dynamics, with an intricate bifurcation sequence as L increases, and for large enough L, extensive 'spatiotemporal chaos' (STC) with a positive density of positive Lyapunov exponents. Consequently, this system is popularly studied as a prototype of complex spatiotemporal dynamics in a partial differential equation (PDE), as visualization of the solutions of this equation in the STC regime (see Figure 1) suggests the presence of structures having characteristic space and time scales, and localized 'events' resembling short-L solutions. This motivates the search for localized models to represent the evolution and interactions of the coherent structures.

Due to the periodic boundary conditions, the POD approach [10] does not yield localized modes, as the POD eigenmodes for such a translationally invariant system are Fourier modes. This observation led to the proposal of localization via wavelet-based models [3, 6]. In previous work [22], we showed how a projection onto wavelet modes captured the space and scale localization of the dynamics remarkably well, and we obtained a characterization of the dynamics in terms of these modes. In the present paper, after reviewing those results in Section 2, we show how wavelets permit the extraction of relatively low-dimensional local models which capture the essentials of the complex dynamics. We emphasize that in contrast to the usual POD paradigm for extracting coherent structures, the wavelet modes are not intended to reproduce the typical shapes of the structures; rather, we use them to isolate features that are localized in both physical and Fourier space. This notion differs from that of spatially localized modes (Anderson localization) discussed in other papers in this issue: for us wavelets provide 'windowing' tools to examine local properties of solutions which are statistically homogeneous and spatially distributed.

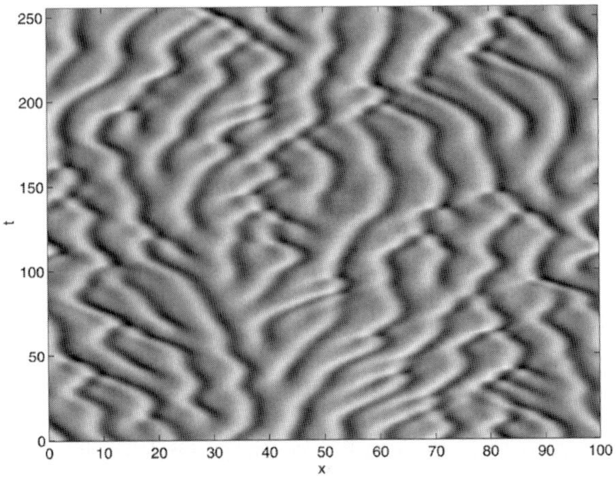

Figure 1. Gray-scale representation of the evolution of a solution of the KS equation (2) on the spatiotemporally chaotic attractor, showing typical local 'events', the interactions of coherent structures at a characteristic scale. Lighter shading indicates local maxima, darker shows minima, and the shading interpolates between extreme values of white at $u = +3.5$ and black at $u = -3.5$.

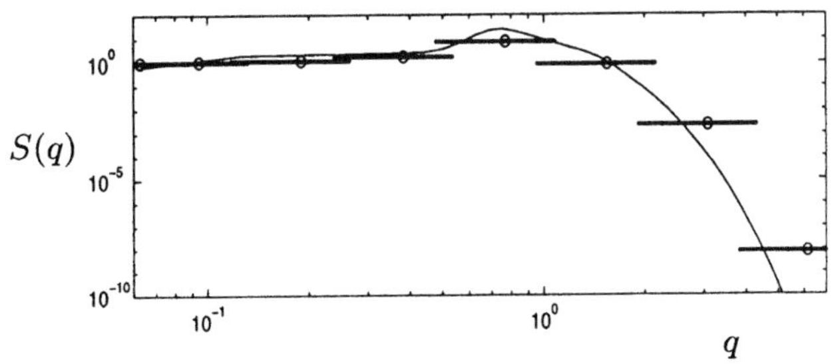

Figure 2. Power spectrum $S(q)$, compared to the energy per individual wavelet at each level j (horizontal lines). The wavelet distribution is plotted with the center and range of 99% of the support in Fourier space. From [22].

2. Characterization of Spatiotemporally Chaotic KS Dynamics

Spatiotemporal chaos (STC) in the KS equation has been extensively characterized and reported in the literature; see, for instance, [11, 17, 22], and the review in [20] (particularly chapters 1 and 4), which contains many additional details and references. Of particular importance in understanding the qualitative dynamics in the extensively chaotic regime is the power spectrum $S(q)$, or mean energy per Fourier mode, shown in Figure 2 [16, 19]. Here the power spectrum is defined in terms of the Fourier modes $\hat{u}_q$ by $S(q) = L\langle \hat{u}_{-q}\hat{u}_q \rangle$, and by invoking ergodicity, we replace ensemble averages $\langle \cdot \rangle$ on the attractor by time averages along a single solution trajectory.

The shape of the power spectrum reveals three characteristic regimes of the overall dynamics at different scales. The spectrum is almost flat in the *large* scale (low-q) region, reminiscent of a thermodynamically equilibrated regime with equipartition of energy. The *active* scales contain most of the energy, and have a pronounced peak for $q = q_m$ near the most linearly unstable mode, $q_m \approx q_0 = 1/\sqrt{2}$; that is, the local interactions occur at length scales near $l_m = 2\pi/q_m \approx 2\pi\sqrt{2}$. There is exponential decay of energy due to strong dissipation at the *small* scales.

2.1. The Wavelet Decomposition

Fourier-based properties such as the power spectrum are continuous in q, giving no separation of scales; we may wish to group ranges of Fourier modes to separate and distinguish between characteristic features of the dynamics at different scales. Furthermore, the Fourier representation is unable to capture any properties that arise from spatial localization, as each Fourier basis function is uniformly supported on the entire domain. This motivates the use of a basis localized in space and scale to detect dynamic features largely due to concentrated events at characteristic length scales, and wavelets are well suited to this purpose [22]. Following [3, 6, 7] we use a set of functions $\{\psi_{jk}\}$, symmetric about their centers $x_{jk} = L2^{-j}(k+1/2)$, indexed by the scale, or wavelet *level*, j, with k denoting the horizontal position within the level (by our choice of convention, $j = 0$ refers to the largest, or coarsest, scale, while increasing j implies 'zooming in' to smaller, or finer, scales). It is frequently convenient to visualize the set of wavelets in terms of a 'wavelet pyramid', with rows arranged in decreasing order of scale (increasing j), and horizontal location within each row representing the central position x_{jk}, as in the schematic cartoons of Figures 4, 5 and 8.

These wavelet functions $\{\psi_{jk}\}$ form an orthonormal basis for zero mean, finite energy periodic functions on $[0, L]$ of periodized, m-th order spline wavelets [15], which are of class C^{m-2}, have $m-1$ vanishing moments, exponential decay in x, and algebraic decay in q as $|q| \to 0$ and $|q| \to \infty$. In terms of this basis, in the following we use the wavelet representation of the solution $u(x,t)$ of the KS equation:

$$u(x,t) = \sum_{j=0}^{J} \sum_{k=0}^{2^j-1} a_{jk}(t)\psi_{jk}(x) = \sum_{\alpha} a_\alpha(t)\psi_\alpha(x). \tag{4}$$

The KS equation in terms of the wavelet coefficients, obtained by Galerkin projection of the governing equation (2) onto the decomposition (4), is then given by

$$\frac{\mathrm{d}}{\mathrm{d}t} a_\alpha(t) = \sum_{\alpha'} l_{\alpha\alpha'} a_{\alpha'} + \sum_{\alpha',\alpha''} n_{\alpha\alpha'\alpha''} a_{\alpha'} a_{\alpha''}, \tag{5}$$

where

$$l_{\alpha\alpha'} = -\int_0^L \psi_\alpha \left(\partial_{xx}\psi_{\alpha'} + \partial_{xxxx}\psi_{\alpha'}\right) \mathrm{d}x, \qquad n_{\alpha\alpha'\alpha''} = -\int_0^L \psi_\alpha \psi_{\alpha'} \partial_x \psi_{\alpha''} \,\mathrm{d}x,$$

and α represents the multi-index (j,k). The time-dependent wavelet coefficients $a_{jk}(t) = a_\alpha(t)$ satisfying (5) reveal the behavior of $u(x,t)$ at scale or level j and position given by k.

The results in this paper on wavelet characterization and models for the KS equation are all reported for $L = 100$, well within the STC regime. For other lengths, the results are qualitatively similar, but the details depend on the interplay between wavelet and intrinsic length scales, specifically, on the relation between the distance $l_j = L2^{-j}$ between adjacent wavelets at level j of the dyadic wavelet decomposition and the characteristic length of the dynamics l_m [20]. For $L = 100$, the peak of the energy spectrum is located in wavelet level $j = 4$, as seen in Figure 2, which shows the energy per individual wavelet by comparison with the Fourier power spectrum, and confirms that wavelets are well-localized in Fourier space. On combining the energy distribution per wavelet with the number 2^j of wavelets at

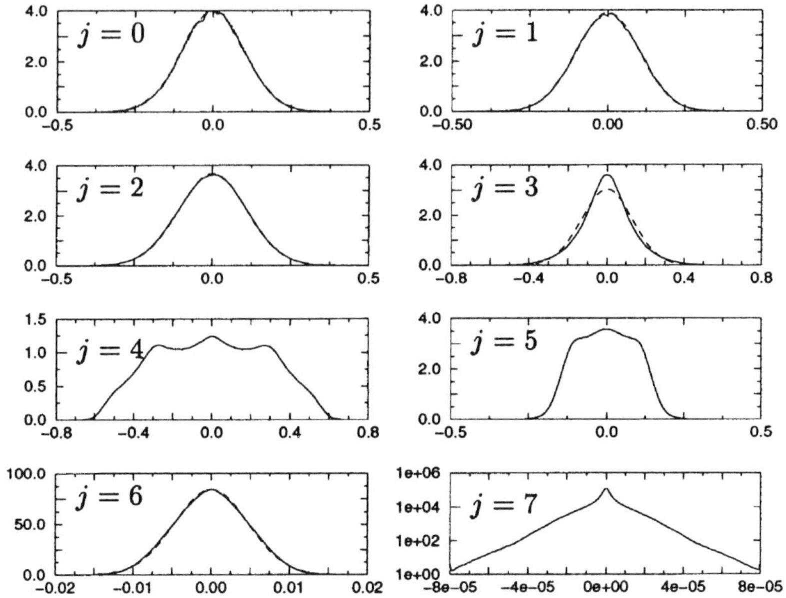

Figure 3. Probability density functions for the wavelet coefficients at each level $j = 0, \ldots, 7$, for $L = 100$. For $j = 0$-3 and 6, a best-fit Gaussian distribution is superposed (dashed line); the logarithmic axis for $j = 7$ indicates a near-exponential distribution. The figure for $j = 0$ also contains the PDF (dotted line) for a long-time simulation of the stochastic model (7), which is nearly coincident with the PDF from the KS simulation.

level j, the fraction of energy per wavelet level may be computed, and allows us to clearly distinguish between the levels with distinct behaviors. In fact, the active, energetic scales, levels $j = 3$, 4 and 5, together account for over 95% of the total energy, with level 4 alone containing about 71%. The large scales, levels 0, 1 and 2, contain under 5% of the total energy, and the small scale levels $j = 6, 7, \ldots$ are strongly damped with less than 0.1% of the energy. Inspection of time series and temporal correlations at the different levels indicates slow large-scale dynamics, correlation times of the order of $\tau \approx 10$ for the active scales – comparable to those of the system as a whole, and more rapid and intermittent behavior at the small scales.

2.2. Probability Distributions of Wavelet Coefficients

The most striking distinctions between the different scales and their dynamical significance are displayed in Figure 3 by the probability distribution functions (PDFs) for different j, which clearly show how the wavelet representation captures both scale and spatially localized information. At large scales, the distributions are Gaussian, and the dynamics resemble slow noise. This is consistent with the conjecture [23] that the large-scale dynamics may be described by a noise-driven Burgers equation with a positive effective viscosity $\nu > 0$,

$$u_t = \nu u_{xx} + \lambda u u_x + f, \tag{6}$$

where the stochastic forcing term may be interpreted as the derivative of Gaussian white noise. In this picture, the (deterministically chaotic) active and small scales simulate the effect of a random forcing on the large scales, as well as renormalizing the viscosity through nonlinear coupling. The form of this effective equation is motivated by symmetry principles, and the asymptotic validity of this description has been confirmed by several analytical and numerical studies [13, 18]; see [20, sec. 4.2.2]. The equation (6) has generated much interest since it was shown [12] that its integral form, known as the KPZ equation, is the simplest equation describing the kinetic roughening of growing interfaces with stochastic local dynamics [2, 8].

For modeling purposes, one can exploit the observed Gaussian dynamics of Figure 3 by replacing the 'noisy' large scales with an autonomously generated stochastic process motivated by the effective description (6). Computations [9, 18] of the parameters in (6) relevant to the KS equation show that for lengths L less than several thousand, the $\lambda u u_x$ term is asymptotically irrelevant to the scaling of solutions; so for the lengths we are interested in, we may linearize (6) by setting $\lambda = 0$. Additionally, the forcing term f has finite correlation time τ_f of order 10 [24], as expected since it is derived from the deterministic active and small scales. This suggests a linear stochastic model in which we use (6) with $\lambda = 0$ in Fourier space, and model the forcing with the simplest Langevin equation form that gives finite-time (exponential) correlations in the long-time limit. Our stochastic description for a large-scale Fourier mode $\hat{v}_q$ driven by colored noise is thus

$$\frac{\mathrm{d}}{\mathrm{d}t}\hat{v}_q = -\nu q^2\, \hat{v}_q + \hat{f}_q, \qquad \frac{\mathrm{d}}{\mathrm{d}t}\hat{f}_q = -\frac{1}{\tau_f}\,\hat{f}_q + \sqrt{D_f}\,\hat{\eta}_q, \tag{7}$$

where $D_f = Dq^2/L\tau_f^2$, and $\hat{\eta}_q$ is zero-mean, unit-variance, Gaussian white noise, with covariance $\langle \hat{\eta}_q(t)\hat{\eta}_{q'}(t')\rangle = \delta_{q+q',0}\delta(t-t')$. An advantage of this description is that the parameters in the effective forced Burgers description have all been determined for the KS equation, so no fitting is required; we use the values $D = 17.9$, $\nu = 7.5$ and $\tau_f = 7.0$ quoted in [9].

In the limit of long times and large scales ($q \to 0$), the spectra and correlations for this modified Ornstein-Uhlenbeck process (Langevin equation with colored noise) agree with theoretical predictions for the forced Burgers/KPZ model (6). With the values for D, ν and τ_f given above, we have simulated our stochastic model (7), and found it to reproduce the large-scale KS behavior extremely well, as shown in the PDF for level $j = 0$ in Figure 3 (our model is also appropriate for large-scale wavelets, since these are superpositions of few Fourier modes). We shall use the model (7) in numerical experiments below.

At the energetic scales $j = 3$, 4 and 5, the wavelet coefficient PDFs are strikingly different from any observed in the Fourier representation. The $j = 3$ level shows a distinct steepening of the distribution relative to a Gaussian, while levels $j = 4$ and (to a slightly lesser extent) $j = 5$ show strongly nonequilibrium PDFs, comparable to the distributions for pointwise values of u [9]: a broad, triple-humped distribution, superposing a peak at 0 and a double-humped peak reminiscent of cellular (sinusoidal-like) solutions. At these scales, the support of the wavelets is comparable to the characteristic intrinsic wavelength l_m of the dynamics, and the wavelet coefficients appear to capture on average the spatially local structures and events at the active scales.

The PDF for level $j = 6$ is again almost Gaussian; it is interesting that such an equilibrium distribution should appear well within the dissipative range, and this property is incompletely understood. At the smallest scales, $j \geq 7$, the effects of strong dissipation are apparent, and the wavelet coefficients display near-exponential PDFs reminiscent of those associated with small-scale intermittency in turbulence: the small-scale coefficients remain near zero most of the time, and undergo occasional excursions of relatively large amplitude driven by events at larger scales, and associated with the passage of 'coherent structures', that is, peaks or troughs of the field u.

2.3. Space Localization

Visual inspection of the collision and birth of peaks prevalent throughout the STC regime, as in Figure 1, suggests that the interactions of coherent structures are spatially localized [22]. Computation of characteristic lengths, such as those associated with the exponential decay of the two-point correlation function or the mutual information, confirms that sufficiently distant spatial locations are statistically independent; comparison of the bulk KS dynamics in the presence of changing boundary conditions also indicates the absence of significant interactions beyond a certain distance.

In [22] we presented an alternative, stronger view of spatial localization, showing with the aid of the wavelet Galerkin form of the KS equation (5) that $\bar{l}_c \approx 25$ is a good estimate of an dynamical interaction distance, capturing the localization of *instantaneous* coupling between modes instead merely of statistical averages on the attractor. Specifically, we observed that the Galerkin coefficients $l_{\alpha\alpha'}$ in (5) describe the coupling between wavelets ψ_α and $\psi_{\alpha'}$ separated by a distance $d_{\alpha\alpha'} = |x_\alpha - x_{\alpha'}|$, and decay due to the spatial exponential decay of $\psi(x)$; similarly for $n_{\alpha\alpha'\alpha''}$. This allows us to manipulate the dynamics, eliminating all coupling beyond an interaction distance l_c by setting $l_{\alpha\alpha'} = 0$ when $d_{\alpha\alpha'} > l_c$, and similarly for the nonlinear term. A range of experiments with varying l_c showed that for large enough l_c, we recover the essential KS dynamics, while decreasing l_c, cutting more and more interactions, significantly alters the dynamics, causing departures from typical events, space and time scales, inducing shifts in the energy distributions to larger scales, and frequent divergence and blowup of the temporal evolution due to disruption of the energy dissipation mechanisms. This led us to estimate a typical *dynamical* interaction length $\bar{l}_c \approx 25$, beyond which cutting interactions significantly disrupts the KS dynamics, and suggested that in low-dimensional 'short' models for the observed spatiotemporally complex behavior as described in Section 3 below, wavelets should be coupled, either to other modes or to external forcing, up to at least a distance $\sim \bar{l}_c$, to ensure sufficient interactions.

2.4. Scale-by-Scale Structure of the Dynamics

Whereas the above statistical results indicate how wavelet spectra and PDFs can discriminate between *averaged* contributions of different scales, we may also use wavelets as an experimental tool to discern the detailed *dynamical* significance of different wavelet levels to the overall spatiotemporally complex dynamics; the ensuing insights will form a further ingredient for our construction of local models.

In principle, our experiments here and in Section 3 focus on the evolution of a set of internal modes of our model, $\{a_\alpha\}, \alpha \in \mathbf{B}$, where $\mathbf{B}$ is a subset of the complete set $\mathbf{A}$ of wavelet indices; for the experiments of this section, $\mathbf{B}$ contains complete wavelet levels, whereas the models of Section 3 use a spatially localized subset, or 'box', of wavelet modes. These internal modes are forced by the external modes $\{b_{\alpha'}\}, \alpha' \notin \mathbf{B}$, so that the a_α satisfy

$$\begin{aligned} \frac{\mathrm{d}}{\mathrm{d}t} a_\alpha &= \sum_{\alpha' \in \mathbf{B}} l_{\alpha\alpha'} a_{\alpha'} + \sum_{\alpha' \notin \mathbf{B}} l_{\alpha\alpha'} b_{\alpha'} + \sum_{\alpha', \alpha'' \in \mathbf{B}} n_{\alpha\alpha'\alpha''} a_{\alpha'} a_{\alpha''} \\ &+ \sum_{\alpha' \in \mathbf{B},\ \alpha'' \notin \mathbf{B}} (n_{\alpha\alpha'\alpha''} + n_{\alpha\alpha''\alpha'}) a_{\alpha'} b_{\alpha''} + \sum_{\alpha', \alpha'' \notin \mathbf{B}} n_{\alpha\alpha'\alpha''} b_{\alpha'} b_{\alpha''}, \quad \alpha \in \mathbf{B}. \end{aligned} \tag{8}$$

In practice, this formulation is equivalent to overwriting modal coefficients at each time step, and this is the method we use for our experiments. Specifically, we integrate the full

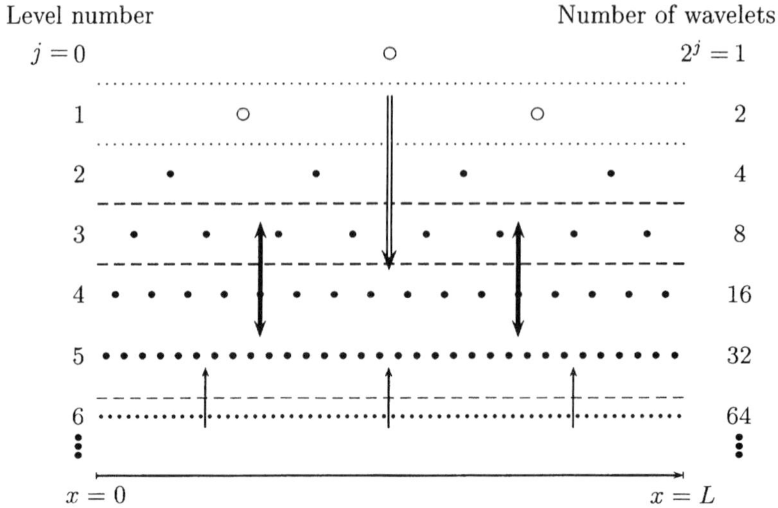

Figure 4. Schematic representation of Section 2.4, in which the interactions of complete wavelet levels and their contributions to the overall dynamics are investigated, particularly the influence of large scales on the active scale modes.

KS equation for the wavelet modes $\{a_\alpha\} = \{a_{jk}\}$ using a wavelet pseudospectral solver. We can then, say, eliminate wavelet level j by setting all the modes $b_{jk'}$ at that level to zero after each time step; alternatively, we can force at particular modes by replacing them with values computed from independent KS integrations, in which case the $\{b_{\alpha'}\}$ satisfy (5), or via an autonomous model of some kind for the $\{b_{\alpha'}\}$ (see [4] for a similar approach to the investigation of two-dimensional Navier–Stokes dynamics). By successively eliminating or driving different levels or combinations of levels, we deduce their respective contributions to the spatiotemporally chaotic dynamics, as suggested in the cartoon of Figure 4; in Section 3 we use a similar approach to driving a local set of modes to evaluate various localized models.

From the results of numerous such experiments for $L = 100$, described in [22] and more fully in [20], we find that the smallest scales, $j \geq 6$, are essentially irrelevant to the dynamics, being slaved to the larger scales, and may be eliminated at little cost except to detailed tracking. The most active, energetic scales $j = 4$ and 5 are crucial for both the characteristic spatial structure and the energy transfer mechanism, and must be included as internal modes in any model. Experiments in which some or all of the large-scale levels $j = 0$–2 are 'turned off', demonstrate that these scales contribute the noisy excitation that prevents the system from settling down in a cellular state and maintains the spatiotemporal disorder, though no one of these levels itself is vital to this purpose. Indeed, the large scales can drive the active scales even in the absence of intermediate levels, indicating that nonlocal energy transfer occurs. Level $j = 3$, intermediate between the large and most active scales, appears to play a major role in sustaining the typical 'events', the interactions of coherent structures through the creation and annihilation of defects.

In an autonomous model for a well-chosen subset of the total set of modes of the KS equation, we shall need to model neglected modes with external forcing terms. It is important to know how much freedom we have in this choice: how faithful does the forcing need to be to the 'true' behavior to induce qualitatively and, if possible, quantitatively correct dynamics? Answers to this question yield much information on the essential ingredients of the complex dynamics, and the robustness of the STC state. In view of spatial translational invariance and the strong scale-dependence of the behavior of the wavelet modes, as before we investigate the effects of forcing on a scale-by-scale basis [20, ch. 5].

As we observed from eliminating individual levels, forcing at levels $j \geq 6$ has negligible effect on the large- and active-scale dynamics, there being little back-propagation of energy; consequently, these small-scale levels will typically be neglected in the models of Section 3. In contrast, driving at or manipulating the most energetic levels $j = 4$ and/or 5, even with correct statistics obtained from an independent integration of the KS equation, sufficiently destroys the essential phase relationships and energy transfer mechanisms to lead to finite-time blowup. At the large-scale levels there is more flexibility, and driving them from an independent KS run certainly yields the 'correct' dynamics and statistics. In fact, we may go further: We have already observed in Section 2.2 that the statistics at the large scales are Gaussian, and may be well-simulated by an autonomously generated linear stochastic process. However, the KS equation is completely deterministic, and we might conceivably expect its characteristic dynamics to depend on correlations between large- and active-scale modes. We have tested this by letting the modes at wavelet levels 0, 1 and 2 each be driven independently by the stochastic process (7), the remaining levels undergoing KS dynamics; the resulting evolution is visually and statistically remarkably close to that of the full KS equation [20]. That is, purely stochastic large-scale evolutions appear to have the same effect as their deterministically derived counterparts in the way they drive chaotic dynamics at the active scales: the dynamical contributions of the large scales in the KS equation are essentially random. As a consequence, in a short model for the spatiotemporally complex dynamics, it is sufficient to use the simple linear random process (7) to simulate the effect of the noisy large scales which are present in a long system, thereby providing the 'heat bath' which keeps the system 'alive' and away from a simple equilibrium.

An interesting test of robustness, relevant to our modeling, is the extent of variation in the large-scale driving tolerated by the dynamics. It turns out that variations in the *rate* of large-scale forcing have little effect on the active-scale dynamics, which compensate for excessively rapid or slow forcing and respond at their intrinsic time scales. This is unsurprising if indeed, as is suggested by other experiments reported above, the sole purpose of the large scales is to provide Gaussian excitation, for then altering large-scale time scales merely varies the rate at which the Gaussian distribution is sampled.

On the other hand, the dynamics are sensitive to the driving *amplitudes*: excessive energy at the large scales induces rapidly traveling structures and shock-like features. Such behavior may be interpreted, at least qualitatively, by the observation that from the perspective of the active levels, high large-scale energy has the effect of changing the local mean, which by the Galilean invariance of the KS equation (2) is equivalent to imposing a drift. The traveling and shock solutions are also reminiscent of those observed in a destabilized version of the KS equation [21], in which the linear dispersion relation is shifted up uniformly through an additional $(\varepsilon^2 - 1)u$ term, $\varepsilon > 1$, on the right-hand side of (2). For the purposes of modeling, it is thus necessary to take more care in simulating the amplitudes than the time scales of the neglected modes.

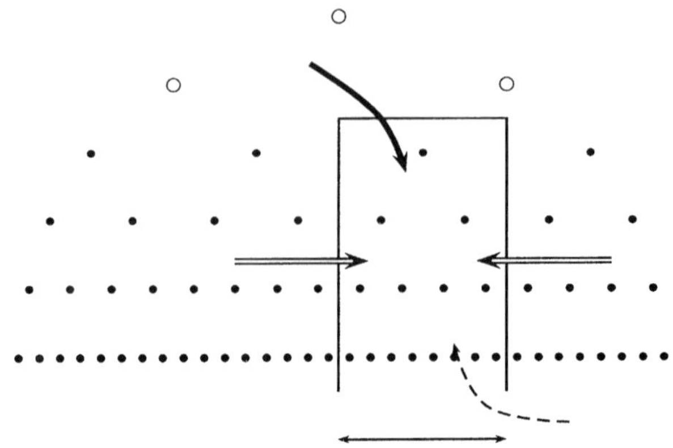

Figure 5. Simplified 'cartoon' representation of our approach to extracting externally forced models for a spatially localized subsystem of the full system, as in Section 3.1.

3. Spatially Localized Models

We now combine the above ingredients into the construction of localized models for spatiotemporally chaotic dynamics. We have shown that the dynamics are localized in space and strongly scale-dependent, and that the well-localized wavelet modes enable a good characterization of the KS dynamics and a detailed understanding of the relevant contributions of respective wavelet levels. As previously discussed, we wish to critically examine the idea of a large, extensively chaotic system in the 'thermodynamic limit' being composed of smaller, weakly interacting subsystems of relatively low dimension, by the explicit construction of such candidate spatially localized subsystems. This motivates us to extract a 'box' of internal wavelet modes $\{a_\alpha\}$ evolving according to (8), using the results on the scale-by-scale structure to choose appropriate external forcing $\{b_{\alpha'}\}$.

As depicted schematically in Figure 5, for our localized model we extract a subset **B** of the full wavelet hierarchy, of wavelets centered on a domain of length $\bar{L} = 2^{-j_0}L$; typically we use $j_0 = 2$ for a full $L = 100$ system, to yield short systems of length $\bar{L} = 25$. With this choice of lengths, we may exploit the results of the previous sections and of [22] on the separation of scales and contribution of different wavelet levels to the $L = 100$ KS dynamics, which carry over to multiples of L by powers of two. We choose these parameters for various reasons:

- We know that $\bar{l}_c \approx 25$ is a good estimate for a characteristic dynamic interaction length, so $\bar{L} = 25$ seems a reasonable length for a localized model.
- Larger $\bar{L}$ implies more complex internal dynamics, and requires a higher-dimensional model; for large enough $\bar{L}$, no external forcing is needed to obtain spatiotemporally complex dynamics, rendering the modeling redundant. In contrast, for the range $\bar{L} \in [20, 30]$, various types of 'simple' attractors exist for the KS equation, including fixed points, standing and traveling waves and heteroclinic cycles [11].
- For given $\bar{L}$, a larger 'full system' length L just adds more large-scale levels, providing slow Gaussian forcing, and contains additional boxes which do not significantly interact with the subsystem we are interested in, being separated from it by a distance greater than $\bar{l}_c$.

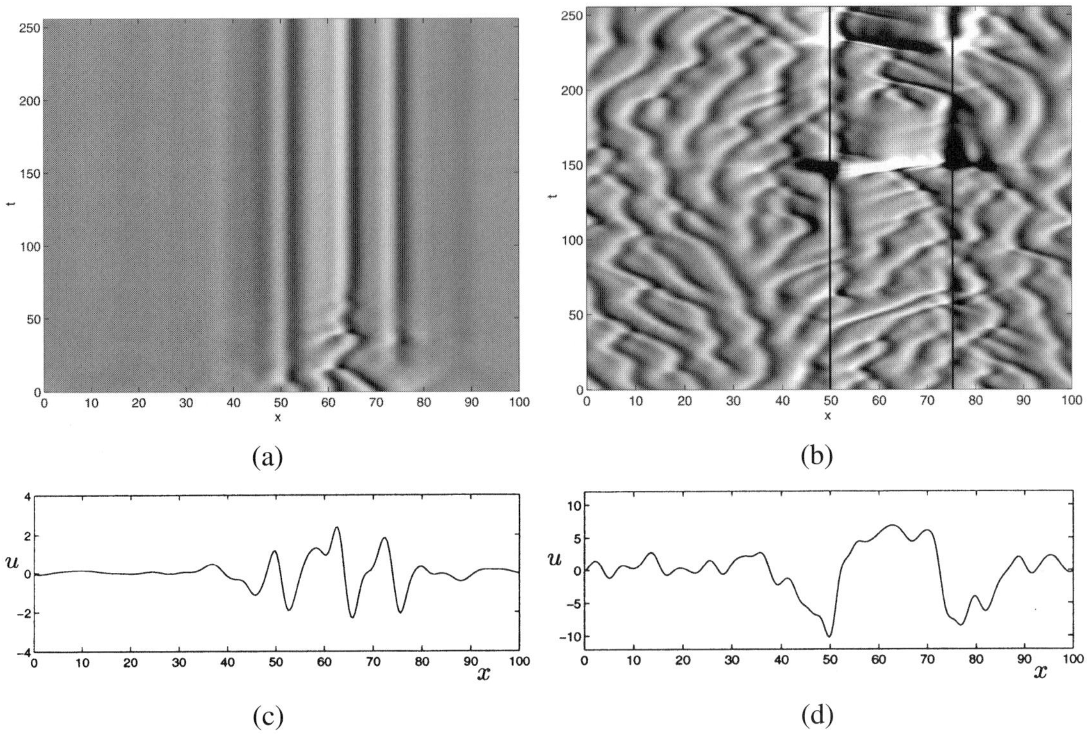

Figure 6. Local, nonperiodized model for the modes in the box of Figure 5. (a) No external forcing implies rigid dynamics; the $t = 150$ cross-section is shown in (c). (b) Forcing at external modes at levels $j = 0$–5 leads to excessive amplitudes and boundary effects, as in the $t = 150$ cross-section shown in (d).

3.1. Localized, Nonperiodic Models with Forcing

In the full KS equation, each local 'box' of wavelet modes interacts with its neighbors and with the large scales, which act as external forcing; and, provided the system size is sufficiently large, the dynamics are locally unaware of the global periodicity, the bulk dynamics being independent of boundary conditions. Thus it may seem reasonable to seek models in which the local box is forced from the large scales and external adjacent modes, without imposing the constraint of periodicity, as schematically portrayed in Figure 5. One would expect that success would depend on feeding in the 'correct' statistics for the active scale external modes, as described above [22]. It turns out, however, that in such models, the translational symmetry of the KS dynamics is broken too drastically, leading to atypical dynamics, as previously observed in preliminary unpublished experiments of J. Elezgaray (referred to in [14]).

In these experiments, the modes $\{a_\alpha\}$ inside the chosen 'box' are evolving within the localized, forced model, being driven by the outside modes $\{b_{\alpha'}\}$ according to (8). Our experimental procedure is described in Section 2.4, and in more detail in [20]; in summary, we integrate a full KS equation for the complete hierarchy of wavelet modes, and at each time step, all the modes outside the short box are either set to zero, or overwritten by values obtained from an independent full KS integration running simultaneously.

Figure 6 shows some representative results. In Figure 6a, the modes inside the box, rooted at wavelet level 2, evolve in the absence of any forcing: $b_{\alpha'} = 0, \alpha' \notin \mathbf{B}$. Such a system rapidly settles down to a steady state, with fixed peaks, as seen in the $t = 150$ cross-section in (c). In the absence of both forcing and periodic boundary conditions, therefore, this 15-dimensional dynamical system (containing wavelet levels $j = 2$–5) remains rigid.

The opposite extreme is shown in Figure 6b, in which the box **B** is driven at all large- and active-scale external modes $\alpha' \notin \mathbf{B}$ with time series $b_{\alpha'}(t)$ obtained from a full independent KS integration (5). The forcing terms thus automatically have the 'correct' statistics at the different levels (note that the figure is reconstructed on $L = 100$ from both the internal modes of the model and the forcing terms). The large amplitudes (particularly at the boundaries of the box) and rapid fronts apparent in Figure 6b, and in the $t = 150$ cross-section in (d), are due to the disruption of the normal feedback mechanisms, so that wavelet coefficients may grow abnormally before they decay. Specifically, the evolution (5) of the external modes $b_{\alpha'}$, which provide the driving time series, is independent of the modes a_α within the box, so that they cannot provide the 'correct' compensation for growth in the internal modes, sometimes serving rather to reinforce unusually large amplitudes (relative to those observed in a full KS run such as that of Figure 1).

The origin of the difficulties may be traced to the fact that the set **B** of modes internal to the model includes ones that are (sufficiently) linearly unstable and autonomously excited, instead of being slaved to the forcing. Consider the evolution of the differences $d_\alpha = a_\alpha - b_\alpha$ between internal modes a_α evolving according to (8), and corresponding driving coefficients b_α satisfying (5): the d_α obey the equation

$$\begin{aligned} \dot{d}_\alpha &= \sum_{\alpha' \in \mathbf{B}} l_{\alpha\alpha'} d_{\alpha'} + \sum_{\alpha', \alpha'' \in \mathbf{B}} n_{\alpha\alpha'\alpha''}(d_{\alpha'} b_{\alpha''} + d_{\alpha''} b_{\alpha'} + d_{\alpha'} d_{\alpha''}) \\ &\quad + \sum_{\alpha' \in \mathbf{B},\ \alpha'' \notin \mathbf{B}} (n_{\alpha\alpha'\alpha''} + n_{\alpha\alpha''\alpha'}) d_{\alpha'} b_{\alpha''}, \end{aligned} \tag{9}$$

which may be rewritten as

$$\dot{d}_\alpha = \sum_{\alpha' \in \mathbf{B}} [l_{\alpha\alpha'} + g_{\alpha\alpha'}(t)] d_{\alpha'} - \tfrac{1}{2} \sum_{\alpha', \alpha'' \in \mathbf{B}} n_{\alpha'\alpha''\alpha} d_{\alpha'} d_{\alpha''}. \tag{10}$$

Here we have used the identity $n_{\alpha'\alpha''\alpha} = -n_{\alpha\alpha'\alpha''} - n_{\alpha\alpha''\alpha'}$ [14], and defined the parametric driving term $g_{\alpha\alpha'}(t) = -\sum_{\alpha''} n_{\alpha'\alpha''\alpha} b_{\alpha''}(t)$, which has vanishing mean since all the $b_{\alpha''}$ have zero time average (see Figure 3).

If only a single internal mode α is driven by all others, then the nonlinear term in (10) vanishes (since $n_{\alpha\alpha\alpha} = 0$) and the difference d_α grows or decays exponentially, on average, depending on the sign of $l_{\alpha\alpha}$, that is, according to the linear stability or instability of mode α. That is, a model in which a single large-scale mode is driven by all others, will blow up, while a small-scale (high-j) mode will be synchronized to the control run.

In the general case in which **B** contains more than one element, the nonlinear term in (10) conserves energy, but transfers it among the internal modes, and the situation is more complicated; stability and instability depend on the operator $A_{\alpha\alpha'}(t) = l_{\alpha\alpha'} + g_{\alpha\alpha'}(t)$. However, we expect that if the internal modes are (sufficiently) linearly unstable, with all the stable modes participating in the driving, then $A_{\alpha\alpha'}(t)$ is, on average, positive definite, and the model energy grows exponentially. Conversely, if all large-scale modes are involved in the driving, and the internal modes $\alpha \in \mathbf{B}$ are all linearly stable, then $l_{\alpha\alpha'}$ (and thus, on average, $A_{\alpha\alpha'}(t)$) is negative definite, the difference decays, and tracking to the control run occurs: sufficiently many low modes uniquely determine the asymptotic dynamics, as small scales are slaved to large scales – in fact this can be shown rigorously under slightly stronger conditions [20].

More generally, if the 'box' contains some linearly unstable and some stable modes, then energy transfer from large to small scales prevents blow-up of the model. However, the possibility of tracking, that is, $d_\alpha \to 0$ for $\alpha \in \mathbf{B}$, depends on the linear part of (10), that is, on $A_{\alpha\alpha'}(t)$. We expect, from (10), that the presence of (sufficiently) unstable modes among the set $\mathbf{B}$ of internal modes implies positive eigenvalues of $A_{\alpha\alpha'}(t)$, and hence, instability of the synchronized state $\{d_\alpha = 0\}$. Our experience is that (for $L = 100$) any modes at wavelet levels $j \leq 3$ satisfy the requirements for instability; the situation for $j = 4$ is less clear. Thus, in models of the form of Figure 5, the modes $a_\alpha, \alpha \in \mathbf{B}$ in the box fail to synchronize to the behavior of the corresponding modes b_α satisfying the full KS equation. For more discussion, and results of numerical experiments confirming all the above scenarios, see [20, secs. 5.3, 6.1].

These results and simulations such as that of Figure 6b pose a serious problem for the construction of non-periodized, externally forced local models: since undesirable results including excessive amplitudes occur even if the forcing is from a full KS integration – that is, the driving necessarily has all the 'correct' statistical properties – we can hardly expect better results from more simplified types of external random fluctuation. The problem of excessive growth seems destined to occur whenever our model is able to support its own dynamics by containing unstable modes. Consequently, these conclusions do not necessarily depend on our wavelet formulation, or even on the KS equation; they may apply more generally to driving and synchronization of a localized model through boundary forcing.

Numerous other experiments including various types of external forcing – for instance, forcing only at certain wavelet levels, or only from a 'halo' of nearest neighboring active-scale modes – essentially confirm the above observations. We find in general that for models of the type depicted in Figure 5, one of two possibilities occurs, broadly corresponding to the two extreme cases of Figures 6a and 6b:

- The first case occurs when the excitation of the internal modes is weak; for instance, when it arises from the large scales, or is due only to a few external modes, as in Figure 7a, in which the modes in the box are forced only at level $j = 3$. In this case, the model dynamics are rigid; there may be a fluctuating peak in the interior, undergoing some typical 'events', but peaks at the boundaries of the model are stationary and excessively constrain the internal dynamics.
- In the second scenario, the internal modes are influenced more strongly by exterior modes, usually when forcing is at the active-scale levels $j = 4$ and/or 5, such as in Figure 7b. In this situation, more realistic dynamics can occur, including motion of the peaks; but there are mismatches at the boundaries and large-scale internal modes tend to be excited excessively due to failure of the standard damping mechanisms, leading to large amplitudes, high local means and rapid fronts.

In some experiments, the conditions for both these anomalies are satisfied, leading to even more disrupted model dynamics. The effect of breaking the translational symmetry, so that the modes at the edges of the box are essentially inequivalent to those in the center, is too drastic, and such local models, it appears, cannot reproduce the spatiotemporally complex dynamics typically observed in the KS equation. Another difficulty with such externally forced models is that, while they may be readily investigated numerically, they appear too complicated to be analytically tractable.

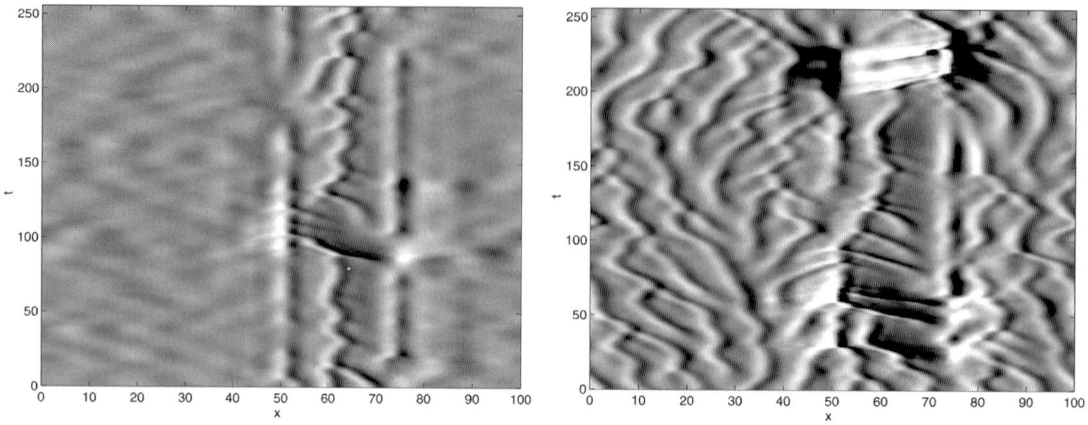

(a) (b)

Figure 7. Nonperiodized model, with modes in the distinguished box forced from the active scales of an independent control run at (a) level $j = 3$, (b) levels $j = 4$ and 5.

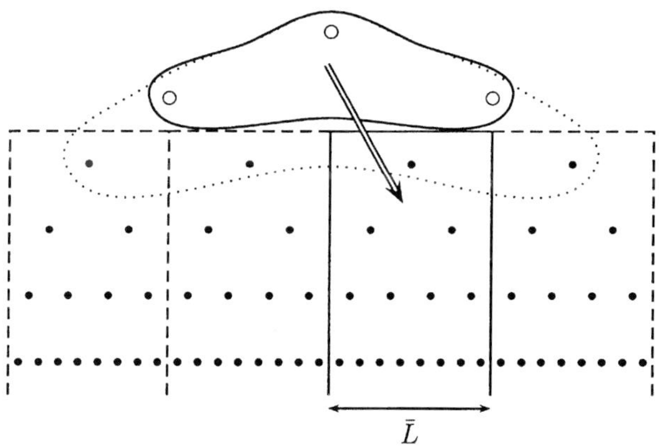

Figure 8. Schematic representation of periodized models of a short subsystem, forced from the largest-scale external and/or internal modes, as in Sections 3.2 and 3.3.

3.2. Periodized Models with Aperiodic Forcing

In the light of the above-mentioned difficulties, we have investigated another class of models, in which active-scale interactions are restored by periodizing. A motivation for imposing periodic boundary conditions is to replace the influence of missing adjacent 'external' modes by distant 'internal' modes [3], with the advantage of eliminating the drastic inequivalence between internal modes at the center and boundaries of the model, and thereby better approximating the infinitesimal translation symmetry of the KS equation (Fourier-based models of short periodic systems, which preserve the translation invariance, have been studied in [5]).

As before, we include a caricature of this approach in Figure 8. In practice, numerical experiments with such models are performed by selecting a 'distinguished box' of length $\bar{L} = 2^{-j_0}L$, and at each time step, setting wavelet coefficients in all adjacent boxes to the corresponding values in the distinguished box (where the centers of 'corresponding' wavelets are separated by a multiple of $\bar{L}$). That is, essentially we only evolve the modes within the distinguished box, subject to large-scale forcing from wavelet levels $j < j_0$.

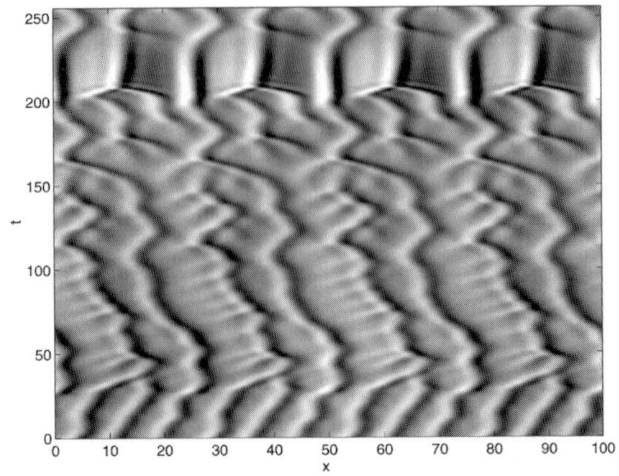

Figure 9. Periodized model, forced by levels $j = 0$ and 1 of an independent KS integration.

The theoretical basis and a consistent notation and formalism for such models, analogous to (8), are introduced in chapter 3 of [20], which also includes the straightforward demonstration that, in the absence of any large-scale forcing at $j < j_0$, a length L system periodized from level j_0 is exactly equivalent to an $\bar{L}$-periodic KS equation. In particular, for $L = 100$, $\bar{L} = 25$, without the largest wavelet levels the periodized model rapidly converges to a trimodal cellular state, known to be the attractor for $\bar{L} = 25$ [11]. If we retain levels $j < j_0$, however, such models provide a systematic approach to interpreting a smaller periodic system as a subsystem of a larger one, by retaining an influence of large scales as driving inputs to short periodic boxes.

We may also write evolution equations for the internal modes in the box, interpreting them as wavelet coefficients for a basis periodized on a domain of length $\bar{L}$. However, the Galerkin evolution equations are not particularly tractable, containing $2^{j_0} - 1$ forcing terms in the equation for each mode. Furthermore, the large-scale forcing term is not $\bar{L}$-periodic, and hence acts differently on the different boxes, as is apparent from the representation of Figure 8. Consequently, an initially $\bar{L}$-periodic system will not remain periodic, unless constrained to be so. This is achieved in our simulations by choosing a distinguished box *a priori*, and after each time step overwriting the coefficients in the other boxes by those in the distinguished box, as described above.

A typical experiment of this kind is shown in Figure 9, in which we force from levels $j = 0$ and 1. We note that the large-scale driving (which is essentially equivalent to Gaussian noise) maintains the periodized subsystem away from the steady state; and that local 'events' reminiscent of those in STC, such as traveling structures and creation and collision of peaks, occur persistently. That is, the large-scale forcing keeps the system 'alive'. However, visual inspection indicates that the dynamics of this system are rather unusual, a conclusion supported by comparison of long-time statistical calculations of the model power spectrum and distributions at different wavelet levels with Figures 2 and 3. The inconsistencies may be related to occasional anomalous large-amplitude shock-like events, observed for instance near $t = 225$ of Figure 9, which are associated with excessive energy at the largest internal scales of the model, similar to those previously seen when forcing excessively at the large scales.

While localized periodic models with aperiodic large-scale forcing may seem natural to capture the effect of the large-scale modes on the active scales, and do appear to reproduce some of the typical KS events, the presence of atypical structures and shocks due to the asymmetry of the forcing term indicates that this type of model is inadequate. $\bar{L}$-periodic forcing leads to a more tractable model, and we now turn to this.

3.3. Periodized Models with Periodic Forcing

We avoid the problems due to aperiodicity, described in Sections 3.1 and 3.2, by averaging the large-scale forcing terms in space, to render them $\bar{L}$-periodic. In practice, this is equivalent to averaging corresponding wavelet coefficients over the boxes, after performing a time step with aperiodic driving; and this latter method is the one we have used for our numerical experiments.

Since all wavelets and driving terms in this model are now $\bar{L}$-periodic, it is possible to write an effective PDE for the solution inside a box such as one of those in Figure 8; the existence of such a relatively simple formulation is a major reason for preferring such periodized models. By writing out explicitly the wavelet Galerkin formulation of the KS equation, considering the evolution of modes within the box subject to periodicity and large-scale periodic forcing, and then 'inverting' the Galerkin procedure, one finds that the solution $\bar{u}(x, t)$ on the box satisfies a KS equation with external forcing (compare (3)),

$$\bar{u}_t = \mathcal{L}\bar{u} + \bar{u}\mathcal{D}\bar{u} + \mathcal{D}(\bar{u}\bar{v}) + \mathcal{L}\bar{v} + \bar{w}, \tag{11}$$

with $\bar{L}$-periodic boundary conditions [20, sec. 3.3]. Here it turns out that $\bar{v} = 0$, since it is the projection of the large-scale modes $j < j_0$ onto the scales $j \geq j_0$: the periodization removes linear and parametric influences of large scales. Similarly, $\bar{w}$ is the projection of the nonlinear term reconstructed from the large scales, onto the small scales. This does not vanish in general (since the subspaces in a multiresolution analysis are not closed under multiplication and differentiation), but there are some special cancellations for symmetric wavelets. This procedure thus yields an effective PDE for a short system, subject to weak averaged forcing $\bar{w}$ from the large scales:

$$\bar{u}_t = \mathcal{L}\bar{u} + \bar{u}\mathcal{D}\bar{u} + \bar{w}. \tag{12}$$

We have performed experiments for $L = 100$ ($\bar{L} = 25$) with such forcing, in which the driving from levels $j = 0$ and 1 is derived from an independent KS integration (see Figure 10a). The atypical dynamics and excessive amplitudes which were prevalent in the aperiodic models have disappeared under periodization, but the large-scale forcing $\bar{w}$, while nonzero, appears too weak under the averaging to sustain complex spatiotemporal activity. In related experiments with sinusoidal or other forcing at the large scales, the system also settled down to a steady cellular state.

The formulation (12) for a short local system modeling a subsystem of a larger spatiotemporally complex system, while showing promise for analytical investigation, appears to be unsuccessful in numerical experiments, at least for the parameter values we have investigated; the periodization seems to eliminate too much of the desirable driving. However, we can use the ideas leading to (11) and (12) to suggest extensions of the model, in which the forcing is periodic and thus does not incur the undesirable effects of symmetry-breaking, but there is still a strong direct influence on the active scale modes. One possibility is to introduce a nonzero function $\bar{v}$ into (11), to simulate the effects of parametric and linear large-scale driving; this may be a promising avenue for future exploration. An alternative, which we explore briefly here, is to simulate the driving effect of the large Gaussian scales on the box, much of which is lost by periodization, by *explicitly driving* the largest-scale internal modes.

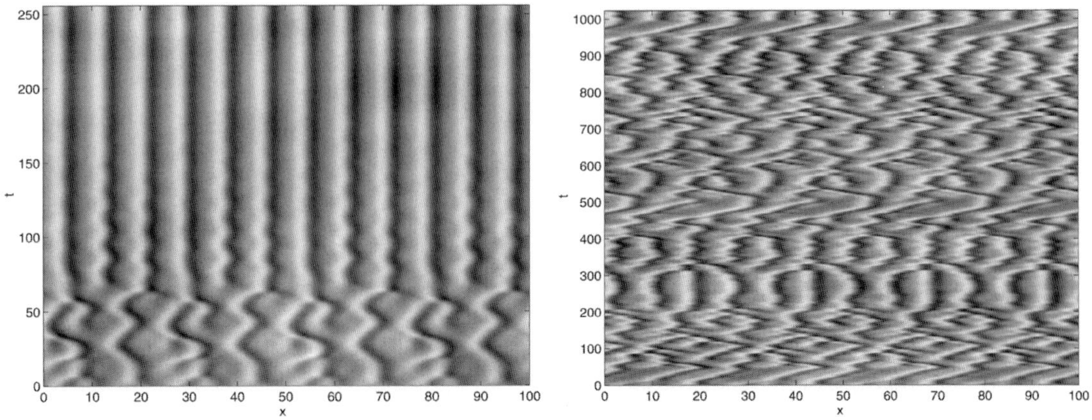

(a) (b)

Figure 10. Periodized model with periodized forcing from an independent KS integration, driving at (a) levels $j = 0$ and 1; (b) levels $j = 0$ and 1, and forcing the lowest internal mode at $j = 2$.

Figure 10b shows the result of an experiment with $\bar{L}$-periodic forcing in which, in addition to driving at levels $j = 0$ and 1 via the averaged $\bar{w}$, the lowest internal mode at $j = 2$ is also driven from an independent KS integration. The dynamics over most of the time interval shown (apart from those times at which the level 2 forcing is unusually small) look remarkably like those of the full KS equation; this is confirmed by long-time statistical calculations of spectra and PDFs. By comparison with Figure 10a, it is apparent that the $j = 2$ excitation has by far the strongest effect, the periodized forcing from levels 0 and 1 exerting little influence on the complex dynamics. It is thus reasonable to ask whether the external forcing $\bar{w}$ from $j = 0$ and 1 is necessary at all; what happens if we just drive at level 2? Simulations confirm that, unsurprisingly, driving the periodized model at level $j = 2$ with a time series taken from the control run is sufficient to maintain complex dynamics and good statistics. *Forcing at only the lowest internal mode of a periodized subsystem captures the effect of the large scales on the short local model.*

3.4. Short Models with Spatiotemporally Complex Dynamics

It is now readily apparent that this short periodized model, in the absence of the large-scale term $\bar{w}$, is exactly a full KS equation (2) on a short $\bar{L}$-periodic domain, in which the lowest level is externally specified. This motivates us to propose an autonomous 'minimal model': small L-periodic KS systems with forcing (where we now use L rather than $\bar{L}$ to emphasize that these are not necessarily subsystems of larger systems). Within this framework, a wide range of experiments is possible, encompassing driving with different amplitudes, time scales and temporal characteristics, as well as variations in the length of the domain (implying differing elementary attractors in the absence of forcing). In the present study, we present the results of only a few such simulations, to demonstrate that such short systems with forcing form a reasonable model for STC in the KS equation.

Our construction of the minimal model for a short box, driven at its largest scale, is confirmed by some of our previous conclusions:

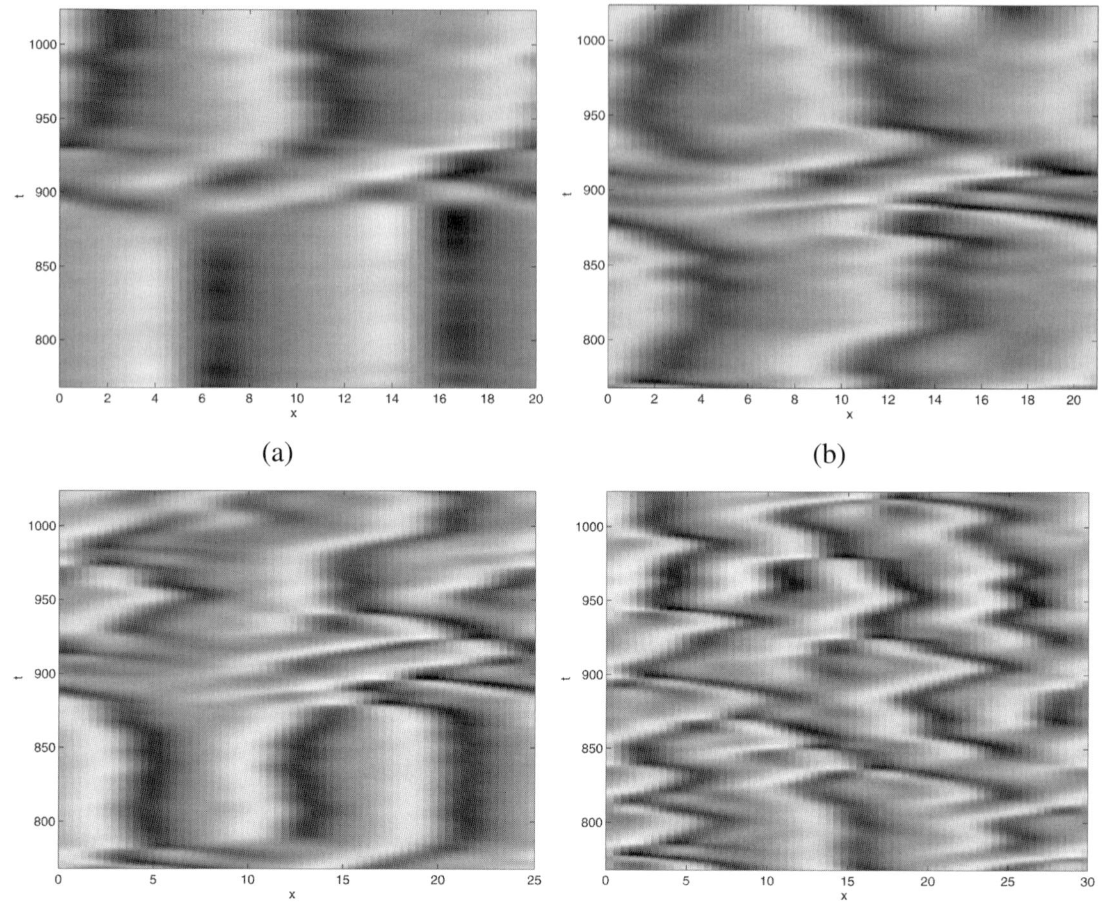

Figure 11. Short systems for a range of L, randomly driven at their lowest wavelet level by a suitably chosen colored Gaussian process: (a) $L = 20$, (b) $L = 21$, (c) $L = 25$, (d) $L = 30$.

- Within the STC regime, the dynamics are spatially localized, but interactions up to a length $l \sim \bar{l}_c \approx 25$ are dynamically relevant.
- The typical interactions of coherent structures occur at the active scale levels, but the large scales provide the excitation essential to sustain the complex spatiotemporal dynamics.
- The strength of the large-scale driving plays an important role; forcing that is too weak may lead to collapse into a cellular state, while unduly powerful driving leads to large amplitudes and rapidly traveling peaks, or even to a shock-like solution.
- The temporal structure and correlation time of the driving has a relatively small influence.
- To avoid excessive growth, the forcing should be spatially L-periodic; but it should be sufficiently strong at the largest scales within the model to drive the characteristic spatiotemporal behavior.

We have performed numerical simulations on short boxes of a range of lengths, in which the excitation of the largest-scale internal mode is derived from the autonomous stochastic

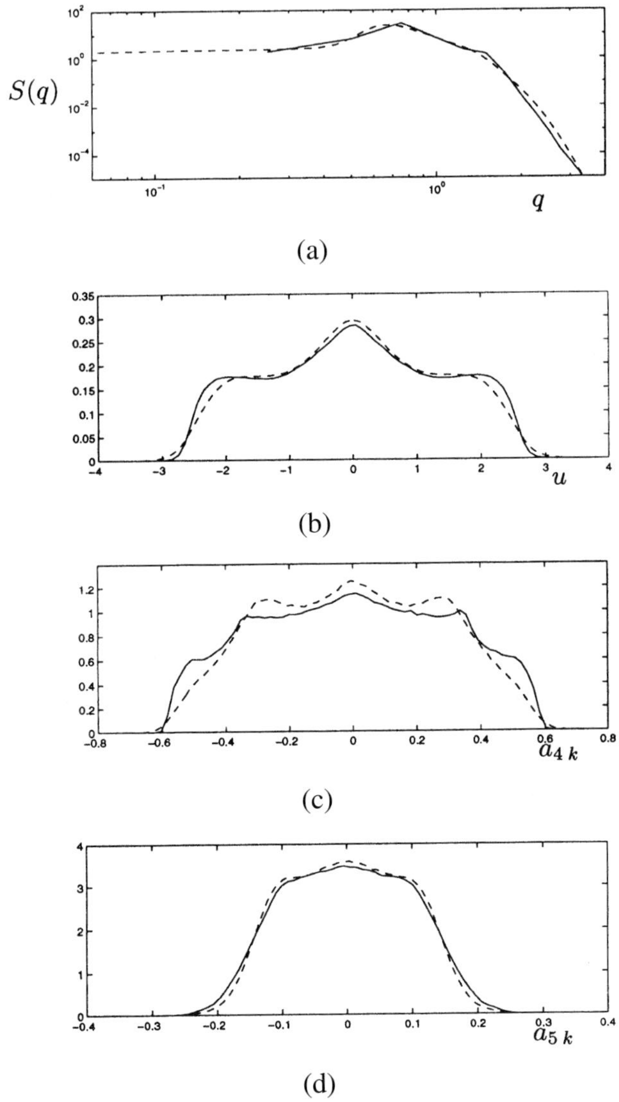

Figure 12. Comparison of statistics for $L = 25$ model driven by colored Gaussian noise at the lowest wavelet level (solid lines), with statistics of full KS equation on $L = 100$ (dashed lines): (a) power spectra (see Figure 2); (b) PDF of pointwise values of the field u; (c) PDF of wavelet level $j = 2$ of model, compared with $j = 4$ for the $L = 100$ full system; and (d) model $j = 3$ PDF, compared with full $j = 5$ PDF (compare Figure 3).

process (7) (Langevin equation with colored noise); as described previously, this form is motivated by the effective forced Burgers equation description of the large-scale dynamics and the fact that this colored Gaussian noise model effectively simulates the statistics and dynamical effects of the large scales.

Figure 11 shows some representative results. It appears from (a) for $L = 20$ that if L is too short, the system is too highly constrained to generate the typical events necessary for STC, and settles into an apparent heteroclinic cycle; this is consistent with the existence of a characteristic dynamical interaction length $\bar{l}_c$. However, the systems with $L \geq 21$ all display sustained chaotic behavior closely reminiscent of the STC state for long KS systems, as borne out by a comparison of the statistics between the $L = 25$ model of Figure 11c and the full KS equation of Figure 1, some of which are shown in Figure 12. Note however that by comparison with the full system, the PDFs of u in (b), and of the level 2 wavelets in (c), display a trend towards a cellular state, evidenced by broader distributions, with the model spending more time at large amplitudes; this effect is readily understood by recalling that an unforced $L = 25$ system tends to the attracting cellular state, so that the driven model will temporarily approach such a state over time intervals during which the stochastic forcing has small amplitude.

It is interesting to note that similar results are obtained if the autonomous forcing at the lowest wavelet level is chosen to be deterministic, sinusoidal in time: the qualitative dynamics resemble those typical for STC in a large KS system almost as often, for a range of L, as in the case of random forcing; and (because the sinusoidal driving does not linger near zero for extended intervals) some of the statistics appear to be even closer to those of the full system than those shown in Figures 12b and 12c. This suggests that the Gaussian statistics at the large scales, and the temporal characteristics of the colored noise process, are inessential to drive active-scale chaotic dynamics. Thus it seems that both deterministic and stochastic forcing at the lowest wavelet level of a short KS system are capable of driving the system sufficiently that it generates complex dynamics similar to those observed in the large-L thermodynamic limit. Such a relatively low-dimensional model of a short system with large scales forced with externally generated processes appears to qualify as a 'minimal box' containing the ingredients for STC in the KS equation, and we are continuing to study such systems to explore the origins and nature of spatiotemporal chaos.

4. Conclusion

The wavelet decomposition has given us insight into the localization of KS dynamics in space and scale. Motivated by these results, we have proposed and investigated various wavelet-based models for short, forced subsystems of an extended spatiotemporally complex system. A major conclusion is that the relevant symmetries, implied by periodic boundary conditions, are essential both in the system and in the forcing, to avoid excessive amplitudes and atypical events. Within the context of periodized subsystems with periodic forcing, driven also at the largest-scale internal modes, we have discovered model systems that appear to emulate the STC of the full KS equation.

We deduce that short, driven systems, provided they have length at least of order of the dynamical interaction length, can sustain spatiotemporal complexity. In this view, the attractor for the appropriately forced short systems is in some sense similar to that in the extensive limit. We postulate an, as yet purely heuristic, description for this idea that the large system may be fully understood in terms of the dynamics of smaller subsystems, by writing

$$\text{`}\mathcal{A}_L \approx \oplus \mathcal{A}_{\bar{L}}\text{'},$$

where $\mathcal{A}_L$ represents the attractor for a length L system; of course, this representation presupposes the, as yet unproven, extensivity of the KS equation. This picture deduced from our experiments of Section 3.4 allows us to speculate that the main difference between 'large' and 'small' systems is the noisy dynamics at the lowest modes which drive the typical events. This noise arises due to the collective interactions of the subsystems; and the threshold to sustained

STC occurs when sufficiently many subsystems interact to maintain the Gaussian large-scale dynamics, which in turn feed into the subsystems and maintain activity within them, in a similar way to our latest experiments.

Acknowledgements

RW would like to thank J. Elezgaray for providing his computer code for some of the wavelet transform calculations. This work was partially supported by DoE grant DE-FG02-95ER25238 and by a Charlotte Elizabeth Procter Fellowship at Princeton University.

References

1. Aubry, N., Holmes, P., Lumley, J. L., and Stone, E., 'The dynamics of coherent structures in the wall region of a turbulent boundary layer', *Journal of Fluid Mechanics* **192**, 1988, 115–173.
2. Barabási, A.-L. and Stanley, H. E., *Fractal Concepts in Surface Growth*, Cambridge University Press, Cambridge, 1995.
3. Berkooz, G., Elezgaray, J., and Holmes, P., 'Coherent structures in random media and wavelets', *Physica D* **61**, 1992, 47–58.
4. Browning, G. L., Henshaw, W. D., and Kreiss, H.-O., 'A numerical investigation of the interaction between the large and small scales of the two-dimensional incompressible Navier–Stokes equations', preprint, April 1998; UCLA CAM Report 98-23, 1998.
5. Dankowicz, H., Holmes, P., Berkooz, G., and Elezgaray, J., 'Local models of spatio-temporally complex fields', *Physica D* **90**, 1996, 387–407.
6. Elezgaray, J., Berkooz, G., and Holmes, P., 'Wavelet analysis of the motion of coherent structures', in *Progress in Wavelet Analysis and Applications*, Y. Meyer and S. Roques (eds.), Editions Frontières, Gif-sur-Yvette, 1993, pp. 471–476.
7. Elezgaray, J., Berkooz, G., and Holmes, P., 'Large-scale statistics of the Kuramoto-Sivashinsky equation: A wavelet-based approach', *Physical Review E* **54**, 1996, 224–230.
8. Halpin-Healy, T. and Zhang, Y.-C., 'Kinetic roughening phenomena, stochastic growth, directed polymers and all that', *Physics Reports* **254**, 1995, 215–414.
9. Hayot, F., Jayaprakash, C., and Josserand, C., 'Long-wavelength properties of the Kuramoto–Sivashinsky equation', *Physical Review E* **47**, 1993, 911–915.
10. Holmes, P., Lumley, J. L., and Berkooz, G., *Turbulence, Coherent Structures, Dynamical Systems and Symmetry*, Cambridge University Press, Cambridge, 1996.
11. Hyman, J. M., Nicolaenko, B., and Zaleski, S., 'Order and complexity in the Kuramoto–Sivashinsky model of weakly turbulent interfaces', *Physica D* **23**, 1986, 265–292.
12. Kardar, M., Parisi, G., and Zhang, Y.-C., 'Dynamic scaling of growing interfaces', *Physical Review Letters* **56**, 1986, 889–892.
13. L'vov, V. S., Lebedev, V. V., Paton, M., and Procaccia, I., 'Proof of scale invariant solutions in the Kardar-Parisi-Zhang and Kuramoto–Sivashinsky equations in 1+1 dimensions: Analytical and numerical results', *Nonlinearity* **6**, 1993, 25–47.
14. Myers, M., Holmes, P., Elezgaray, J., and Berkooz, G., 'Wavelet projections of the Kuramoto–Sivashinsky equation I. Heteroclinic cycles and modulated traveling waves for short systems', *Physica D* **86**, 1995, 396–427.
15. Perrier, V. and Basdevant, C., 'Periodical wavelet analysis, a tool for inhomogeneous field investigation. Theory and algorithms', *La Recherche Aérospatiale* **1989**(3), 1989, 53–67.
16. Pomeau, Y., Pumir, A., and Pelce, P., 'Intrinsic stochasticity with many degrees of freedom', *Journal of Statistical Physics* **37**, 1984, 39–49.
17. Pumir, A., 'Statistical properties of an equation describing fluid interfaces', *Journal de Physique* **46**, 1985, 511–522.
18. Sneppen, K., Krug, J., Jensen, M., Jayaprakash, C., and Bohr, T., 'Dynamic scaling and crossover analysis for the Kuramoto–Sivashinsky equation', *Physical Review A* **46**, 1992, R7351–R7354.

19. Toh, S., 'Statistical model with localized structures describing the spatio-temporal chaos of Kuramoto–Sivashinsky equation', *Journal of the Physical Society of Japan* **56**, 1987, 949–962.
20. Wittenberg, R. W., 'Local dynamics and spatiotemporal chaos. The Kuramoto–Sivashinsky equation: A case study', Ph.D. Thesis, Princeton University, 1998.
21. Wittenberg, R. W., 'Dissipativity, analyticity and shocks in the (de)stabilized Kuramoto–Sivashinsky equation', 2000, in preparation.
22. Wittenberg, R. W. and Holmes, P., 'Scale and space localization in the Kuramoto–Sivashinsky equation', *Chaos* **9**, 1999, 452–465.
23. Yakhot, V., 'Large-scale properties of unstable systems governed by the Kuramoto–Sivashinski equation', *Physical Review A* **24**, 1981, 642–644.
24. Zaleski, S., 'A stochastic model for the large scale dynamics of some fluctuating interfaces', *Physica D* **34**, 1989, 427–438.

Nonlinear Dynamics **25:** 133–156, 2001.

Mode Localization in Dynamics and Buckling of Linear Imperfect Continuous Structures

ANGELO LUONGO
Department of Structural Engineering, University of L'Aquila, 67040 Monteluco Roio, Italy

(Received: 21 December 1998; accepted: 10 August 1999)

Abstract. Localization phenomena in one-dimensional imperfect continuous structures are analyzed, both in dynamics and buckling. By using simple models, fundamental concepts about localization are introduced and similarities between dynamics and buckling localization are highlighted. In particular, it is shown that strong localization of the normal modes is due to turning points in which purely imaginary characteristic exponents assume a non zero real part; in contrast, if turning points do not occur, only weak localization can exist. The possibility of a disturbance propagating along the structure is also discussed. A perturbation method is then illustrated, which generalizes the classical WKB method; this allows the differential problem to be transformed into a sequence of algebraic problems in which the spatial variable appears as a parameter. Applications of the method are worked out for beams and strings on elastic soil. All these structures are found to have nearly-defective system matrices, so their characteristic exponents are highly sensitive to imperfections.

Keywords: Localization, dynamics, buckling, perturbation methods, turning point, defective system.

1. Introduction

The mechanical behavior of high modal density structures can be considerably influenced by structural imperfections, even in the linear field. In particular, the normal modes of oscillations or the buckling modes, which in the absence of irregularities extend to the whole system, can localize in a restricted region of the structure. Thus, if an external disturbance is applied at a point, its effects cannot propagate very far but are confined to the neighborhood of the source.

Localization has been extensively analyzed for discrete systems. Pioneering studies were originally pursued in physics, by investigating disordered chains of atoms (see [1, 2] for a review) and many years later in mechanics, by analyzing chains of pendulums [3]. Extensive reviews on the subject are given in [4, 5].

In dynamics, attention has been mainly focused on discrete one-dimensional structures, such as open or closed chains of pendulums [3, 6–10] or bi-pendulums [11]. Periodic continuous systems such as chains of beams, multi-span beams and taut strings or cables with lumped masses have been analyzed through finite elements or Galerkin techniques [7, 8, 12–15] or by exact transfer matrices [16–21]. Axially vibrating beams with or without distributed elastic constraints have been considered in [22, 23] by following a stochastic or deterministic approach, respectively. Nonlinear problems have been studied only recently [24, 25]; in particular the existence of localized nonlinear normal modes has been proved in [26–28]. In [26] it has been observed that the origin of nonlinear mode localization in lumped-mass periodic systems is the dependence of the frequencies of substructures on their vibration amplitude; thus nonlinearities cause mistuning and localization takes place even in perfectly periodic systems.

In contrast, localization in buckling has received little attention in the linear field [29] and considerable attention in the postcritical nonlinear analysis, where the phenomenon is explained to be a consequence of the interaction of a cluster of nearly-simultaneous modes. Many studies have been devoted to this subject; a review is presented in [30]. However, similarly to the dynamic problem, localization in buckling of perfect nonlinear structures has strict analogies with localization in buckling of imperfect linear structures, as it was observed in [31].

The localization phenomenon in linear continuous one-dimensional systems is considered here, both in dynamics and buckling. The problem is discussed with the following aims: (1) to improve the understanding of the mechanical phenomenon; (2) to emphasize that strong localization is closely connected with the occurrence of *turning points* [32], in absence of which only weak localization can occur; (3) to show that localization of vibrations and buckling patterns have common aspects; (4) to show that continuous structures undergoing strong localization are nearly-defective systems.

The paper is organized as follows. In Section 2 basic concepts on localization are introduced by referring to continuous systems, both in dynamics and buckling. In Section 3 a perturbation method for a general imperfect linear system with high modal density is illustrated. Three examples are worked out in Section 4: namely, the free vibrations of a string and of a beam resting on a Winkler soil, and the buckling pattern of a compressed beam on elastic soil. In Section 5, it is shown that, since the system matrices of these structures are nearly-defective (i.e. they have nearly-coincident characteristic exponents with associated nearly-parallel eigenvectors), their imperfection sensitivity is high. Therefore, small defects cause large variations in the characteristic exponents, thus possibly exalting the tendency towards localization.

2. Basis Concepts

Basic concepts about localization in continuous systems, both in dynamics and buckling, are illustrated and examples are given.

2.1. Localization of Vibrations

Let us consider a continuous one-dimensional structure with spatially constant mechanical properties (perfect system) vibrating at an undamped natural frequency. Let us assume first that under suitable boundary conditions the (spatial) modal shape is sinusoidal, i.e. a couple of characteristic exponents of the differential equation of motion is purely imaginary. If some spatially-dependent imperfection is introduced in the system, such characteristic exponents are perturbed from their nominal value, and assume new values that depend on the site. If they have a non zero real part throughout most of the structure, and if the structure is sufficiently long, the modal shape localizes. The points at which the characteristic exponents change from imaginary to real or complex conjugate are called *turning points*. If more general boundary conditions are prescribed, even the perfect system has real or complex characteristic exponents contributing to the eigenfunctions. However, the exponents are independent of the site and significantly affect the solution only in narrow boundary layers, except perhaps for the first few modes. When sufficiently large imperfections are introduced, the real parts exponents depend on the site and, through the mechanism previously illustrated, can enforce localization to occur somewhere in the dominion. In order to separate the effects of boundary conditions from that of imperfections, reference will be made to systems having purely sinusoidal eigenfunctions.

The occurrence of turning points represents the simplest mechanism which triggers localization. However, it is *not the only way* for a normal mode to localize. As an example, the axially vibrating beam studied in [22] can undergo localization albeit its characteristic exponents are always purely imaginary, independently from the mass and/or stiffness imperfection shape. For such kind of structures, localization is due to the fact the amplitude of the oscillating modal shape decays exponentially in the space, as a result of the changes of the mechanical properties along the structure. Thus, if the imperfections are stepwise variable, the normal modes are made of pieces of sinusoids with different amplitudes and phases. If the amplitudes tend to zero far away from a certain section of the structure, then the mode is localized. This circumstance always occurs in a sufficiently long structure and it is strictly connected with the random nature of the imperfections (see, e.g., [2], where the Furstemberg theorem concerning the limit of noncommuting random variables is used to explain the phenomenon).

The two mechanisms of localization illustrated are quite different and should not be confused. Moreover, while the first one (when it can occur) leads very directly to localization, the second one usually calls for longer structures and/or larger random disorder. For this reason the two phenomena could be referred to as *strong localization* and *weak localization*, thus more sharply defining a difference already discussed by Pierre [9], although from a different point of view. According to that paper, weak localization has little interest in engineering structures; therefore, attention is mainly paid here to strong localization. Hence, the word 'localization' will be used as shortening of 'strong localization'.

To illustrate the main aspects of the phenomenon, reference is made to a specific problem, already analyzed in [23]. The main findings are resumed here and new results are illustrated.

Consider a string weakly taut by a small force T, having mass per unit length m_0, resting on a Winkler soil of nominally constant stiffness $k_f = k_0$. The perfect system has sinusoidal modes with close frequencies. The problem is to determine the modes and the frequencies of the imperfect system when small defects affect the stiffness of the springs, i.e. when $k_f = k_f(x)$ depends on the abscissa x [23]. The normal modes $w(x)$ of a fixed-fixed string satisfy the following homogeneous boundary value problem

$$\begin{aligned} &\beta^2 w'' + (\lambda - \chi(x))w = 0, \quad x \in [-1/2, 1/2], \\ &w(-1/2) = 0, \quad w(1/2) = 0, \end{aligned} \tag{1}$$

in which $\beta = (T/k_0 l^2)^{1/2} \ll 1$ is the coupling parameter, responsible for the high modal density of the perfect system, $\lambda = m_0\omega^2/k_0$ is the nondimensional frequency and $\chi(x) = k_f(x)/k_0$ is the nondimensional stiffness of the imperfect springs.

Equation (1a) can easily be integrated if a single defect consisting in a rectangular stiffness well centered at the midspan $x = 0$ is present. In this case $\chi(x) = 1 - \varepsilon$ for $0 \leq |x| < \Delta$ and $\chi(x) = 1$ for $\Delta < |x| \leq 1/2$, where ε and 2Δ are the depth and the width of the well, respectively. Correspondingly the solution $w(x)$ has two branches, $w_i(x) = \exp(\theta_i x)$ and $w_e(x) = \exp(\theta_e x)$, internal and external to the well, respectively. Two cases must be distinguished:

(a) When $\lambda > 1$, both $w_i(x)$ and $w_e(x)$ are sinusoidal functions, so the associated normal mode is extended, as for the perfect structure. The a_e/a_i ratio between the amplitudes of the two branches is found to be always larger than 1, so that no weak localization occurs. However,

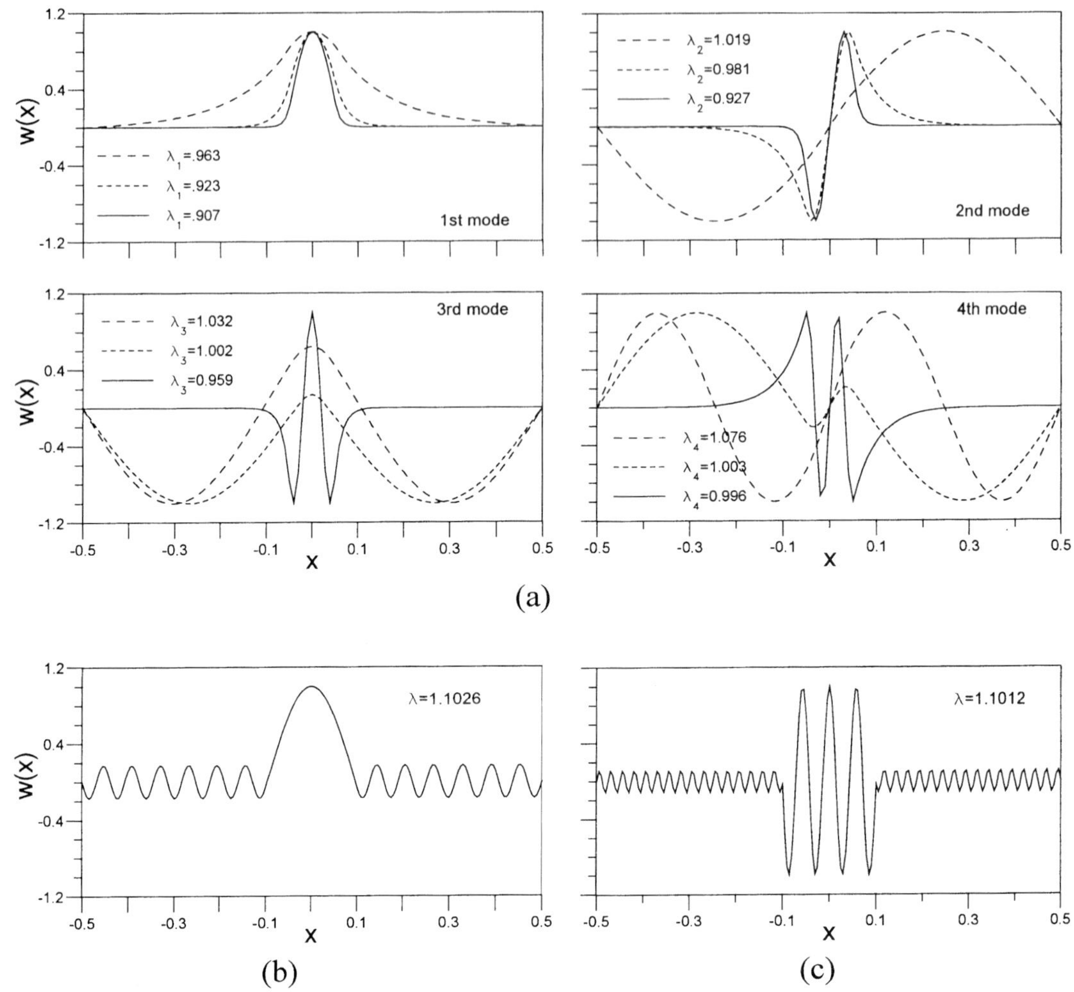

Figure 1. Normal modes for the string on elastic soil with rectangular stiffness well (a) or peak (b) centered at midspan; (a) $\varepsilon = 0.1$, $\Delta = 0.05$, - - - $\beta = 0.02236$, $\cdots$ $\beta = 0.00707$, — $\beta = 0.00316$; (b) $\varepsilon = -0.1$, $\Delta = 0.1$, $\beta = 0.00316$; (c) $\varepsilon = -0.1$, $\Delta = 0.1$, $\beta = 3.16 \times 10^{-4}$.

if $\varepsilon < 0$ (i.e. if a peak of soil stiffness is considered instead of a well), $a_e/a_i < 1$ and weak localization can take place.

(b) When $1-\varepsilon < \lambda < 1$, $w_i(x)$ is sinusoidal while $w_e(x)$ is exponential, so that the mode is localized near the defect; in this case the character of the solution changes at the point $x = \Delta$, which is a turning point. As can easily be checked from Equation (1a), the following relation between the characteristic exponents θ_i (purely imaginary) and θ_e (real) holds: $|\theta_i|^2 + \theta_e^2 = \mu$, where $\mu = \varepsilon/\beta^2$ is the imperfection/coupling ratio. The relation shows that $|\theta_i|$ and θ_e range between zero and $\sqrt{\mu}$. Thus, since the first modes are associated with smaller values of $|\theta_i|$, they also have greater values of θ_e, therefore lower modes are more strongly localized than higher modes. Because $|\theta_i|$ is limited, the number of localized modes is limited as well, and increases with μ. If μ is very small no localized modes can occur. Therefore, *in a finite length system*, a threshold defect must exist in order to localize a given number of eigenfunctions; on the other hand, *if the length is sufficiently large*, any small defect can localize the first modes.

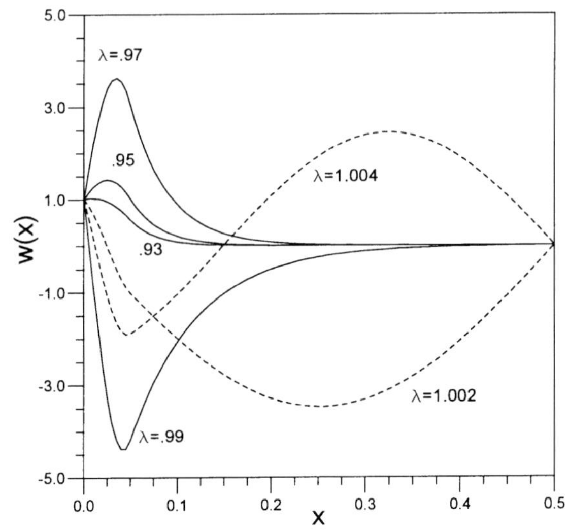

Figure 2. Time-harmonic response of the second string in Figure 1a when the midspan is driven with frequency λ; $\Delta = 0.05$, $\varepsilon = 0.1$, $\beta = 0.00707$.

Some normal modes for different values of the parameters are plotted in Figure 1. Figure 1a shows the effects of the occurrence of turning points. It is seen that, for a fixed imperfection, the number of localized modes increases when β decreases. Thus, while the first mode is localized for all β's considered, the fourth mode localizes only for the smallest β. In Figures 1b and 1c some weakly localized mode produced by a peak of stiffness are shown. They occasionally occur in sequences of nonlocalized modes. To observe localization of more modes it would be necessary to introduce random diffused imperfections.

Examine now the possibility of a disturbance propagating along the string. To this end, assume that the midspan $x = 0$ is harmonically driven by a device with a known frequency λ and unitary amplitude. By accounting for the antisymmetry of the problem, the amplitude $w(x)$ of the time-harmonic response is determined by integrating Equation (1a) in the interval $[0, 1/2]$ with the boundary conditions $w(0) = 1$, $w(1/2) = 0$. The response is plotted in Figure 2 for various λ; the string is the second of those considered in Figure 1a. It is apparent that, when $\lambda < 1$, due to the presence of a turning point at $x = \Delta$, the response is localized around the driven point; when $\lambda > 1$, since no turning points occur, the response is extended to the whole string. It should be noted that resonance occurs when λ is equal to 0.981, 1.003, ... i.e. when the frequency of the excitation is equal to a natural frequency of an antisymmetric mode of the double length string (see Figure 1a). Thus, when λ is equal to 0.97 or 0.99, the response is essentially in the first mode, while when λ is equal to 1.002 or 1.004 the second mode prevails. However, since the first mode is localized, the first disturbances cannot propagate; in contrast, since the second mode is extended, the last disturbances propagate.

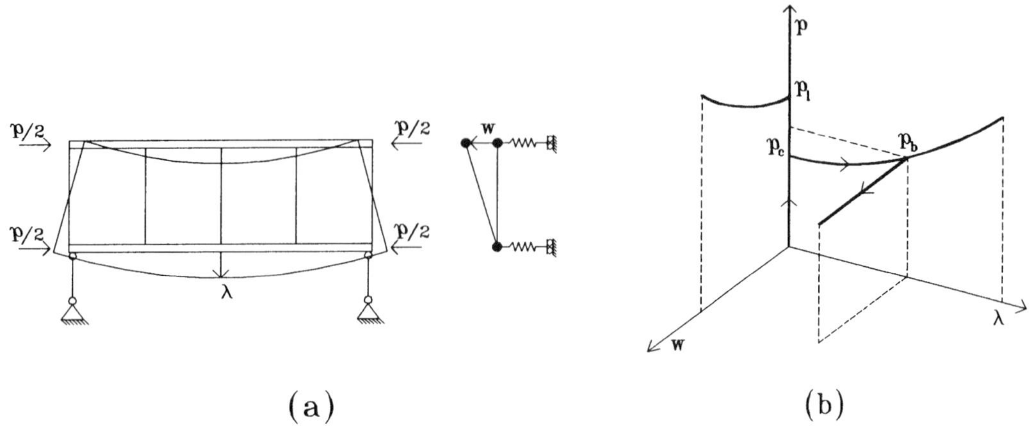

Figure 3. Truss structure elastically restrained; (a) model, (b) equilibrium paths.

2.2. Localization of Buckling Patterns

It is well known from the theory of elastic stability that simple compressed structures such as long beams on elastic soil, long plates, thin-walled beams or spheres under hydrostatic pressure exhibit a large number of nearly coincident buckling modes, i.e. they have dense spectra. This is due to the fact that a small parameter β – typically an aspect ratio – appears in the eigenvalue equation. As a consequence, if imperfections were taken into account in the analysis, localization should occur. To prove that, a simple model similar to the previous one could be considered, consisting in a uniformly compressed beam on imperfect elastic soil. However, in spite of the formal analogy with the dynamical problem, localization in buckling does not seem to have been studied in depth. In fact, the imperfections that are usually accounted for in elastic stability are typically small curvatures or eccentricities of the forces (the so-called 'initial imperfections'), instead of small spatial variations of the mechanical characteristics, as thicknesses or material properties. Since the former imperfections add only known terms to the boundary value problem relevant to the perfect system, they leave its spectral properties unaltered and localization cannot occur. In contrast, the latter imperfections leave the boundary value problem homogeneous and therefore modify the eigensolutions, eventually allowing localization.

On the other hand, a large amount of literature exists on the subject of nonlinear interaction among nearly simultaneous buckling modes. In particular, it has been shown that buckling patterns in the postcritical range can localize in one or more regions, in contrast with the periodic character of the critical modes [33–37]. A simple model studied by the author in [31] clarifies the fact the mechanism which produces localization in nonlinear perfect systems is the same as that which produces localization in linear imperfect systems. In other words, nonlinearities have the same effects as imperfections.

The example given in [31] concerns a simply supported planar truss-beam, having continuous top and bottom longerons which are restrained against out-of-plane displacements by distributed elastic springs (Figure 3a). When the structure is compressed, it can buckle in an overall (Eulerian) mode or in several out-of-plane local modes whose wavelength is small compared with the span. The system behaves as a compressed thin-walled beam that in fact buckles in an overall (Eulerian or flexural-torsional) mode or in a local mode in which the component plates are bent in many longitudinal waves. It is assumed here that the lower critical load multiplier p_c is associated with the Eulerian mode and the higher critical load p_l with the local mode (Figure 3b). When the load p is increased, the structure first buckles in its plane at $p = p_c$ (primary bifurcation), then it buckles out-of-plane at $p = p_b$ (secondary bifurcation); as a result of the interaction it ensues that $p_b < p_l$. The goal of the analysis is to evaluate the secondary bifurcation load p_b along the non trivial bifurcated equilibrium path, together with the associated modal shape. It is assumed that p_l is slightly higher than p_c, so the problem can be linearized in the nondimensional amplitude λ of the overall mode, which is taken as being the control parameter.

The problem is governed by the classical equilibrium equation of the compressed beam resting on Winkler soil. Along the trivial fundamental path the axial force in the two longerons is constant; along the bifurcated path, due to the bending associated with the primary bifurcation, an additional spatially variable traction or compression exists in the two beams. For the most compressed beam, the boundary value problem reads

$$\begin{aligned} &\beta^4 w'''' + 2\beta^2[(p + \lambda f(x))w'/p_l]' + w = 0, \qquad x \in [0, \pi], \\ &w(0) = w''(0) = 0, \qquad w(\pi) = w''(\pi) = 0. \end{aligned} \tag{2}$$

In Equation (2a) $\beta = \pi(EI/k_0 l^4)^{1/4}$, where EI is the out-of-plane flexural stiffness of the longerons, l is their length and k_0 is the stiffness of the springs; since β equals the ratio between the wavelengths of the local and overall modes it is, by hypothesis, a small parameter. In addition, in Equation (2a), $f(x) = \sin x$ accounts for the additional variable compression whose maximum value is assumed to be (approximately) proportional to the overall deflection λ.

Along the trivial path $\lambda \equiv 0$, and p represents the unknown eigenvalue. The problem (2) admits the solution $p = p_l$ and $w = \sin(nx)$, with $n \cong \beta^{-1}$. Other eigenvalues slightly higher than p_l correspond to sinusoidal modes with wavenumbers $n \pm k$, with $k \ll n$, so that the system has high modal density.

Along the bifurcated path $\lambda \neq 0$, p depends on λ; since the Eulerian bifurcation is symmetric and λ is a small quantity, $p = p_c + O(\lambda^2)$. Equations (2) are therefore an eigenvalue problem in the displacement λ. By posing $p_c = 1$, $p_l = 1 + \varepsilon$, where $\varepsilon > 0$ is a small parameter, and retaining only first-order quantities, Equations (2) become

$$\begin{aligned} &\beta^4 w'''' + 2\beta^2[(1 - \varepsilon)w'' + \lambda(f(x)w')'] + w = 0, \qquad x \in [0, \pi], \\ &w(0) = w''(0) = 0, \qquad w(\pi) = w''(\pi) = 0. \end{aligned} \tag{3}$$

The parameter ε, measuring the mistuning between the two critical loads p_c and p_l, represents a 'defect' in the broad sense. In fact, when $\varepsilon = 0$, i.e. when the two critical loads are tuned, Equation (3) admits the solution $\lambda = 0$ and $w = \sin(nx)$: there is therefore no reduction (the so-called erosion [30]) in the local critical load. In contrast, if $\varepsilon \neq 0$, λ can be shown to be of order ε [31], so that $p_b = 1 + O(\varepsilon^2) < p_l$, and erosion takes place.

It should be noted that, unlike the example dealt with before and in the remainder of the paper, the parameter ε does not affect the part of the differential operator depending on x. Thus, when $\varepsilon = 0$, Equation (3a) does not reduce to a constant coefficient equation, if $\lambda = 0$ is not posed as well. However, by scaling the eigenvalue at the order ε, i.e. by posing $\lambda = \varepsilon\sigma$, with $\sigma = O(1)$, this condition is satisfied. In this form Equation (3) appears as a perturbation of order ε of a singular operator with constant coefficients. The perturbation assumes the meaning of an imperfection, due to the nonlinearities, of the geometrical stiffness of the beam.

Since the perfect system has high modal density, it can be predicted that this defect produces localization, as do defects in the true sense of the word.

To evaluate the critical load and the bifurcation mode it is necessary to integrate Equation (3a); however exact integration is not straightforward, as it was in the previous example. An asymptotic solution has been found in [31] by using the multiple scale perturbation method to obtain an amplitude modulating equation. This equation turned out to be exactly of the same form of Equation (1a), with a sinusoidal coefficient. By applying the WKB method [32], amplitude modulating solutions qualitatively similar to the ones in Figure 1a were obtained, still exhibiting turning points, although no more in *a priori* known positions. The localized buckling patterns therefore appear like those in Figure 10 below.

To summarize, in this system nonlinearities play the same role as imperfections in linear problems and, together with the high modal density of the structure, cause localization of the buckling patterns. As has been explained in Section 1, a similar result was found in [26] for a discrete dynamical system.

3. A General Perturbation Method for Continuous Systems

The examples analyzed before have shown that, in order to evaluate the normal modes of an imperfect continuous system and to check the possible occurrence of localization, it is necessary to integrate a differential equation whose coefficients are weakly dependent on the abscissa x. Since exact integration is often impossible, it would be desirable to take advantage of the presence in the equation of a small parameter to obtain asymptotic solutions. However, it is known that series expansions in terms of the imperfection parameter ε are not suited to the scope [6, 7]. In fact, due to the presence of small divisors, asymptotic expansions obtained by perturbing a unique eigenfunction of the perfect system are not uniformly valid. On the other hand, perturbations of a cluster of eigenfunctions of the perfect system lead to a new eigenvalue problem in the unknown amplitudes [38], thus practically nullifying the convenience of using a perturbation method. A more suitable method has been used by Pierre and co-workers [6, 8, 14] in which the structural parameter β, instead of the imperfection parameter ε, has been adopted as the perturbation parameter. This method can be successfully applied for discrete structures but not for a continuous structure, as will be explained ahead.

A general perturbation method for the analysis of one-dimensional linear continuous structures with high modal density, affected by small imperfections, is illustrated here. The method applies to boundary value problems that can be put in the form

$$\begin{aligned} \beta\frac{\mathrm{d}w}{\mathrm{d}x} &= L(x,\varepsilon,\lambda)w, \qquad x \in [0,1], \\ B_0 w(0) &= w_0, \qquad B_1\, w(1) = w_1, \end{aligned} \tag{4}$$

where β and ε are small parameters, $\lambda = O(1)$ is a control parameter, w is the N-state-vector containing generalized displacements and stresses, L is the $N \times N$ system matrix, B_0 and B_1 are Boolean $(N/2) \times N$ matrices, and w_0 and w_1 are prescribed values of the state variables at the ends of the structure.

The parameter λ has the meaning of a time-frequency in dynamics and a load multiplier (or a nondimensional displacement) in buckling problems. It is the unknown eigenvalue of the differential problem when the boundary conditions are homogeneous ($w_0 = w_1 = 0$) or assumes known values when non zero displacements and/or stresses are prescribed as external disturbances at the ends of the structure. The parameter β is a structural parameter, depending on the mechanical characteristics of the perfect structure, and is responsible for the high modal density. The parameter ε is the imperfection parameter, measuring the amplitude of the defects. When $\varepsilon = 0$ the structure is perfect and, by hypothesis, has mechanical properties independent of the abscissa x, so that the matrix $L^* = L(x, 0, \lambda)$ has constant coefficients. By expanding L around $\varepsilon = 0$ it follows

$$L(x, \varepsilon, \lambda) = L^*(\lambda) + \varepsilon L_\varepsilon(x, \lambda) + O(\varepsilon^2). \tag{5}$$

From now on the dependence on λ of all the variables will be taken as understood.

Consider first the perfect system ($\varepsilon = 0$). Equation (4a) admits solutions of the type

$$w(x, \beta) = u^* \exp\left(\frac{\theta^*}{\beta} x\right), \tag{6}$$

where u^* and θ^* are generally complex solutions of the $N \times N$ algebraic eigenvalue problem

$$(L^* - \theta^*)u^* = 0. \tag{7}$$

The eigenvalues θ^* are the characteristic exponents of the differential equation (4a). If L^* admits a complete set of eigenvectors u^*, the general solution to Equation (4a) can be built as a linear combination of the solutions (6). If homogeneous boundary conditions are enforced, a transcendental equation in λ^* is obtained. By hypothesis let

$$\lambda_k^* = 1 + O(\beta^h), \quad h > 1 \tag{8}$$

be the eigenvalues of interest for k ranging over some interval. When β is small the system has high modal density, increasing with the exponent h. β^h is a measure of the spread in the eigenvalues of the perfect system. From Equation (6), if the associated characteristic exponents θ_k^* have modulus $|\theta_k^*| = O(1)$, the eigenfunctions are fast varying; if $|\theta_k^*| = O(\beta)$, they are slowly varying.

Consider now the imperfect system and look for an asymptotic solution. Since two small parameters β and ε are present in the equation, attention will be focused on the case in which the imperfections have amplitude greater than, or equal to, the bandwidth of the eigenvalues λ_k^*, i.e., from Equation (8), when $\varepsilon \geq O(\beta^h)$.

Series expansions could be performed in terms of either β or ε, by fixing one parameter and expanding with respect to the other. However, if β is fixed and $\varepsilon \to 0$, a large number of nearly-simultaneous modes u_k^* has to be considered to express the generating solution, as has already been discussed. In contrast, if ε is fixed and $\beta \to 0$, the nearly-degenerate condition of the generating system is destroyed and a perturbation solution can be pursued in a simpler way. However, since β affects the highest derivative in Equation (4), a singular perturbation problem arises and standard method cannot be applied.

A similar question occurs when discrete periodic structures are considered, like chains of weakly coupled one d.o.f. subsystems. Pierre and co-workers [6, 8, 14] have shown that asymptotic expansions in terms of the coupling parameter β can be successfully employed to determine the eigenfunctions of the perturbed system and to explain the phenomenon of localization. Their generating equation has been obtained by posing $\beta = 0$ in the governing equations, i.e. by considering the uncoupled subsystems as a generating mechanical system. The same operation cannot be performed when continuous systems are dealt with, since kinematic compatibility would be violated. The method presented here enables this difficulty to be overcome.

Since Equation (4a) has variable coefficients, a solution is still sought in the form (6), in which, however, u and θ are assumed to depend on x, i.e.

$$w(x,\beta,\varepsilon) = u(x,\beta,\varepsilon)\exp\left\{\frac{1}{\beta}\int\theta(x,\beta,\varepsilon)\,\mathrm{d}x\right\}. \tag{9}$$

Moreover, u and θ are expanded in series of the perturbation parameter β around $\beta = 0$

$$\begin{aligned} u(x,\beta,\varepsilon) &= \sum_{j=0}^{\infty}\beta^j u_j(x,\varepsilon), \\ \theta(x,\beta,\varepsilon) &= \sum_{j=0}^{\infty}\beta^j \theta_j(x,\varepsilon). \end{aligned} \tag{10}$$

By substituting Equations (9) and (10) in Equation (4a) and zeroing the terms with the same powers of β, the following perturbation equations are drawn

$$\begin{aligned} &(L-\theta_0)u_0 = 0, \\ &(L-\theta_0)u_1 = -u_0' - \theta_1 u_0, \\ &(L-\theta_0)u_2 = -u_1' - \theta_1 u_1 - \theta_2 u_0, \\ &\ldots, \end{aligned} \tag{11}$$

where all quantities depend on x and ε, in addition to λ, and a prime denotes differentiation with respect to x.

It should be noted that the original differential problem has been transformed in a sequence of algebraic problems in which the abscissa x appears as a parameter. Equation (11a) is the generating equation; it is still an eigenvalue problem, like Equation (7), however it furnishes spatially variable characteristic exponents θ_0.

It is supposed that the system matrix L has N distinct eigenvalues; let θ_0 be one of these eigenvalues, with associated right eigenvector u_0 and left eigenvector v_0 (i.e. $L^H v_0 = \bar{\theta} v_0$, where the overbar denotes the complex conjugate and the superscript H the transpose conjugate). By substituting u_0 and θ_0 in Equation (11b) a nonhomogeneous linear equation with singular coefficient matrix is obtained. The solvability condition requires the right-hand member to be orthogonal to the left eigenvector v_0 from which the unknown θ_1 is drawn

$$\theta_1 = -v_0^H u_0' / v_0^H u_0. \tag{12}$$

Equation (11b) can then be solved to within an arbitrary constant and the procedure be continued to higher orders.

After having determined N solutions in this way, the general solution is obtained as a linear combination. By enforcing homogeneous boundary conditions the eigenvalues λ_k and the eigenfunctions $w_k(x)$ ($k = 1, 2, \ldots$) are evaluated. If nonhomogeneous boundary conditions are instead prescribed, for any given λ different from the eigenvalues λ_k the response $w(x)$ is calculated. However, in both cases, the method can furnish solutions that are not uniformly valid in the whole dominion, due to the occurrence of turning points [32]; the problem will be illustrated by reference to applications.

4. Illustrative Examples

Vibrations and buckling patterns of three imperfect one-dimensional structures resting on Winkler soil are analyzed by using the perturbation method developed in Section 3. The tendency towards localization is studied.

4.1. Free and Forced Vibrations of a Weakly Taut Long String on Elastic Soil

The string on elastic soil is considered again, with imperfections affecting the foundation stiffness only. The ends of the string are fixed (free vibrations) or are subjected to time-harmonic transversal displacements (forced vibrations). Harmonic motions of frequency ω are governed by the following boundary value problem

$$Tw'' - k_f(x)w + m_0\omega^2 w = 0,$$

$$w(0) = w_A, \qquad w(l) = w_B, \tag{13}$$

where

$$k_f(x) = k_0(1 + \varepsilon f(x)), \tag{14}$$

in which $f(x)$ accounts for the shape of the imperfections, normalized as $\max(|f(x)|) = 1$.

By introducing the following nondimensional quantities

$$\tilde{w} = \frac{w}{l}, \qquad \tilde{x} = \frac{\pi x}{l}, \qquad \beta = \frac{\pi}{l}\sqrt{\frac{T}{k_0}}, \qquad \lambda = \frac{m_0\omega^2}{k_0} \tag{15}$$

and omitting for simplicity the tilde, Equation (13a) can be recast in the normal form (4a)

$$\beta\begin{Bmatrix} w_1' \\ w_2' \end{Bmatrix} = \left(\begin{pmatrix} 0 & 1 \\ 1-\lambda & 0 \end{pmatrix} + \varepsilon\begin{pmatrix} 0 & 0 \\ f(x) & 0 \end{pmatrix}\right)\begin{Bmatrix} w_1 \\ w_2 \end{Bmatrix}, \tag{16}$$

in which w_1 is the deflection and w_2 the transversal component of the internal force in the deformed configuration. In addition $\beta \ll 1$ if $T \ll k_0 l^2$, i.e. if the stress is small and/or the length is great.

The perfect system ($\varepsilon = 0$) is first considered. The eigenvalues of the matrix L^* are $\theta^* = \pm i\sqrt{\lambda - 1}$. By enforcing the boundary conditions $w(0) = w(\pi) = 0$, $\lambda_k^* = 1 + k^2\beta^2$ is found and, consequently, $\theta_k^* = \pm ik\beta$ ($k = 1, 2, \ldots$) (Figure 4a). Thus the system has high modal density with exponent $h = 2$ (see Equation (8)) and sinusoidal eigenfunctions with characteristic exponents $|\theta_k^*| = O(\beta)$, i.e. slowly varying solutions.

The imperfect system is then studied. Imperfections are assumed to have amplitudes $\varepsilon \geq O(\beta^2)$. The generating Equation (11a) admits the following eigenvalues

$$\theta_0(x) = \pm\sqrt{\varepsilon(f(x) - \sigma)}, \tag{17}$$

where

$$\lambda = 1 + \varepsilon\sigma, \qquad \sigma = O(1) \tag{18}$$

has been posed. The two characteristic exponents θ_0 are real for any x for which $f(x) > \sigma$ (i.e. for $k_f(x)/k_0 > \lambda$) and imaginary for any x for which $f(x) < \sigma$ (i.e. for $k_f(x)/k_0 < \lambda$).

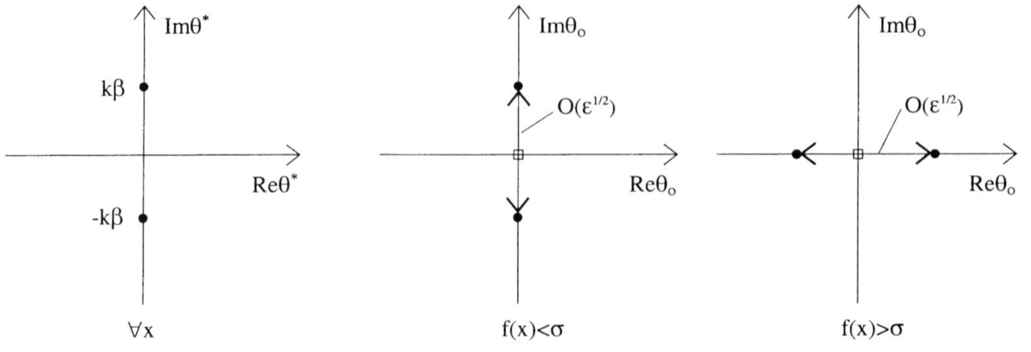

Figure 4. Characteristic exponents for the string on elastic soil; (a) perfect system ($\varepsilon = 0$), (b) imperfect system ($\varepsilon \neq 0$); □ exponents at turning points.

The paths of the eigenvalues on the complex plane for x varying are shown in Figure 4b. When $f(x) = \sigma$ the eigenvalues coalesce at the origin and x is a turning point. It should be noted that $0 \leq |\theta_0(x)| \leq O(\varepsilon^{1/2})$, so, for mode numbers $k = O(1)$, the imperfections produce corrections of the characteristic exponents θ^* of the same order as, or larger than, θ^* , since $\varepsilon \geq O(\beta^2)$. The corresponding eigenfunctions are therefore strongly imperfection sensitive.

Right and left eigenvectors of L are

$$u_0 = \{1 \;\; \theta_0\}^T, \quad v_0 = \{\bar{\theta}_0 \;\; 1\}^T \tag{19}$$

and Equation (12) yields

$$\theta_1(x) = -\frac{1}{2}\frac{\theta_0'(x)}{\theta_0(x)} = \frac{\mathrm{d}}{\mathrm{d}x} \ln \theta_0^{-1/2}(x). \tag{20}$$

By using Equations (17), (19a) and (20) in Equations (9) and (10) the general solution

$$w(x) = \frac{1}{\sqrt[4]{\sigma - f(x)}}\{c_1 \sin \psi(x) + c_2 \cos \psi(x)\} \quad \text{if } f(x) < \sigma,$$

$$w(x) = \frac{1}{\sqrt[4]{f(x) - \sigma}}\{c_3 \exp \psi(x) + c_4 \exp(-\psi(x))\} \quad \text{if } f(x) > \sigma \tag{21}$$

is obtained, with

$$\psi(x) = \int \sqrt{\mu |f(x) - \sigma|}\, \mathrm{d}x. \tag{22}$$

In Equations (21) c_i are arbitrary constants and $\mu = \varepsilon/\beta^2 \geq O(1)$ is the imperfection/coupling ratio. From the boundary conditions a linear system in the constant c_i with coefficient dependent on σ is obtained. The truncated asymptotic solution (21) is known as Liouville–Green or WKB approximation [32].

It should be noted that the solution (21) is singular at the turning points. Since their position depends on σ, they are unknown in a free vibration problem, but are *a priori* known in a forced vibration problem. To eliminate the singularity, polynomial (inner) solutions must be found around the singularities and successively matched with the (outer) solutions (21) that are valid far from the singular points. The matching procedure allows the arbitrary constant to be reduced to two and renders the boundary value problem well posed. The technique is illustrated in [32] and an application relative to a problem with a unique turning point has been developed by the author in [23]. However, for general imperfections, several turning points occur and the solution has oscillatory branches (Equation (21a)) and exponential branches (Equation (21b)) that alternate. Since the exponential decaying is proportional to $\mu^{1/2}$ (Equation (22)), if μ is sufficiently large the eigenfunctions localize around the stiffness gaps. If μ is not large enough, no turning points occur and the modal shapes remain oscillating and extended. This is always true for higher modes, for which $\sigma > \max(f(x)) \leq 1$, even when the lower modes are localized.

When many turning points are present, the matching analytic procedure becomes impracticable and it is necessary to resort to numerical methods. However, even in these cases, the perturbation method allows the acquisition of qualitative information. In particular it permits determination of the conditions under which disturbances prescribed at one end of the string cannot propagate to the other end, remaining localized near the source.

In Figure 5 the lowest modes of the string for different values of the parameters are shown. They have been obtained numerically by direct application of the Galerkin procedure. A single halfwave sinusoidal imperfection is considered first (Figures 5a and 5b), for which $k_f(x)/k_0$ ranges in [0.9, 1]. For the lower value of μ (Figure 5a) the first two modes (associated with eigenvalues $\lambda < 1$) exhibit two turning points while the third mode (with $\lambda > 1$) is oscillatory. For the higher value of μ (Figure 5b) all the three modes have exponential branches near the ends. It is observed that the modes tend to localize around $x = \pi/2$, i.e. around the point in which the stiffness of the soil is minimum. Localization would be magnified if μ were further increased. However, by choosing another shape of the imperfections, localization just manifests itself for the same values of μ. In Figures 5c and 5d the same string but on an elastic soil with two gaps of stiffness, each located near one end (cosinusoidal imperfection), is considered. The stiffness $k_f(x)/k_0$ now ranges in [0.9, 1.1]. For the lower μ (Figure 5c), the first two modes (with $\lambda < 1.1$) are localized around the two gaps, that vibrate in-phase or out-of-phase, while the higher modes (with $\lambda > 1.1$) are extended. For the higher μ (Figure 5d), the first eight modes have turning points, the first four of which, more strongly localized, are shown. It should be noted in this last case, that symmetric and antisymmetric localized modes have nearly-coincident frequencies, since the coupling between the two gaps is very small. In conclusion, the numerical results show that localization strongly depends on the shape $f(x)$ of the imperfections. It is more pronounced for the second imperfection (Figures 5c and 5d) compared with the first imperfection (Figures 5a and 5b) because, due to the narrower gaps, the turning points are closer each other and, due to the greater difference $f(x) - \sigma$ out of the gaps, the exponential decaying is higher.

In Figure 6, the response of the string to an harmonic transversal excitation of frequency λ of the left support is shown; the parameters are the same as in Figures 5c and 5d, respectively. It is seen that, when $\lambda < 1.1$, the response is localized around the left gap, although the localized modes involve motion of both gaps. This is due to the fact that nearly-simultaneous localized symmetric and antisymmetric modes give practically equal contribution to the response, so that motion disappears in the half of the string opposite to the driven end and disturbances cannot propagate. When $\lambda > 1.1$, since the modes with close frequencies are extended, the disturbances propagate.

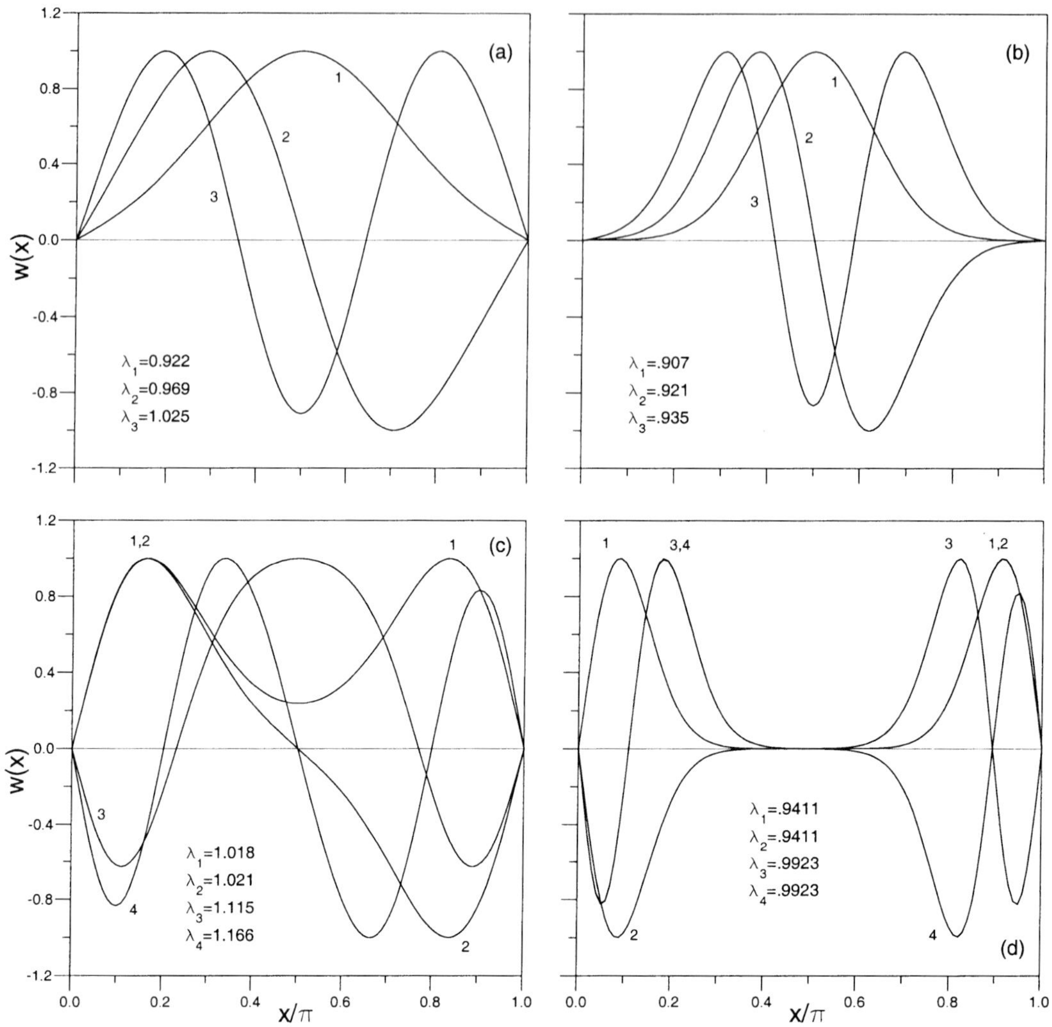

Figure 5. Normal modes for the string on elastic soil; (a, b) sinusoidal imperfections $f(x) = -\sin(x)$; (c, d) cosinusoidal imperfections $f(x) = -\cos(2x)$; $\varepsilon = 0.1$; (a, c) $\beta = 0.1$, (b, d) $\beta = 10^{3/2}$.

4.2. Free Vibrations of a Flexible Long Beam on Elastic Soil

A problem similar to the foregoing is studied; here the string is replaced by a simply supported beam having small flexural stiffness EI. The relevant eigenvalue problem is

$$\begin{aligned} &EIw'''' + k_f(x)w - m_0\omega^2 w = 0, \\ &w(0) = w(l) = 0, \quad w''(0) = w''(l) = 0 \end{aligned} \tag{23}$$

that, after having introduced the nondimensional quantities

$$w = \frac{w}{l}, \quad \tilde{x} = \frac{\pi x}{l}, \quad \beta = \frac{\pi}{l}\sqrt[4]{\frac{EI}{k_0}}, \quad \lambda = \frac{m_0\omega^2}{k_0}, \tag{24}$$

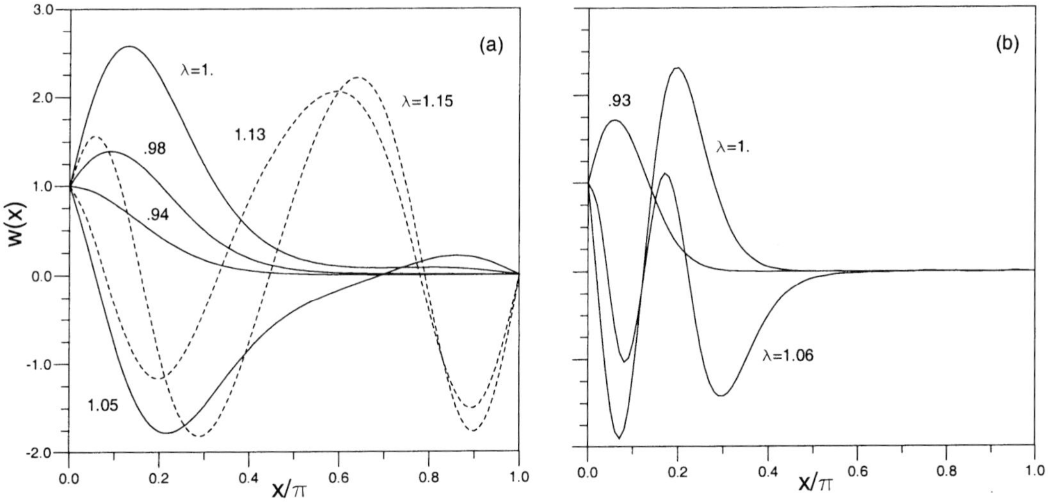

Figure 6. Time-harmonic response of the strings in Figures 5c and 5d when the left support is driven with frequency λ.

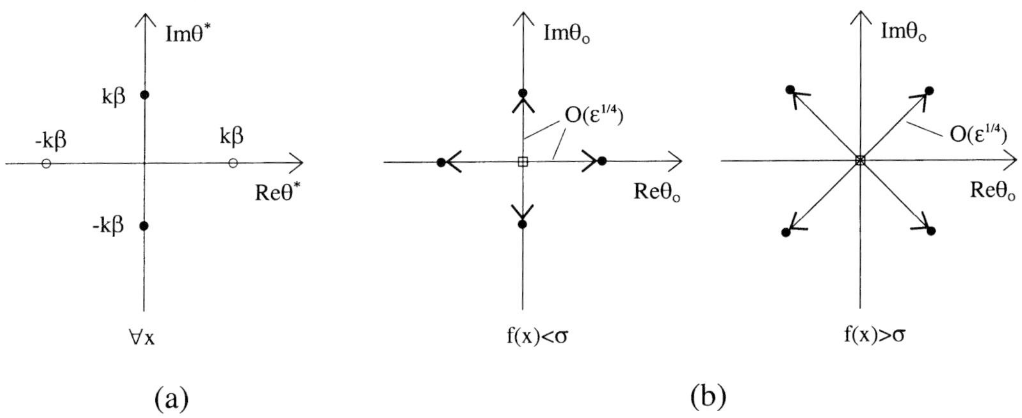

Figure 7. Characteristic exponents for the beam on elastic soil in the dynamic problem; (a) perfect system ($\varepsilon = 0$), (b) imperfect system ($\varepsilon \neq 0$); □ exponents at turning points.

reads

$$\beta \begin{Bmatrix} w_1' \\ w_2' \\ w_3' \\ w_4' \end{Bmatrix} = \left(\begin{Bmatrix} 0 & 1 & 0 & 0 \\ 0 & 0 & 1 & 0 \\ 0 & 0 & 0 & 1 \\ \lambda - 1 & 0 & 0 & 0 \end{Bmatrix} + \varepsilon \begin{pmatrix} 0 & 0 & 0 & 0 \\ 0 & 0 & 0 & 0 \\ 0 & 0 & 0 & 0 \\ -f(x) & 0 & 0 & 0 \end{pmatrix} \right) \begin{Bmatrix} w_1 \\ w_2 \\ w_3 \\ w_4 \end{Bmatrix}. \tag{25}$$

The characteristic exponents of the perfect system are $\theta^*_{1,2} = \pm i(\lambda - 1)^{1/4}$ and $\theta^*_{3,4} = \pm(\lambda - 1)^{1/4}$. From the boundary conditions $\lambda^*_k = 1 + k^4\beta^4 (k = 1, 2, \ldots)$ is found, i.e. $h = 4$ in Equation (8); therefore $\theta^*_{1,2} = \pm ik\beta$ and $\theta^*_{3,4} = \pm k\beta$ (Figure 7a). The associated eigenfunctions are sinusoids, since the real exponents give no contribution; it results $|\theta^*_k| = O(\beta)$, as in the previous example.

The imperfect system is now examined with $\varepsilon \geq O(\beta^4)$. The generating equation furnishes the following characteristic exponents

$$\begin{aligned}\theta_{0_{1,2}}(x) &= \pm i\varphi(x), \quad \theta_{0_{3,4}}(x) = \pm\varphi(x) \quad \text{if } f(x) < \sigma,\\ \theta_{0_{1,2}}(x) &= \pm\frac{\sqrt{2}}{2}(1+i)\varphi(x), \quad \theta_{0_{3,4}}(x) = \pm\frac{\sqrt{2}}{2}(1-i)\varphi(x) \quad \text{if } f(x) > \sigma,\end{aligned} \tag{26}$$

with

$$\varphi(x) = \sqrt[4]{\varepsilon|\sigma - f(x)|} \tag{27}$$

and the right and left eigenvectors

$$u_0 = \{1 \;\; \theta_0 \;\; \theta_0^2 \;\; \theta_0^3\}^T, \quad v_0 = \{\bar{\theta}_0^3 \;\; \bar{\theta}_0^2 \;\; \bar{\theta}_0 \;\; 1\}^T. \tag{28}$$

For different values of x, the characteristic exponents $\theta_0(x)$ are (a) two purely imaginary and two real or (b) two couples of complex conjugate numbers, with opposite real parts (Figure 7b). When $f(x) = \sigma$ the four eigenvalues coalesce at the origin and x is a turning point. Since $0 \leq |\theta_0(x)| \leq O(\varepsilon^{1/4})$, the imperfections modify the characteristic exponents of the perfect system by quantities of the same order as θ^* or larger; thus the eigenfunctions are strongly sensitive to the imperfections.

By following the procedure illustrated above the general solution

$$\begin{aligned}w(x) &= \frac{1}{[\sigma - f(x)]^{3/8}}[c_1 \exp\psi_1(x) + c_2\exp(-\psi_1(x))\\ &\quad + c_3 \sin\psi_1(x) + c_4\cos\psi_1(x)] \quad \text{if } f(x) < \sigma,\\ w(x) &= \frac{1}{[f(x) - \sigma]^{3/8}}\{\exp\psi_2(x)[c_5\sin\psi_2(x) + c_6\cos\psi_2(x)]\\ &\quad + \exp(-\psi_2(x))[c_7\sin\psi_2(x) + c_8\cos\psi_2(x)]\} \quad \text{if } f(x) > \sigma,\end{aligned} \tag{29}$$

is found, where

$$\psi_1(x) = \int \sqrt[4]{\mu|f(x) - \sigma|}\,\mathrm{d}x, \quad \psi_2(x) = \frac{\sqrt{2}}{2}\psi_1(x) \tag{30}$$

and $\mu = \varepsilon/\beta^4 \geq O(1)$ have been posed. The solution is singular at the turning points.

The first three modes of vibration of a beam, numerically obtained by the Galerkin method, are shown in Figure 8 for the same imperfections considered in Figure 5 for the string. Although the values of μ in Figures 8a and 8b are ten times larger than those in Figures 5a and 5b, the distortion of the modes is less important. This circumstance is explained by the fact that, by fixing ε and μ, i.e. by assuming that the values of β^2 and β^4 for the string and for the beam, respectively, are equal, the first system has higher modal density, due to the different powers of the number of mode k. The qualitative behavior is similar to that of the string, except in the neighborhood of the ends, where the modes are strongly exponentially modulated sinusoids. A zoom of the boundary region is illustrated in Figure 8c when μ is increased. When the cosinusoidal imperfection is considered for the lower μ (Figure 8d), the first two modes localize around the two gaps. The first mode is antisymmetric while the second is symmetric.

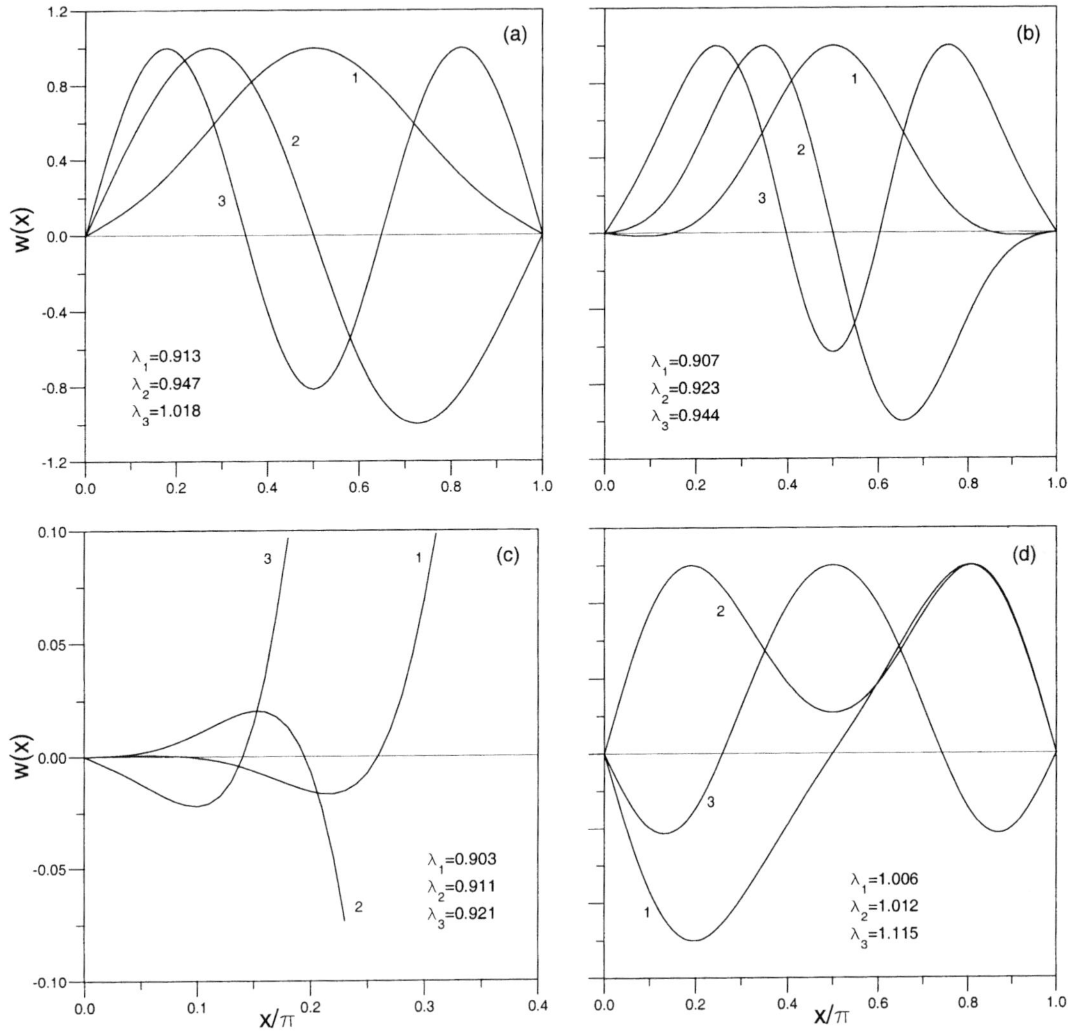

Figure 8. Normal modes of vibration for the beam on elastic soil; (a, b, c) sinusoidal imperfections $f(x) = -\sin(x)$, (d) cosinusoidal imperfections $f(x) = -\cos(2x)$; $\varepsilon = 0.1$; (a, d) $\beta = 10^{-3/4}$, (b) $\beta = 0.1$, (c) $\beta = 10^{-5/4}$.

4.3. Buckling of a Long Flexible Beam on Elastic Soil

The buckling of a long beam with small flexural rigidity EI, simply supported at the ends, resting on Winkler soil, is considered. The beam is compressed by an axial force λN_0 applied at one end and small longitudinal forces $\varepsilon p(x)$ distributed along the axis, having the character of defects. The soil, instead, is considered to be perfect, with constant stiffness k_0. The relevant eigenvalue problem reads

$$\begin{aligned} &EIw'''' + (N(x)w')' + k_0 w = 0, \\ &w(0) = w(l) = 0, \quad w''(0) = w''(l) = 0. \end{aligned} \tag{31}$$

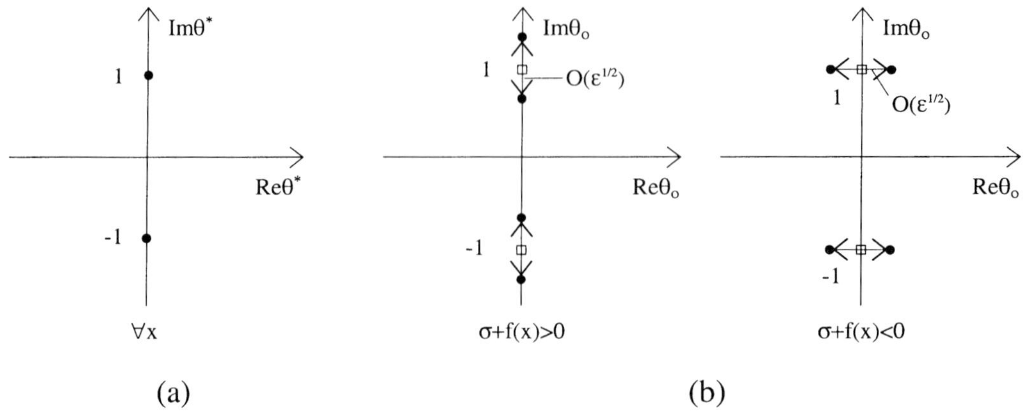

Figure 9. Characteristic exponents for the beam on elastic soil in the buckling problem; (a) perfect system ($\varepsilon = 0$), (b) imperfect system ($\varepsilon \neq 0$); □ exponents at turning points.

The axial stress at the abscissa x can be expressed as

$$N(x) = [\lambda + \varepsilon f(x)]N_0, \tag{32}$$

where $f(x)$ accounts for the shape of the imperfections. By assuming N_0 to be the critical load of the uniformly compressed beam, i.e. $N_0 = 2\sqrt{k_0\,EI}$, and introducing the nondimensional quantities

$$\tilde{w} = \frac{w}{l}, \quad \tilde{x} = \frac{\pi x}{l}, \quad \beta = \frac{\pi}{l}\sqrt[4]{\frac{EI}{k_0}}. \tag{33}$$

Equation (31a) is rewritten as

$$\beta \begin{Bmatrix} w_1' \\ w_2' \\ w_3' \\ w_4' \end{Bmatrix} = \left(\begin{pmatrix} 0 & 1 & 0 & 0 \\ 0 & 0 & 1 & 0 \\ 0 & 0 & 0 & 1 \\ -1 & 0 & -2\lambda & 0 \end{pmatrix} + \varepsilon \begin{pmatrix} 0 & 0 & 0 & 0 \\ 0 & 0 & 0 & 0 \\ 0 & 0 & 0 & 0 \\ 0 & 0 & -2f(x) & 0 \end{pmatrix} \right) \begin{Bmatrix} w_1 \\ w_2 \\ w_3 \\ w_4 \end{Bmatrix}, \tag{34}$$

where the tilde has been omitted and $\beta \ll 1$ if $EI \ll k_0 l^4$.

When the imperfections are absent the beam is uniformly compressed. The eigenvalues of the matrix L^* are

$$\theta^*_{1,2,3,4} = \pm i \left[\sqrt{\frac{\lambda+1}{2}} \pm \sqrt{\frac{\lambda-1}{2}} \right]. \tag{35}$$

When β is small, it follows from the boundary conditions that $\lambda_k^* \cong 1 + 2k^2\beta^2 (k = 0, 1, \ldots)$. Consequently, $\theta^* = \pm i(1 \pm k\beta)$ and the associated eigenfunctions are sinusoids with wavenumbers $n \pm k$, with $n \cong \beta^{-1} \gg 1$. The lower critical load $\lambda_c^* = 1$ is associated with $k = 0$ for which the characteristic exponents coalesce in twos (Figure 9a); the associated critical mode has n halfwaves. The higher critical loads have value close to 1 and are associated with sinusoids having $n \pm k$ halfwaves, where $k \ll n$. The system has high modal density with exponent $h = 2$; the eigenfunctions are fast varying, since $|\theta_k^*| = O(1)$.

The imperfect system with $\varepsilon \geq O(\beta^2)$ is now considered. Equations (11a) admit the following eigenvalues

$$\theta_{0_{1,2,3,4}}(x) = \pm i\left[\sqrt{1+\frac{1}{2}\varepsilon|\sigma+f(x)|} \pm \sqrt{\frac{1}{2}\varepsilon|\sigma+f(x)|}\right] \quad \text{if } \sigma+f(x)>0,$$

$$\theta_{0_{1,2,3,4}}(x) = \pm\left[i\sqrt{1-\frac{1}{2}\varepsilon|\sigma+f(x)|} \pm \sqrt{\frac{1}{2}\varepsilon|\sigma+f(x)|}\right] \quad \text{if } \sigma+f(x)<0, \tag{36}$$

associated with the right and left eigenvectors

$$u_0 = \{1 \;\; \theta_0 \;\; \theta_0^2 \;\; \theta_0^3\}^T, \quad v_0 = \{-\bar{\theta}_0 \;\; -1 \;\; \bar{\theta}_0^3 \;\; \bar{\theta}_0^2\}^T. \tag{37}$$

For different values of x the characteristic exponents can be (a) purely imaginary or (b) two couples of complex conjugate, symmetric with respect the imaginary axis (Figure 9b). When $f(x) = -\sigma$ the four eigenvalues coalesce in twos and x is a turning point. It should be noted that $1 - O(\varepsilon^{1/2}) \leq |\theta_0(x)| \leq 1 + O(\varepsilon^{1/2})$, so that the imperfections produce only a small correction of order $\varepsilon^{1/2}$ of the characteristic exponents of the perfect system, unlike the systems studied before.

The solution can be remarkably simplified if terms of order ε in Equations (36) are neglected in comparison with terms of order $\varepsilon^{1/2}$, i.e. if the following expressions for θ_0 are used

$$\theta_{0_{1,2,3,4}}(x) = \pm i\left[1 \pm \sqrt{\frac{1}{2}\varepsilon|\sigma+f(x)|}\right] + O(\varepsilon) \quad \text{if } \sigma+f(x)>0,$$

$$\theta_{0_{1,2,3,4}}(x) = \pm\left[i \pm \sqrt{\frac{1}{2}\varepsilon|\sigma+f(x)|}\right] + O(\varepsilon) \quad \text{if } \sigma+f(x)<0. \tag{38}$$

After having determined $\theta_1(x)$ from Equation (12), the following solution for $w(x)$ is obtained

$$\begin{aligned} w(x) &= \frac{1}{\sqrt[4]{|\sigma+f(x)|}}\{\sin nx[c_1 \sin\psi(x) + c_2\cos\psi(x)] \\ &\quad + \cos nx[c_3\sin\psi(x) + c_4\cos\psi(x)]\} \quad \text{if } \sigma+f(x)>0, \\ w(x) &= \frac{1}{\sqrt[4]{|\sigma+f(x)|}}\{\sin nx[c_5\exp\psi(x) + c_6\exp(-\psi(x))] \\ &\quad + \cos nx[c_7\exp\psi(x) + c_8\exp(-\psi(x))]\} \quad \text{if } \sigma+f(x)<0, \end{aligned} \tag{39}$$

where

$$\psi(x) = \int\sqrt{\mu\frac{1}{2}|f(x)+\sigma|}\,\mathrm{d}x \tag{40}$$

and $\mu = \varepsilon/\beta^2 \geq O(1)$. It should be noted that Equation (39) furnishes sinusoidal functions with large wavenumber n, slowly modulated with the same law, sinusoidal or exponential, determined for the string (Equation (21)). The asymptotic solution is singular at the turning points.

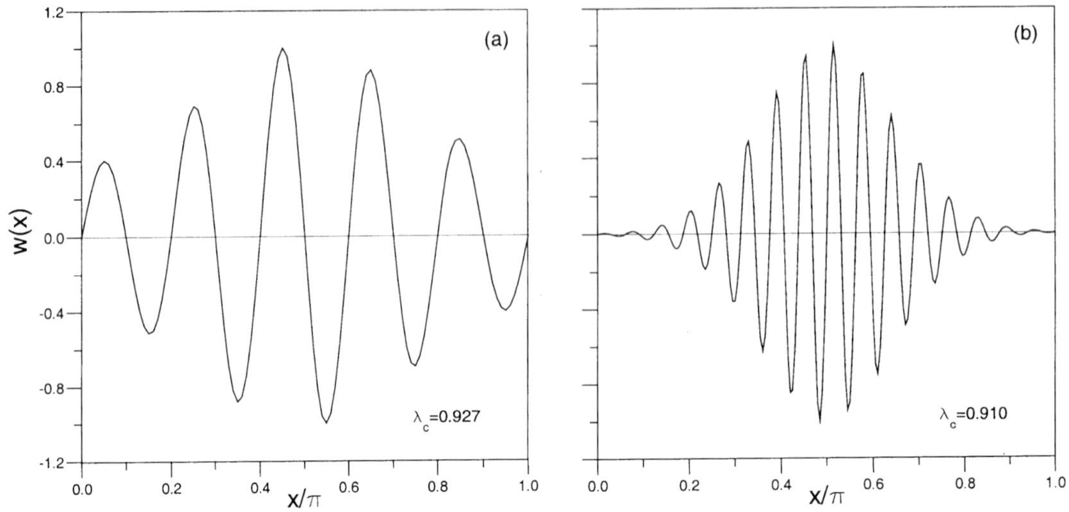

Figure 10. Critical buckling mode for the beam on elastic soil subjected to additional compression forces; $\varepsilon = 0.1$, $f(x) = \sin(x)$; (a) $\beta = 0.1$, (b) $\beta = 10^{-3/2}$.

Figure 10 shows the buckling modes of two compressed beams, obtained numerically. Due to the increment of compression caused by the disturbance forces, the critical multiplier λ_c is smaller than 1; its lower bound is $1-\varepsilon$, and it corresponds to a constant imperfection shape $f(x) = 1$, i.e. to disturbance forces concentrated at the ends. Therefore, from Equation (18), $-1 < \sigma < 0$ and so $\sigma + f(x)$ is positive in the central region (sinusoidal modulation) and negative near the supports (exponential modulation). Thus, if μ is sufficiently large, the buckling mode localizes around the middle point of the beam, i.e. around the point of maximum stress.

5. Remarks on the Characteristic Exponents Sensitivity

It is useful to reconsider the generating equation (11a) and comment on the order of magnitude of the characteristic exponents θ^* and θ_0 of perfect and imperfect systems, respectively.

By using Equation (5), the algebraic eigenvalue problem (11a) reads

$$(L^* + \varepsilon L_\varepsilon - \theta_0)u_0 = 0. \tag{41}$$

This can be considered as an order-ε perturbation of the eigenvalue problem (7), relative to the perfect system. Thus, it might be thought convenient to introduce the series expansions

$$\theta_0 = \theta^* + \varepsilon\theta_\varepsilon + O(\varepsilon^2), \qquad u_0 = u^* + \varepsilon u_\varepsilon + O(\varepsilon^2) \tag{42}$$

and evaluate $\theta_\varepsilon, u_\varepsilon, \ldots$ through a standard perturbation method. However, this procedure would lead to incorrect results. In fact, in all the examples considered in the previous section, it has been observed that small imperfections of order ε modify the eigenvalues θ^* of the system matrix L^* by quantities of order greater than ε. In particular, the characteristic exponents in the dynamic problems have been modified by terms of order $\varepsilon^{1/2}$ and $\varepsilon^{1/4}$ (i.e. by corrections of the same order as the unmodified values θ^*; see Figures 4 and 7) and in the buckling problem by terms of order $\varepsilon^{1/2}$ (while the unmodified values are of order 1; see Figure 9).

In the perturbation theory of the eigenvalues it is known that these fractional powers appear when the unperturbed matrix is defective (i.e. does not have a complete set of eigenvectors) or nearly-defective [39–41]. Thus, defective matrices exhibit high sensitivity to imperfections.

It can be checked that the unperturbed matrices L^* are in fact nearly-defective, because they have nearly-parallel eigenvectors. In particular, the matrices are themselves perturbations of order ε of exactly defective matrices L_{id}. In fact, if Equation (18) is substituted in the differential Equations (16), (25) or (34) and the term $\varepsilon\sigma$ is properly shifted in the perturbation matrix $\varepsilon L_\varepsilon$, the resulting order-one matrix L_{id} is found to be defective. In the first two problems, L_{id} is a Jordan block of order 2 or 4, admitting therefore a unique eigenvector associated with the multiple zero eigenvalue; in the buckling problem, L_{id} admits two double eigenvalues $\theta_{\mathrm{id}} = \pm i$, each associated with a unique eigenvector.

It has been shown by the author in [42] that the most appropriate method to study perturbations of nearly-defective matrices L^* is to build an exactly defective ideal matrix L_{id}, close in some sense to the original matrix L^*, and then to perturb directly L_{id}, by applying the algorithm developed in [41]. Here the ideal matrix L_{id} appears in a natural way in the problem. An example of application of the method is given in the Appendix.

These considerations permit the localization phenomenon to be considered in a different perspective. Namely, since nearly defective continuous structures are strongly sensitive to imperfections, small perturbations cause large variations of the characteristic exponents; in particular, imaginary exponents can become complex and consequently the associated periodic modes localize.

6. Conclusions

Localization phenomena in one-dimensional imperfect continuous structures have been analyzed, both in dynamics and buckling. A general perturbation method has been developed which generalizes the classical WKB method. The method has been applied to analyze the dynamics of a string and the dynamics and buckling of a beam on elastic soil. All these structures have been found to have nearly-defective system matrices, so their characteristic exponents are highly sensitive to imperfections.

The following conclusions are drawn.

1. In engineering structures, the most important mechanism of localization calls for the occurrence of turning points at which the eigenfunctions change from oscillating to exponential or *vice versa*.
2. Localization in continuous systems of finite length occurs when the ratio between the amplitude of the imperfections and the coupling comes to have sufficiently large values. Only the first modes localize and their number increases when the imperfection/coupling ratio increases.
3. In the dynamic problem, harmonic disturbances cannot propagate through the structure if their frequency is close to a localized mode natural frequency; similarly, in the buckling problem, disturbances cannot propagate if the axial load is close to a critical load associated with a localized buckling pattern.
4. Interaction among nearly-simultaneous buckling modes of perfect structures in the nonlinear field can lead to localized buckling patterns. The phenomenon reveals close analogies with the localization of linear imperfect systems.

5. Asymptotic solutions can be built up by applying the perturbation method illustrated; this transforms the differential problem into a sequence of algebraic problems in which the spatial variable appears as a parameter.
6. Localization of vibrations of a string (or beam) on elastic soil is due to characteristic exponents changing from purely imaginary to real (from purely imaginary to complex conjugates, with the real and imaginary parts equal in modulus); localization of the buckling mode of a beam on elastic soil is due to characteristic exponents changing from purely imaginary to complex conjugates with small real part.
7. Systems with nearly defective matrices are highly sensitive; in fact, small perturbations of order ε produce modifications of the characteristic exponents of order $\varepsilon^{1/2}$ or larger. This effect exalts localization.

Appendix: An Example of Solution of a Nearly Defective Eigenvalue Problem

As an example of application of the method outlined in Section 5, the eigenvalue problem (41), relative to the buckling problem dealt with in Section 4.3, is solved. The ideal defective matrix $L_{\rm id}$ is introduced and the perturbation matrix L_ε is redefined as follows

$$L_{\rm id} = \begin{pmatrix} 0 & 1 & 0 & 0 \\ 0 & 0 & 1 & 0 \\ 0 & 0 & 0 & 1 \\ -1 & 0 & -2 & 0 \end{pmatrix}, \quad L_\varepsilon = \begin{pmatrix} 0 & 0 & 0 & 0 \\ 0 & 0 & 0 & 0 \\ 0 & 0 & 0 & 0 \\ 0 & 0 & -2(\sigma + f) & 0 \end{pmatrix} \tag{43}$$

in such a way that $L = L_{\rm id} + \varepsilon L_\varepsilon$ (see Equation (36)). By introducing the series expansions [41]

$$\begin{aligned} \theta_0 &= \theta_{\rm id} + \varepsilon^{1/2}\theta_\varepsilon + \varepsilon\theta_{\varepsilon\varepsilon} + O(\varepsilon^{3/2}), \\ u_0 &= i_{\rm id} + \varepsilon^{1/2}u_\varepsilon + \varepsilon u_{\varepsilon\varepsilon} + O(\varepsilon^{3/2}), \end{aligned} \tag{44}$$

the following perturbation equations are obtained

$$\begin{aligned} (L_{\rm id} - \theta_{\rm id})u_{\rm id} &= 0, \\ (L_{\rm id} - \theta_{\rm id})u_\varepsilon &= \theta_\varepsilon u_{\rm id}, \\ (L_{\rm id} - \theta_{\rm id})u_{\varepsilon\varepsilon} &= -L_\varepsilon u_{\rm id} + \theta_\varepsilon u_\varepsilon + \theta_{\varepsilon\varepsilon} u_{\rm id}. \end{aligned} \tag{45}$$

The right and left eigenvectors of $L_{\rm id}$ associated with $\theta_{\rm id} = \pm i$ are

$$u_{\rm id} = \{1 \ \ \pm i \ \ -1 \ \ \mp i\}^T, \quad v_{\rm id} = \{1 \ \ \pm i \ \ 1 \ \ \pm i\}^T. \tag{46}$$

The perturbation equation (45b) admits the following particular solution

$$u_\varepsilon = \theta_\varepsilon\{0 \ \ 1 \ \ \pm 2i \ \ -3\}^T \tag{47}$$

in which θ_ε is undetermined. By imposing the solvability condition to Equation (45c), two solutions for θ_ε are obtained

$$\theta_\varepsilon = \pm\sqrt{-\frac{1}{2}\,(f + \sigma)}. \tag{48}$$

The series expansions (44) coincide with the exact solution determined in Section 4.3 and then expanded up to $\varepsilon^{1/2}$-order terms.

References

1. Matsuda, H. and Ishii, K., 'Localization of normal modes and energy transport in the disordered harmonic chain', *Supplement of the Progress of Theoretical Physics* **45**, 1970, 56–86.
2. Ishii, K., 'Localization of eigenstates and transport phenomena in the one-dimensional disordered system', *Supplement of the Progress of Theoretical Physics* **53**, 1973, 77–138.
3. Hodges, C. H., 'Confinement of vibration by structural irregularity', *Journal of Sound and Vibration* **82**, 1982, 411–424.
4. Li, D. and Benaroya, H., 'Dynamics of periodic and near-periodic structures', *Applied Mechanics Review* **45**(11), 1982, 447–459.
5. Benaroya, H., 'Localization and effects of irregularities in structures', *Applied Mechanics Review* **49**, 1996, 57–135.
6. Pierre, C. and Dowell, E. H., 'Localization of vibrations by structural irregularity', *Journal of Sound and Vibration* **114**(3), 1987, 549–564.
7. Bendiksen, O., 'Mode localization phenomena in large space structures', *AIAA Journal* **25**(9), 1987, 1241–1248.
8. Pierre, C., 'Mode localization and eigenvalue loci veering phenomena in disordered structures', *Journal of Sound and Vibration* **126**(3), 1988, 485–502.
9. Pierre, C., 'Weak and strong vibration localization in disordered structures: A statistical investigation', *Journal of Sound and Vibration* **139**(1), 1990, 111–132.
10. Wu, G., 'Free vibration of cyclic assemblies with a single disordered component', *Journal of Sound and Vibration* **165**(3), 1993, 567–570.
11. Ottarsson, G. and Pierre, C., 'A transfer matrix approach to free vibration localization in mistuned blade assemblies', *Journal of Sound and Vibration* **197**(5), 1996, 589–618.
12. Hodges, C. H. and Woodhouse, J., 'Vibration isolation from irregularity in a nearly periodic structure: Theory and measurements', *Journal of Acoustical Society of America* **79**(9), 1983, 894–905.
13. Cornwell, P. J. and Bendiksen, O., 'Localization of vibrations in large space reflectors', *AIAA Journal* **27**(2), 1989, 219–226.
14. Pierre, C. and Cha, P. D., 'Strong mode localization in nearly periodic disordered structures', *AIAA Journal* **27**(2), 1989, 227–241.
15. Al-Jawi, A. A., Ulsoy, A. G., and Pierre, C., 'Vibration localization in dual-span, axially moving beams', *Journal of Sound and Vibration* **179**(2), 1995, 243–266 (part I), 267–287 (part II).
16. Keane, A. J. and Price, W. G., 'On the vibrations of mono-coupled periodic and near-periodic structures', *Journal of Sound and Vibration* **128**(3), 1989, 423–450.
17. Cai, G. Q. and Lin, Y. K., 'Localization of wave propagation in disordered periodic structures', *AIAA Journal* **29**, 1990, 450–456.
18. Lin, Y. K. and Cai, G. Q., 'Disordered periodic structures', in *Computational Stochastic Mechanics*, C. A. Brebbia and P. Spanos (eds.), Elsevier, Amsterdam, 1991, pp. 777–788.
19. Cha, P. D. and Pierre, C., 'Vibration localization by disorder in assemblies of monocoupled multimode component systems', *Journal of Applied Mechanics* **58**, 1991, 1072–1081.
20. Cheng, S. P. and Perkins, N. C., 'Closed-form vibration analysis of sagged cables/mass suspension', ASME Paper 92WA/APM-7, 1992.
21. Bouzit, D. and Pierre, C., 'Localization of vibration in disordered multi-span beams with damping', *Journal of Sound and Vibration* **187**(4), 1995, 625–648.
22. Scott, J. M. F., 'The statistics of waves propagating in a one-dimensional random medium', *Proceedings of Royal Society of London* **A398**, 1985, 341–363.
23. Luongo, A., 'Mode localization by structural imperfections in one-dimensional continuous systems', *Journal of Sound and Vibration* **155**(2), 1992, 249–271.
24. Tjavaras, A. A. and Triantafyllou, M. S., 'Non-linear response of two disordered pendula', *Journal of Sound and Vibration* **190**(1), 1996, 65–76.

25. Zevin, A. A., 'Localization of periodic oscillations in discrete non-linear systems', *Journal of Sound and Vibration* **193**(4), 1996, 847–862.
26. Vakakis, A., Nayfeh, T., and King, M., 'A multiple-scales analysis of nonlinear, localized modes in a cyclic periodic system', *Transactions of the ASME* **60**, 1993, 388–397.
27. Emaci, E., Nayfeh, T. A. and Vakakis, A. F. ,'Numerical and experimental study of nonlinear localization in a flexible structure with vibro-impacts', *Journal of Applied Mathematics and Mechanics (ZAMM)* **77**(7), 1997, 527–541.
28. Nayfeh, T. A., Emaci, E. and Vakakis, A. F., 'Application of nonlinear localization to the optimization of a vibration isolation system', *AIAA Journal* **35**(8), 1997, 1378–1386.
29. Pierre, C., 'Curve veering and mode localization in a buckling problem', *Journal of Applied Mathematics and Physics (ZAMP)* **40**, 1989, 758–761.
30. Gioncu, V., 'General theory of coupled instabilities', *Thin Walled Structures* **19**(2–4), 1994, 81–127.
31. Luongo, A., 'On the amplitude modulation and localization phenomena in interactive buckling problems', *International Journal of Solids and Structures* **27**(15), 1991, 1943–1954.
32. Nayfeh, A. H., *Perturbation Methods*, Wiley, New York, 1973.
33. Tvegaard, V. and Needleman, A., 'On the localization of buckling patterns', *Journal of Applied Mechanics* **47**, 1980, 613–619.
34. Tvegaard, V. and Needleman, A., 'On the development of localized buckling patterns', in *Collapse, The Buckling of Structures in Theory and Practice*, J. M. T. Thompson and G. W. Hunt (eds.), Cambridge University Press, Cambridge, 1983, pp. 1–17.
35. Potier-Ferry, M., 'Amplitude modulation, phase modulation and localization of buckling patterns', in *Collapse, The Buckling of Structures in Theory and Practice*, J. M. T. Thompson and G. W. Hunt (eds.), Cambridge University Press, Cambridge, 1983, pp. 149–159.
36. Luongo, A. and Pignataro, M. ,'Multiple interaction and localization phenomena in the postbuckling of compressed thin-walled members', *AIAA Journal* **26**(11), 1988, 1395–1402.
37. Hunt, G. W., Bolt, H. M., and Thompson, J. M. T., 'Structural localization phenomena and the dynamical phase-space analogy', *Proceedings of Royal Society London* **A425**, 1989, 245–267.
38. Luongo, A., 'Perturbation methods for the eigenvalue analysis of imperfect systems with high modal density', Parts I and II, *Accademia Peloritana dei Pericolanti*, Messina (Italy), Vol. LXVIII, 401–423, 1991, 424–447 [in Italian].
39. Wilkinson, J. H., *The Algebraic Eigenvalue Problem*, Oxford University Press, London, 1965.
40. Kato, T., *A Short Introduction to Perturbation Theory for Linear Operators*, Springer-Verlag, New York, 1982.
41. Luongo, A., 'Eigensolutions sensitivity for nonsymmetric matrices with repeated eigenvalues', *AIAA Journal* **31**(7), 1994, 1321–1328.
42. Luongo, A., 'Eigensolutions of perturbed nearly-defective matrices', *Journal of Sound and Vibration* **185**(3), 1995, 377–395.

Nonlinear Dynamics **25:** 157–182, 2001.

Dynamics of Relative Phases: Generalised Multibreathers

TAEHOON AHN*, ROBERT S. MACKAY**, and JACQUES-A. SEPULCHRE***
The Nonlinear Centre, Department of Applied Mathematics and Theoretical Physics, Silver Street, University of Cambridge, Cambridge CB3 9EW, U.K.

(Received: 15 March 1999; accepted: 10 August 1999)

Abstract. For small Hamiltonian perturbation of a Hamiltonian system of arbitrary number of degrees of freedom with a normally non-degenerate submanifold of periodic orbits we construct a nearby submanifold and an 'effective Hamiltonian' on it such that the difference between the two Hamiltonian vector fields is small. The effective Hamiltonian is independent of one coordinate, the 'overall phase', and hence the corresponding action is preserved. Unlike standard averaging approaches, critical points of our effective Hamiltonian subject to given action correspond to exact periodic solutions. We prove there has to be at least a certain number of these critical points given by global topological principles. The linearisation of the effective Hamiltonian about critical points is proved to give the linearised dynamics for the full system to leading order in the perturbation. Hence in the case of distinct eigenvalues which move at non-zero speed with ε, the linear stability type of the periodic orbit can be read off from the effective Hamiltonian. Our principal application is to networks of oscillators or rotors where many such submanifolds of periodic orbits occur at the uncoupled limit – simply excite a number $N \geq 2$ of the units in rational frequency ratio and put the others on equilibria, subject to a non-resonance condition. The resulting exact periodic solutions for weak coupling are known as multibreathers. We call the approximate solutions given by the effective Hamiltonian dynamics, 'generalised multibreathers'. They correspond to solutions which look periodic on a short time scale but the relative phases of the excited units may evolve slowly. Extensions are sketched to travelling breathers and energy exchange between degrees of freedom.

Keywords: Breathers, phases, Hamiltonian, dynamics.

1. Introduction

Since the seminal papers of Takeno et al. [1, 2], several papers have dealt with the mathematical analysis of 'discrete breathers': spatially localised time-periodic solutions of networks of weakly coupled oscillators [3–11]. Discrete breathers can be constructed by continuation of solutions from the decoupled limit, for which some units are oscillating and the remainder are at rest. A discrete breather resulting from one oscillating unit at the uncoupled limit we call a 'single breather'; those resulting from more than one oscillating unit we call 'multibreathers'.

The present paper firstly provides a variation on the general Hamiltonian proof of existence of 'multibreathers' that was promised in [4], given independently in [5], and later worked out in [11]. Secondly, it derives their linear stability types more generally than [5]. Thirdly, and most significantly, it constructs approximate solutions which look like multibreathers but whose relative phases evolve slowly and are found to be governed by an effective Hamiltonian.

* On leave from: Nonlinear and Complex Systems Laboratory, Department of Mathematics, Basic Science Research Institute, POSTECH, Pohang, Korea 790-784, E-mail: ath@postech.ac.kr.
** Author for correspondence. Address from 1 March 2000: Mathematics Institute, University of Warwick, Coventry CV4 7AL, U.K., E-mail: mackay@maths.warwick.ac.uk.
*** New permanent address: Institut Non-Linéaire de Nice-Sophia, 1361, route des Lucioles, 06560 Valbonne, France, E-mail: sepulchr@inln.cnrs.fr.

Thus we enlarge the concept of 'multibreather' to 'generalised multibreather' solutions which are not necessarily exactly time-periodic, but which we expect to be of physical importance.

The simplest example of system to which our results apply is a chain of weakly coupled pendula in which at the uncoupled limit N pendula, not necessarily neighbours, are excited with approximately equal frequencies, non-resonant with the frequencies of linearised motion of the remaining pendula about stable equilibrium. Our method constructs a $\mathbb{T}^N \times \mathbb{D}^N$ in the phase space of the full system, which is approximately invariant and on which the vector field is close to that of an effective Hamiltonian H^{eff} on $\mathbb{T}^N \times \mathbb{D}^N$. The coordinates on the $\mathbb{T}^N \times \mathbb{D}^N$ can be thought of as a global phase, $N-1$ relative phases, a global action and $N-1$ relative momenta. The effective Hamiltonian does not depend on the global phase, thus the global action A is preserved and for each A, $H^{\text{eff}}_A = H^{\text{eff}}(A, \cdot)$ describes slow evolution of the relative phases and their momenta on a $\mathbb{T}^{N-1} \times \mathbb{D}^{N-1}$.

The idea of constructing an effective Hamiltonian for approximate dynamics of slow variables is old. For example, Chirikov [12] constructed an effective Hamiltonian for the dynamics of resonant phases $m \cdot \theta$ of a Hamiltonian system $H(I, \theta)$ on $T^*\mathbb{T}^N$ near the resonant surfaces where $m \cdot \partial H/\partial I$ is small, for one or more $m \in \mathbb{Z}^N \setminus \{0\}$. Earlier, Poincaré [13] introduced an effective Hamiltonian for the dynamics of the Laplace vectors – eccentricities and inclinations with their associated angles, for the planetary ellipses in the solar system by averaging over the relatively rapid orbital motion (for a history, see [14]).

Similarly, Whitham [15] introduced an averaged Lagrangian to describe the evolution of the envelope of oscillatory solutions of some Lagrangian PDEs, and under the name of 'collective coordinates' this idea was applied to slow evolution of solitons and breathers for perturbations of integrable PDEs by various people, e.g. [16, 17].

Our approach is weaker than that of Poincaré and Chirikov, in that we require periodic reference dynamics whereas they allow quasiperiodic. But our approach is stronger than theirs in two respects:

(i) critical points of our H^{eff}, including degenerate ones, are in one-to-one correspondence with exactly time-periodic solutions,
(ii) we allow the zone where the reference dynamics is good to be a neighbourhood of a torus of lower dimension than the total number of degrees of freedom, and obtain an effective Hamiltonian for just the reduced number of degrees of freedom, subject to a non-resonance condition on the normal dynamics.

Moser [18] obtained an effective Hamiltonian for the motion of relative phases on perturbing a vector field on a manifold consisting entirely of periodic solutions of constant period, but only to first order in the perturbation. Thus to deduce existence of periodic orbits for the perturbed system he requires non-degenerate critical points of his effective Hamiltonian. Also he did not consider the questions of perturbation of a vector field with only a submanifold of periodic solutions, nor of uniformity of the estimates with respect to system size.

Persistence of some number of periodic solutions from a submanifold of periodic solutions was obtained in [19], but no properties of the non-equilibrium solutions of the effective Hamiltonian are considered there, nor is uniformity with respect to system size considered.

In Section 2, we specify the kind of Hamiltonian under study. In Section 3 we give a method to find multibreather solutions. Section 4 presents the main result which enables one to describe approximate dynamics of generalised multibreathers. In Section 5, we investigate the relation between the linear stability of equilibria of the approximate dynamics and of the multibreathers. In Section 6 we apply the method to a range of situations in oscillator networks. Conclusions and discussions end the paper in Section 7.

2. Relative Momenta and Relative Phases for Multibreathers

We consider the following type of C^2 Hamiltonian describing networks of coupled units (which may be either oscillators or rotors)

$$H_\varepsilon(\underline{p}, \underline{q}) = \sum_{s\in S} H_s(p_s, q_s) + \varepsilon H^1(\underline{p}, \underline{q}), \quad (\underline{p}, \underline{q}) = (p_s, q_s)_{s\in S} \in (\mathbb{R}^2)^S, \tag{1}$$

where S is a countable index set and ε is a coupling constant. For each $s \in S$ we assume that $H_s(p_s, q_s)$ is a 1 DOF Hamiltonian describing an oscillator or rotor and has an elliptic equilibrium (p_s^e, q_s^e). Without loss of generality, we set $(p_s^e, q_s^e) = (0, 0)$ and $H_s(p_s^e, q_s^e) = 0$. We endow the phase space with the l_2 norm

$$||(\underline{p}, \underline{q})||_2 := \sqrt{\sum_{s\in S} |(p_s, q_s)|^2},$$

so that the Hamiltonian in (1) is convergent.

At $\varepsilon = 0$ we choose a finite subset $S' \subset S$ of size $N + 1 \geq 2$ and consider periodic solutions for which the units belonging to S' exhibit periodic motions with frequencies in the irreducible ratio $\underline{m} \in \mathbb{Z}^{N+1}$ and the remaining units are on elliptic equilibria. These solutions give multibreathers. The state of the units belonging to S' can be represented by action-angle variables $(I_s, \psi_s \bmod 1)$. We divide the variables as follows:

$$(\underline{p}, \underline{q}) = (\underline{I}, \underline{\psi}, p_r, q_r),$$

where $\underline{I} = (I_s)_{s\in S'}$, $\underline{\psi} = (\psi_s)_{s\in S'}$, and r runs through $S'' = S \setminus S'$. The multibreathers of period T and ratio $\underline{m}$ at $\varepsilon = 0$ are given by

$$\underline{I}(t) = I_0, \quad \underline{\psi}(t) = ft\,\underline{m} + \psi_0, \quad (p_r(t), q_r(t)) = (\underline{0}, \underline{0}),$$

where

$$f = \frac{1}{T}, \quad (\underline{0}, \underline{0}) := (0, 0)_{s\in S''},$$

and I_0 is defined so that

$$\frac{dH_s}{dI_s}(I_{0s}) = fm_s, \quad \forall s \in S'.$$

Thus the multibreathers with given period form an invariant $(N + 1)$-torus $\mathbb{T}^{N+1}$. If the units belonging to S' are locally anharmonic, i.e. they satisfy $d^2H_s/dI_s^2 \neq 0$ (actually, it would suffice to have all but one of the units in S' anharmonic), then this $\mathbb{T}^{N+1}$ is part of an invariant set $E \times \mathbb{T}^{N+1}$ for some interval E in $\mathbb{R}$, filled by periodic orbits of the same frequency ratio with possibly different period. The projection of $E \times \mathbb{T}^{N+1}$ to the action space is a resonant curve $\underline{I} = \underline{I}^{\mathrm{res}}(A)$ parametrised by the *area* $A = \underline{m} \cdot \underline{I}$, satisfying

$$\frac{\partial H_0}{\partial \underline{I}}(\underline{I}^{\mathrm{res}}(A), \underline{0}, \underline{0}) = f(A)\,\underline{m},$$

for some function $f(A)$. It is easy to see that on $E \times \mathbb{T}^{N+1}$

$$\dot{\underline{I}} = 0, \quad \dot{\underline{\psi}} = f(A)\,\underline{m}, \quad \dot{A} = 0, \quad \dot{p}_r = 0, \quad \dot{q}_r = 0$$

and

$$f(A) = \frac{\mathrm{d}}{\mathrm{d}A} H_0(\underline{I}^{\mathrm{res}}(A), \underline{0}, \underline{0}).$$

Since $m \in \mathbb{Z}^{N+1}$ is irreducible, there exists $M \in GL(N+1, \mathbb{Z})$ such that $Me_1 = m$ where e_1 is the first standard basis vector (see appendix A of [20], for example). This induces a canonical change of coordinates for $(\underline{I}, \underline{\psi})$ given by

$$\underline{\psi} = M \begin{bmatrix} \varphi \\ \underline{\theta} \end{bmatrix}, \quad \begin{bmatrix} A \\ \underline{J} \end{bmatrix} = M^t \underline{I}.$$

Then the new variables $(A, \varphi) \in \mathbb{R} \times \mathbb{R}/\mathbb{Z}$ are called the *area* and *overall phase* and $(\underline{J}, \underline{\theta}) \in \mathbb{R}^N \times \mathbb{R}^N/\mathbb{Z}^N$ are called the *relative momenta* and *relative phases*. The resonance curve $\underline{I} = \underline{I}^{\mathrm{res}}(A)$ is transformed into $\{(A, \underline{J}^{\mathrm{res}}(A)) : A \in E\}$. In the new variables we write the Hamiltonian as

$$H_\varepsilon = H^0(A, \underline{J}, p_r, q_r) + \varepsilon H^1(A, \varphi, \underline{J}, \underline{\theta}, p_r, q_r),$$

which satisfies

$$\begin{aligned} D_A H^0(A, \underline{J}^{\mathrm{res}}(A), \underline{0}, \underline{0}) &= f(A), \\ D_{\underline{J}} H^0(A, \underline{J}^{\mathrm{res}}(A), \underline{0}, \underline{0}) &= 0. \end{aligned}$$

The multibreathers at $\varepsilon = 0$ are represented by

$$\begin{aligned} A(t) &= A, \quad \varphi(t) = f(A)t + \varphi_0, \\ \underline{J}(t) &= \underline{J}^{\mathrm{res}}(A), \quad \underline{\theta}(t) = \underline{\theta}, \quad (p_r(t), q_r(t)) = (\underline{0}, \underline{0}), \end{aligned} \tag{2}$$

for arbitrary $(A, \varphi_0, \underline{\theta})$, which parametrise the invariant $E \times \mathbb{T}^{N+1}$. Note that $(\varphi_0, \underline{\theta})$ are periodic variables of period 1. Since φ_0 is just the initial phase along the periodic orbits we have an $E \times \mathbb{T}^N$ of multibreathers, labelled by $(A, \underline{\theta})$. The phase space, in a neighbourhood of these multibreathers is as follows:

$$\Gamma = \{(A, \varphi_0, \underline{J}, \underline{\theta}, p_r, q_r) \in E \times \mathbb{T}^1 \times \Omega \times \mathbb{T}^N \times \mathbb{R}^{2S''}\},$$

where $\Omega \subset \mathbb{R}^N$ is a neighbourhood of $\underline{J}^{\mathrm{res}}(A) \in \mathbb{R}^N$ and $||(p_r, q_r)||_2$ is finite. The tangent space of Γ denoted by Λ is a Banach space $\mathbb{R}^2 \times \mathbb{R}^{2N} \times \mathbb{R}^{2S''}$ with l_2 norm.

When the coupling constant ε is different from zero, there is generically no nearby invariant $E \times \mathbb{T}^{N+1}$ consisting entirely of periodic orbits of bounded period, because for generic Hamiltonian systems periodic orbits of given energy and period less than a given bound are isolated. Nonetheless, one can prove that at least N periodic solutions survive small perturbation (and at least 2^{N-1} if all non-degenerate). See [3–5, 11, 21] for various cases. We will present a variation on the methods of [4, 5, 11], which is essential for the extension to our generalised multibreathers.

3. Periodic Orbits of the System under Constraints

The starting point is a variational formulation of the problem of finding a periodic orbit. Let Z be a space of functions

$$z(s) = (A(s), \varphi(s), \underline{J}(s), \underline{\theta}(s), p_r(s), q_r(s)),$$

with the periodicity condition

$$z(s+1) = z(s) + (0, 1, \underline{0}, \underline{0}, \underline{0}, \underline{0}),$$

known as a 'loop space'. More precisely, we choose the loop space to be $Z = C^1(S^1, \Gamma)$ where $S^1 = \mathbb{R}/\mathbb{Z}$ is a circle of length 1. The tangent space $T_z Z$, $z \in Z$, is isomorphic to the Banach space $C^1(S^1, \Lambda)$ with the norm

$$||\xi||_2^2 = ||A||_{C^1}^2 + ||\varphi||_{C^1}^2 + ||\underline{J}||_{C^1}^2 + ||\underline{\theta}||_{C^1}^2 + \sum_{s \in S'} ||(p_s, q_s)||_{C^1}^2.$$

We define the *symplectic area* of a loop z by

$$\mathcal{A}(z) = \oint A \,\mathrm{d}\varphi + \underline{J} \cdot \mathrm{d}\underline{\theta} + p_r \cdot \mathrm{d}q_r.$$

By the Poincaré–Cartan variational principle, $z(ft)$ is a periodic orbit of H of symplectic area $a \in \mathbb{R}$ if and only if $(z, f) \in Z \times \mathbb{R}$ is a critical point of the function from $Z \times \mathbb{R}$ to $\mathbb{R}$ defined by:

$$S_a(z, f) = \oint H \circ z \,\mathrm{d}s - f\left(\mathcal{A}(z) - a\right). \tag{3}$$

This amounts to expressing the periodic orbits of H as critical points of the *averaged energy*

$$\mathcal{H}(z) = \oint H \circ z \,\mathrm{d}s$$

in the space Z subject to fixed area $\mathcal{A}(z) = a$. It can be easily checked that the Lagrange multiplier f associated with the latter constraint is just the frequency of the periodic orbit (time frequency, not angular frequency, as we did not insert a factor $1/2\pi$ in the definition of $\mathcal{A}$). Let us note that since the system is autonomous, these periodic solutions are determined up to an arbitrary phase that we can fix by choosing $\varphi(0) = \varphi_0$.

Now, for $\varepsilon = 0$, $a \in E$, and fixed φ_0, we have critical points $(z_{a,\underline{\theta}}, f(a))$ of S_a corresponding to rescaled solutions in time given in (2) for each $\underline{\theta} \in \mathbb{T}^N$. These critical points are degenerate, so not guaranteed to persist to $\varepsilon \neq 0$. The non-trivial homology of the N-torus, however, obliges there to remain at least some number of nearby critical points for ε small, corresponding to particular choices of relative phase $\underline{\theta} \in \mathbb{T}^N$, thus giving some multibreathers for $\varepsilon > 0$. To prove this, our approach is to stationarize S subject to additional constraints

$$\int_0^1 \underline{\theta}(s) \,\mathrm{d}s = \bar{\underline{\theta}}, \qquad \int_0^1 \underline{J}(s) \,\mathrm{d}s = \bar{\underline{J}},$$

which is a non-degenerate variational problem, and then examine for which of the resulting critical points the constraints can be removed. It would suffice to constrain only the relative phases, not the relative momenta, but the above extension is essential to derive generalised multibreather solutions later. Let us introduce Lagrange multipliers $\underline{u}, \underline{v} \in \mathbb{R}^N$ and $k \in \mathbb{R}$ which represent artificial constraint forces and consider the *constrained energy* $W : Z \times \mathbb{R} \times \mathbb{R}^{2N} \times \mathbb{R} \to \mathbb{R}$

$$W_{a,\varphi_0,\bar{\underline{J}},\bar{\underline{\theta}},\varepsilon}(z, f, \underline{u}, \underline{v}, k) = \mathcal{H}(z) - f(\mathcal{A}(z) - a) - \underline{v} \cdot \left(\int_0^1 \underline{J}\, \mathrm{d}s - \bar{\underline{J}} \right) - \underline{u} \cdot \left(\int_0^1 \underline{\theta}\, \mathrm{d}s - \bar{\underline{\theta}} \right) - k \left(\int_0^1 (\varphi(s) - s - \varphi_0)\, \mathrm{d}s \right).$$

For the last term of W we lift φ to $\mathbb{R}$; it is introduced simply to eliminate the global phase shift degeneracy due to the autonomous nature of the system.

The constrained energy W is differentiable. The derivative DW is a mapping from $Z \times \mathbb{R}^{2N+2}$ to the dual $(TZ \times \mathbb{R}^{2N+2})^*$, the bounded linear maps from $C^1(S^1, \Lambda) \times \mathbb{R}^{2N+2}$ to $\mathbb{R}$. However, DW can equally well be regarded as a mapping from $Z \times \mathbb{R}^{2N+2}$ to $C^0(S^1, \Lambda) \times \mathbb{R}^{2N+2}$, and this is the point of view we shall take. Let us write

$$w = (z, f, \underline{u}, \underline{v}, k) \quad \text{and} \quad \alpha = (a, \varphi_0, \bar{\underline{J}}, \bar{\underline{\theta}}).$$

Taking a general variation $\delta w = (\delta z, \delta f, \delta \underline{u}, \delta \underline{v}, \delta k) \in C^1(S^1, \Lambda) \times \mathbb{R}^{2N+2}$ and using integration by parts, DW at w is given by

$$\begin{aligned} DW_{w,\alpha,\varepsilon}(\delta w) = & \int_0^1 \big[(D_A H(z) - f\dot{\varphi})\delta A + (D_\varphi H(z) + f\dot{A} - k)\, \delta\varphi \\ & + (D_{\underline{J}} H(z) - f\dot{\underline{\theta}} - \underline{v}) \cdot \delta \underline{J} + (D_{\underline{\theta}} H(z) + f\dot{\underline{J}} - \underline{u}) \cdot \delta\underline{\theta} \\ & + (D_{p_r} H(z) - f\dot{q}_r)\, \delta p_r + (D_{q_r} H(z) + f\dot{p}_r)\, \delta q_r \big]\, \mathrm{d}s \\ & - \delta f\, (\mathcal{A}(z) - a) - \delta k \int_0^1 (\varphi(s) - s - \varphi_0)\, \mathrm{d}s \\ & - \delta\underline{u} \cdot \int_0^1 (\underline{\theta} - \bar{\underline{\theta}})\, \mathrm{d}s - \delta\underline{v} \cdot \int_0^1 (\underline{J} - \bar{\underline{J}})\, \mathrm{d}s. \end{aligned} \tag{4}$$

Note that $DW_{w,\alpha,\varepsilon}(\delta w)$, $\delta w \in C^1(S^1, \Lambda) \times \mathbb{R}^{2N+2}$, is well-defined because $C^0(S^1, \Lambda) \times \mathbb{R}^{2N+2} \subset (TZ \times \mathbb{R}^{2N+2})^*$. The Euler–Lagrange equations for critical points are:

$$\begin{aligned} & f\dot{A} + D_\varphi H(z) - k = 0, \\ & f\dot{\varphi} - D_A H(z) = 0, \quad a = \mathcal{A}(z), \\ & f\dot{\underline{J}} + D_{\underline{\theta}} H(z) - \underline{u} = 0, \quad \bar{\underline{J}} = \int_0^1 \underline{J}(s)\, \mathrm{d}s, \end{aligned}$$

$$f\underline{\dot\theta} - D_{\underline{J}}H(z) + \underline{v} = 0, \quad \underline{\bar\theta} = \int_0^1 \underline{\theta}(s)\,\mathrm{d}s,$$

$$f\dot p_r + D_{q_r}H(z) = 0, \quad \varphi_0 = \int_0^1 \varphi(s)\,\mathrm{d}s - \frac{1}{2},$$

$$f\dot q_r - D_{p_r}H(z) = 0. \tag{5}$$

If $w_{\alpha,\varepsilon} = (z_{\alpha,\varepsilon}(s), f_{\alpha,\varepsilon}, \underline{u}_{\alpha,\varepsilon}, \underline{v}_{\alpha,\varepsilon}, k_{\alpha,\varepsilon})$ is a critical point of $W_{\alpha,\varepsilon}$ then, by taking constant functions for the variations with respect to $\underline{\theta}$, $\underline{J}$, φ, A respectively, the Lagrange multipliers are given by

$$\underline{u}_{\alpha,\varepsilon} = \int_0^1 D_{\underline{\theta}}H(z_{\alpha,\varepsilon}(s))\,\mathrm{d}s, \quad \underline{v}_{\alpha,\varepsilon} = \int_0^1 D_{\underline{J}}H(z_{\alpha,\varepsilon}(s))\,\mathrm{d}s,$$

$$k_{\alpha,\varepsilon} = \int_0^1 D_{\varphi}H(z_{\alpha,\varepsilon}(s))\,\mathrm{d}s, \quad f_{\alpha,\varepsilon} = \int_0^1 D_A H(z_{\alpha,\varepsilon}(s))\,\mathrm{d}s. \tag{6}$$

Sometimes, we will denote $w_{\alpha,\varepsilon}$ by $w_\varepsilon(\alpha)$.

We see that (z, f) is a critical point of S_a if and only if $(z, f, 0, 0, 0)$ is a critical point of $W_{\alpha,\varepsilon}$, $\alpha = (a, \varphi_0, \underline{\bar J}, \underline{\bar\theta})$ for some φ_0, $\underline{\bar J}$, $\underline{\bar\theta}$. So we obtain multibreathers by finding α and critical points of $W_{\alpha,\varepsilon}$ for which $\underline{u}, \underline{v}, k = 0$. For $\varepsilon = 0$ and any $\alpha = (a, \varphi_0, \underline{\bar J}, \underline{\bar\theta})$, $W_{\alpha,0}$ has a critical point $w_{\alpha,0}$ given by

$$A(s) = a, \quad \varphi(s) = s + \varphi_0, \quad \underline{J}(s) = \underline{\bar J}, \quad \underline{\theta}(s) = \underline{\bar\theta}, \quad p_r(s) = p_r^e, \quad q_r(s) = q_r^e,$$
$$f = D_A H^0(a, \underline{\bar J}, p_r^e, q_r^e), \quad \underline{u} = 0, \quad \underline{v} = D_{\underline{J}}H^0(a, \underline{\bar J}, p_r^e, q_r^e), \quad k = 0. \tag{7}$$

It has $\underline{u}, \underline{v}, k = 0$ if and only if $\underline{\bar J} = J^{\mathrm{res}}(a)$. The following proposition states that $w_{\alpha,0}$ is a non-degenerate critical point of the constrained energy $W_{\alpha,0}$ provided a non-resonance condition holds:

$$2\pi n D_A H^0(a, \underline{\bar J}, p_r^e, q_r^e) \notin \overline{\bigcup_{s\in S''} \operatorname{spec}\left\{\begin{pmatrix} 0 & -1 \\ 1 & 0 \end{pmatrix} D^2_{p_s q_s} H_s(p_s^e, q_s^e)\right\}} \quad \text{for all } n \in \mathbb{Z}. \tag{8}$$

So, it persists to some $w_{\alpha,\varepsilon}$ for ε small and $\underline{\bar J} \in \Omega$.

PROPOSITION 1. *Suppose that the non-resonance condition* (8) *holds on* $(a, \underline{J}) \in E \times \Omega$. *Then* $w_{\alpha,0}$, $\alpha \in E \times \mathbb{T}^1 \times \Omega \times \mathbb{T}^N$ *is a non-degenerate critical point of* $W_{\alpha,0}$ *and there is* $\varepsilon_0 > 0$ *such that there is a* 1*-parameter family of critical points* $w_{\alpha,\varepsilon}$ *of* $W_{\alpha,\varepsilon}$ *for* $0 \le \varepsilon \le \varepsilon_0$ *for every* $\alpha \in E \times \mathbb{T}^1 \times \Omega \times \mathbb{T}^N$.

Proof. The proof is based on the implicit function theorem, which applies if DW is C^1 and its derivative $D^2W_{\alpha,0}$ with respect to $Z \times \mathbb{R}^{2N+2}$ has bounded inverse.

It is easy to check that DW is C^1, as a function from $Z \times \mathbb{R}^{2N+2}$ to $(TZ \times \mathbb{R}^{2N+2})^*$, or $C^0(S^1, \Lambda) \times \mathbb{R}^{2N+2}$ since $H \in C^2$. Its derivative $D^2W_{\alpha,0}$ at $w_{\alpha,0}$ is given by

$$D^2W_{\alpha,0}(w_{\alpha,0})(\delta w, \Delta w) = \int_0^1 \left[(D_A D_z H^0(z)\,\Delta z - \Delta f - f\,\Delta\dot\varphi)\,\delta A\right.$$

$$+ (f\Delta\dot{A} - \Delta k)\,\delta\varphi + (D_{\underline{J}} D_z H^0(z)\,\Delta z - f\Delta\dot{\underline{\theta}} - \Delta\underline{v})\cdot\delta\underline{J}$$
$$+ (f\Delta\dot{\underline{J}} - \Delta\underline{u})\cdot\delta\underline{\theta} + (D_{p_r} D_z H^0(z)\,\Delta z - f\delta\dot{q}_r)\cdot\delta p_r$$
$$+ (D_{q_r} D_z H^0(z)\,\Delta z + f\delta\dot{p}_r)\cdot\delta q_r\big]\,\mathrm{d}s - \delta f\, D\mathcal{A}(z)\,\Delta z$$
$$-\,\delta k\int_0^1 \Delta\varphi\,\mathrm{d}s - \delta\underline{u}\cdot\int_0^1 \Delta\underline{\theta}\,ds - \delta\underline{v}\cdot\int_0^1 \Delta\underline{J}\,\mathrm{d}s,$$

where $\delta w, \Delta w \in C^1(S^1,\Lambda)\times\mathbb{R}^{2N+2}$. $D^2 W_{\alpha,0}(w_{\alpha,0})$ has a bounded inverse if and only if given $\pi \in C^0(S^1,\Lambda)\times\mathbb{R}^{2N+2}$, there is a unique $\Delta w \in C^1(S^1,\Lambda)\times\mathbb{R}^{2N+2}$ such that

$$D^2 W_{\alpha,0}(w_{\alpha,0})(\delta w, \Delta w) = \pi(\delta w) \tag{9}$$

for all $\delta w \in C^1(S^1,\Lambda)\times\mathbb{R}^{2N+2}$, and the ratio $||\Delta w||/||\pi||$ is bounded. Equation (9) is equivalent to

$$f\Delta\dot{A} - \Delta k = \pi_\varphi, \tag{10}$$

$$f\Delta\dot{\varphi} + \Delta f - D_A H^0(z)\Delta A - D_A D_{\underline{J}} H^0(z)\,\Delta\underline{J} = -\pi_A, \tag{11}$$

$$f\Delta\dot{\underline{J}} - \Delta\underline{u} = \pi_{\underline{\theta}}, \tag{12}$$

$$f\Delta\dot{\underline{\theta}} - D_{\underline{J}} D_A H^0(z)\,\Delta A - D^2_{\underline{J}} H^0(z)\,\Delta\underline{J} - \Delta\underline{v} = -\pi_{\underline{J}}, \tag{13}$$

$$f\Delta\dot{p}_s + D_{q_s} D_{p_s} H_s(0,0)\Delta p_s + D^2_{q_s} H_s(0,0)\Delta q_s = \pi_{q_s}, \tag{14}$$

$$f\Delta\dot{q}_s - D^2_{p_s} H_s(0,0)\Delta p_s - D_{p_s} D_{q_s} H_s(0,0)\Delta q_s = -\pi_{p_s} \quad \text{for } s \in S'', \tag{15}$$

$$\int_0^1 (\Delta A + a\Delta\dot{\varphi} + \bar{\underline{J}}\cdot\Delta\dot{\underline{\theta}})\,\mathrm{d}s = -\pi_f, \tag{16}$$

$$\int_0^1 \Delta\varphi\,\mathrm{d}s = -\pi_k, \quad \int_0^1 \Delta\underline{J}\,\mathrm{d}s = -\pi_{\underline{v}}, \quad \int_0^1 \Delta\underline{\theta}\,\mathrm{d}s = -\pi_{\underline{u}}, \tag{17}$$

where the subscripts on π represents the components of π.

For every $s \in S''$, (14) and (15) are solved for Δp_s and Δq_s as follows. Let us write

$$x = \begin{pmatrix}\Delta p_s\\ \Delta q_s\end{pmatrix}, \quad B = \begin{pmatrix} D_{q_s} D_{p_s} H_s(0,0) & D^2_{q_s} H_s(0,0)\\ -D^2_{p_s} H_s(0,0) & -D_{p_s} D_{q_s} H_s(0,0)\end{pmatrix}, \quad y = \begin{pmatrix}\pi_{q_s}\\ -\pi_{p_s}\end{pmatrix}$$

and rewrite (14) and (15) as

$$f\dot{x} + Bx = y.$$

The solutions are given by

$$x(s) = \exp\left(-\frac{1}{f}Bs\right)x(0) + \frac{1}{f}\int_0^s \exp\left(-\frac{1}{f}B(s-s')\right)y(s')\,\mathrm{d}s', \tag{18}$$

which satisfies $x(1) = x(0)$ if and only if

$$x(0) = \left(I - \exp\left(-\frac{1}{f}B\right)\right)^{-1}\frac{1}{f}\int_0^1 \exp\left(-\frac{1}{f}B(1-s')\right)y(s')\,\mathrm{d}s'. \tag{19}$$

The resonance condition provides that $(I - \exp(-B/f))$ has a bounded inverse. Thus $x(s)$ is uniquely determined by (18) and (19), and the ratio $||x||/||y||$ is bounded. Integrating (10), ΔA and Δk are solved by

$$\Delta k = -\int_0^1 \pi_\varphi(s)\,\mathrm{d}s, \qquad \Delta A(s) = \frac{1}{f}\int_0^s \pi_\varphi(s')\,\mathrm{d}s' + \Delta A(0) + \frac{s}{f}\Delta k,$$

where $\Delta A(0) = -\pi_f$ from (16). Similarly, $\Delta \underline{J}$, $\Delta \underline{u}$ are solved from (12) and (17), and $\Delta\underline{\theta}$, $\Delta\underline{v}$ are solved from (13) and (17). Finally, Δf and $\Delta\varphi$ are solved from (11) and (17). One can easily check that the ratio $||\Delta w||/||\pi||$ is bounded. □

As already remarked, from the Euler–Lagrange Equation (5) it can be seen that if $\underline{u}_{\alpha,\varepsilon} = 0$, $\underline{v}_{\alpha,\varepsilon} = 0$ and $k_{\alpha,\varepsilon} = 0$, then $z_{\alpha,\varepsilon}(ft)$ is a periodic solution of the free system. We prove existence of periodic solutions for H_ε by considering the function

$$H_\varepsilon^{\mathrm{eff}}(\alpha) = \oint H_\varepsilon \circ z_{\alpha,\varepsilon}\,\mathrm{d}s, \qquad \alpha = (a, \varphi_0, \underline{\bar{J}}, \underline{\bar{\theta}}).$$

The following lemma states that $\underline{u}_{\alpha,\varepsilon}$, $\underline{v}_{\alpha,\varepsilon}$ are given by derivatives of H^{eff}, H^{eff} is independent of φ_0, and $k_{\alpha,\varepsilon}$ is automatically zero.

LEMMA 1. *If $(z_{\alpha,\varepsilon}, f_{\alpha,\varepsilon}, \underline{u}_{\alpha,\varepsilon}, \underline{v}_{\alpha,\varepsilon}, k_{\alpha,\varepsilon})$ is a critical point of $W_{\alpha,\varepsilon}$, $\alpha = (a, \varphi_0, \underline{\bar{J}}, \underline{\bar{\theta}})$, then*

$$\underline{u}_{\alpha,\varepsilon} = D_{\underline{\bar{\theta}}}H_\varepsilon^{\mathrm{eff}}(\alpha), \qquad \underline{v}_{\alpha,\varepsilon} = D_{\underline{\bar{J}}}H_\varepsilon^{\mathrm{eff}}(\alpha),$$

$D_{\varphi_0}H_\varepsilon^{\mathrm{eff}}(\alpha) = 0$ and $k_{\alpha,\varepsilon} = 0$.

Proof. Let $w_{\alpha,\varepsilon} = (z_{\alpha,\varepsilon}, f_{\alpha,\varepsilon}, \underline{u}_{\alpha,\varepsilon}, \underline{v}_{\alpha,\varepsilon}, k_{\alpha,\varepsilon})$ be a critical point of $W_{\alpha,\varepsilon}$ and consider a variation of α, $\alpha' = (a, \varphi_0, \underline{\bar{J}}, \underline{\bar{\theta}} + \Delta\underline{\bar{\theta}})$ for $|\Delta\underline{\bar{\theta}}|$ small. Then we have

$$\lim_{|\Delta\underline{\bar{\theta}}|\to 0} \frac{W_{\alpha,\varepsilon}(z_{\alpha',\varepsilon}, f_{\alpha,\varepsilon}, \underline{u}_{\alpha,\varepsilon}, \underline{v}_{\alpha,\varepsilon}, k_{\alpha,\varepsilon}) - W_{\alpha,\varepsilon}(w_{\alpha,\varepsilon})}{|\Delta\underline{\bar{\theta}}|} = 0. \tag{20}$$

On the other hand, $z_{\alpha',\varepsilon}$ is a component of the critical point $w_{\alpha',\varepsilon}$ of $W_{\alpha',\varepsilon}$ which satisfies

$$\mathcal{A}(z_{\alpha',\varepsilon}) = a, \quad \int_0^1 \underline{J}_{\alpha',\varepsilon}\,\mathrm{d}s = \underline{\bar{J}}, \quad \int_0^1 \underline{\theta}_{\alpha',\varepsilon}\,\mathrm{d}s = \underline{\bar{\theta}} + \Delta\underline{\bar{\theta}}, \quad \int_0^1 (\varphi_{\alpha',\varepsilon} - s - \varphi_0)\,\mathrm{d}s = 0.$$

Thus (20) implies

$$\lim_{|\Delta\underline{\bar{\theta}}|\to 0} \frac{\mathcal{H}(z_{\alpha',\varepsilon}) - \mathcal{H}(z_{\alpha,\varepsilon}) - \underline{u}_{\alpha,\varepsilon}\cdot\Delta\underline{\bar{\theta}}}{|\Delta\underline{\bar{\theta}}|} = 0,$$

or

$$D_{\underline{\bar{\theta}}} H_{\varepsilon}^{\text{eff}}(\alpha) = \underline{u}_{\alpha,\varepsilon}.$$

Similarly, we have

$$\begin{aligned} D_{\underline{\bar{J}}} H_{\varepsilon}^{\text{eff}}(\alpha) &= \underline{v}_{\alpha,\varepsilon}, \\ D_{\varphi_0} H_{\varepsilon}^{\text{eff}}(\alpha) &= k_{\alpha,\varepsilon}. \end{aligned}$$

It is easily checked that $(z_{\alpha,\varepsilon}(s+\Delta\varphi), f_{\alpha,\varepsilon}, \underline{u}_{\alpha,\varepsilon}, \underline{v}_{\alpha,\varepsilon}, k_{\alpha,\varepsilon})$ is a critical point of $W_{\alpha',\varepsilon}$, where $\alpha' = (a, \varphi_0 + \Delta\varphi, \underline{\bar{J}}, \underline{\bar{\theta}})$. Consequently, we have

$$\frac{\partial}{\partial \varphi_0} z_{\alpha,\varepsilon}(s) = \frac{\partial}{\partial s} z_{\alpha,\varepsilon}(s). \tag{21}$$

Then

$$\begin{aligned} \frac{\partial}{\partial \varphi_0} H_{\varepsilon}^{\text{eff}}(\alpha) &= \frac{\partial}{\partial \varphi_0} \int_0^1 H(z_{\alpha,\varepsilon}(s))\, \mathrm{d}s = \int_0^1 DH(z_{\alpha,\varepsilon}(s)) \frac{\partial}{\partial \varphi_0} z_{\alpha,\varepsilon}(s)\, \mathrm{d}s \\ &= \int_0^1 DH(z_{\alpha,\varepsilon}(s)) \frac{\partial}{\partial s} z_{\alpha,\varepsilon}(s)\, \mathrm{d}s = \int_0^1 \frac{\mathrm{d}}{\mathrm{d}s} H(z_{\alpha,\varepsilon}(s))\, \mathrm{d}s = 0. \end{aligned}$$

Thus $H_{\varepsilon}^{\text{eff}}$ is independent of φ_0 and $k_{\alpha,\varepsilon} = 0$. □

So periodic orbits of area a close to one of those of the unperturbed $E \times \mathbb{T}^N$ for ε small are in one-to-one correspondence with the critical points of $H_{a,\varepsilon}^{\text{eff}}(\underline{\bar{J}}, \underline{\bar{\theta}}) := H_{\varepsilon}^{\text{eff}}(a, \varphi_0, \underline{\bar{J}}, \underline{\bar{\theta}})$. Now under the anharmonic assumption that all but one excited unit is locally anharmonic we have that $D_{\underline{\bar{J}}}^2 H_{a,0}^{\text{eff}}$ is invertible at $\underline{\bar{J}} = \underline{J}^{\text{res}}(a)$, and so the same is true for ε small. But $D_{\underline{\bar{J}}} H_{a,0}^{\text{eff}} = \underline{0}$ at $\underline{\bar{J}} = \underline{J}^{\text{res}}(a)$ for all $\underline{\bar{\theta}}$, so by the implicit function theorem for ε small there exists a locally unique $\underline{\bar{J}}_{a,\varepsilon}(\underline{\bar{\theta}})$ such that

$$D_{\underline{\bar{J}}} H_{a,\varepsilon}^{\text{eff}}(\underline{\bar{J}}_{a,\varepsilon}(\underline{\bar{\theta}}), \underline{\bar{\theta}}) = \underline{0}.$$

Furthermore $\underline{\bar{J}}_{a,\varepsilon}$ is C^1 in $\underline{\bar{\theta}}$. Let

$$V_{a,\varepsilon}^{\text{eff}}(\underline{\bar{\theta}}) = H_{a,\varepsilon}^{\text{eff}}(\underline{\bar{J}}_{a,\varepsilon}(\underline{\bar{\theta}}), \underline{\bar{\theta}}),$$

which we call an 'effective potential'. This is nearly the same as the 'Melnikov-type' potential referred to in [4] and used in [11], and the 'effective action' of [5], except that here we use constant area rather than constant period. V^{eff} could have been derived directly without introducing the constraint $\underline{\bar{J}}$, but we require H^{eff} in the following section, not just V^{eff}.

THEOREM 1. *For all $a \in E$, and ε small enough, H_ε has at least N periodic orbits of area a near $E \times \mathbb{T}^{N+1}$ and at least 2^{N-1} if all are non-degenerate.*

Proof. $V^{\text{eff}}_{a,\varepsilon} : \mathbb{T}^N \to \mathbb{R}$ is C^2, so we obtain at least N critical points by the theory of Ljusternik-Schnirelman category [22]. If all critical points are non-degenerate (as is usually the case) then we obtain at least 2^{N-1} critical points by Morse theory [23]. We will show later that non-degenerate critical points of $H^{\text{eff}}_{a,\varepsilon}$ correspond to non-degenerate periodic orbits (only double Floquet multipliers $+1$), thus completing the proof. □

Note that the critical points do not necessarily depend continuously on ε: for example, it could be that some annihilate in pairs while other pairs are created, or that a continuous branch of critical points for $\varepsilon > 0$ does not tend to a unique limit as $\varepsilon \to 0$. The only constraint is that for each ε small there are at least N critical points. If, however,

$$V^{\text{eff}}_{a,\varepsilon}(\underline{\bar{\theta}}) = C + g(\varepsilon)\, V^0_a(\underline{\bar{\theta}}) + o(g(\varepsilon)) \quad \text{as } \varepsilon \to 0$$

for some C^2 function V^0_a, and function $g(\varepsilon) \to 0$ as $\varepsilon \to 0$, with $g(\varepsilon) \neq 0$ for $\varepsilon \neq 0$, then each non-degenerate critical point of V^0_a persists to a critical point of $V^{\text{eff}}_{a,\varepsilon}$ depending smoothly on ε (cf. theorem 4 of [11]). Furthermore, if V^0_a is a Morse function (all critical points non-degenerate) with K critical points ($K \geq 2^{N-1}$ by Morse theory) then the set of critical points of $V^{\text{eff}}_{a,\varepsilon}$ for ε small non-zero consists of precisely K points continuing the critical points of V^0_a and they depend smoothly on ε. This is directly analogous to the variational approach to Melnikov's method for finding 'subharmonic' periodic orbits in a periodically forced integrable system (cf. [24] for the case of homoclinic orbits). In the common case where the coupling between the phases is first order in ε, then $g(\varepsilon)$ can be taken to be ε and V^0_a is given simply by evaluating $\int H_\varepsilon$ on the unperturbed orbits.

Note that weaker conditions on critical points of V^0_a than non-degeneracy may suffice to control the location of some critical points of $V^{\text{eff}}_{a,\varepsilon}$, for example a strict local minimum or maximum, cf. [25].

If we start from an initial condition $z_{\alpha,\varepsilon}(0)$ where $\underline{u}$ and $\underline{v}$ do not vanish, the solution can not be periodic of nearby period. Indeed, the averaged values $\underline{\bar{J}}$ and $\underline{\bar{\theta}}$ over one revolution of the overall phase φ will slowly vary in time. In the next section we will prove that approximate solutions of the free system can be constructed by considering $H^{\text{eff}}_\varepsilon$ as an $N+1$ DOF Hamiltonian system for $\alpha = (a, \varphi_0, \underline{\bar{J}}, \underline{\bar{\theta}})$.

4. Approximate Dynamics of the Relative Variables

In the free system the rate of change of the relative variables is approximately given by the opposite of the constraint forces. Let us consider the following dynamical system:

$$\begin{aligned}
\dot{a} &= 0, \\
\dot{\varphi}_0 &= f_\varepsilon(\alpha), \\
\dot{\underline{\bar{J}}} &= -\underline{u}_\varepsilon(\alpha), \\
\dot{\underline{\bar{\theta}}} &= \underline{v}_\varepsilon(\alpha),
\end{aligned} \tag{22}$$

where $\alpha = (a, \varphi_0, \underline{\bar{J}}, \underline{\bar{\theta}})$. Actually, the right-hand side does not depend on φ_0 as was proved in Lemma 1. We will show that the dynamics of (22) gives approximate solutions of the free system. First we prove that (22) is Hamiltonian with the Hamiltonian function

$$H^{\text{eff}}_\varepsilon(\alpha) = \oint H_\varepsilon(z_{\alpha,\varepsilon}(s))\, \mathrm{d}s. \tag{23}$$

PROPOSITION 2. *The dynamical system* (22) *is a Hamiltonian system with the Hamiltonian* $H_\varepsilon^{\text{eff}}$.

Proof. It is proved in Lemma 1 that

$$\underline{u}_\varepsilon(\alpha) = D_{\underline{\bar\theta}} H_\varepsilon^{\text{eff}}(\alpha), \quad \underline{v}_\varepsilon(\alpha) = D_{\underline{\bar J}} H_\varepsilon^{\text{eff}}(\alpha)$$

and

$$D_{\varphi_0} H_\varepsilon^{\text{eff}}(\alpha) = 0.$$

By the same argument as in Lemma 1,

$$f_\varepsilon(\alpha) = D_a H_\varepsilon^{\text{eff}}(\alpha).$$

Therefore, (22) is Hamiltonian with the Hamiltonian function $H_\varepsilon^{\text{eff}}(\alpha)$. □

In the system (22), a is constant in time. For fixed $a \in E$, the relevant dynamics is for the relative variables $(\underline{\bar J}, \underline{\bar\theta})$

$$\begin{aligned} \dot{\underline{\bar J}} &= -D_{\underline{\bar\theta}} H_{a,\varepsilon}^{\text{eff}}(\underline{\bar J}, \underline{\bar\theta}), \\ \dot{\underline{\bar\theta}} &= D_{\underline{\bar J}} H_{a,\varepsilon}^{\text{eff}}(\underline{\bar J}, \underline{\bar\theta}) \end{aligned} \tag{24}$$

which is also a Hamiltonian system with the Hamiltonian $H_{a,\varepsilon}^{\text{eff}}(\underline{\bar J}, \underline{\bar\theta}) = H_\varepsilon^{\text{eff}}(\alpha)$. Now we consider a surface

$$\mathcal{P}_\varepsilon = \{z_{\alpha,\varepsilon}(0) : \alpha \in E \times \mathbb{T}^1 \times \Omega \times \mathbb{T}^N\},$$

for $0 \le \varepsilon \le \varepsilon_0$. This surface is almost invariant under H_ε, and the dynamics on it is given approximately by (22). The following theorem states that the errors depend only on $\varepsilon|\underline{u}|$ and $\varepsilon|\underline{v}|$, thus they are especially small near a critical point of $H_{a,\varepsilon}^{\text{eff}}$, and the critical point corresponds to an exact periodic orbit of H_ε as was pointed out in the previous section.

THEOREM 2. *Let* $\alpha(t) = (a(t), \varphi_0(t), \underline{\bar J}(t), \underline{\bar\theta}(t))$ *be a solution of* (22)*. Then for* $0 \le \varepsilon \le \varepsilon_0$, $z_{\alpha(t),\varepsilon}(0)$ *is an approximate solution of* H_ε *in the sense that the deviation of its rate of change from the equation of motion is* $\mathcal{O}(\varepsilon|\underline{u}|, \varepsilon|\underline{v}|)$*. Furthermore, for fixed* $a \in E$*, equilibrium points* $(\underline{\bar J}_\varepsilon, \underline{\bar\theta}_\varepsilon)$ *correspond to exact periodic orbits of* H_ε *with period* $1/f_\varepsilon(a, \underline{\bar J}_\varepsilon, \underline{\bar\theta}_\varepsilon)$.

Proof. Evaluate $(\mathrm{d}/\mathrm{d}t) z_{\alpha(t),\varepsilon}(0)$ with (21) to obtain

$$\frac{\mathrm{d}}{\mathrm{d}t} z_{\alpha(t),\varepsilon}(0) = f\frac{\partial z}{\partial s} - \frac{\partial z}{\partial \underline{\bar J}}\underline{u} + \frac{\partial z}{\partial \underline{\bar\theta}}\underline{v}. \tag{25}$$

From the Euler–Lagrange equation and the fact $k_{\alpha,\varepsilon} = 0$, $\partial z/\partial s$ satisfies

$$\begin{aligned} f\frac{\partial A}{\partial s} &= -D_\varphi H(z), \\ f\frac{\partial \varphi}{\partial s} &= D_A H(z), \\ f\frac{\partial \underline{J}}{\partial s} &= -D_{\underline{\theta}} H(z) + \underline{u}, \end{aligned}$$

$$f\frac{\partial \underline{\theta}}{\partial s} = D_{\underline{J}} H(z) - \underline{v},$$

$$f\frac{\partial p_r}{\partial s} = -D_{q_r} H(z),$$

$$f\frac{\partial q_r}{\partial s} = D_{p_r} H(z). \tag{26}$$

From (25) and (26)

$$\frac{\mathrm{d}}{\mathrm{d}t}\begin{bmatrix} A \\ \varphi \\ \underline{J} \\ \underline{\theta} \\ p_r \\ q_r \end{bmatrix} = X_{H_\varepsilon} + \begin{bmatrix} 0 \\ 0 \\ \underline{u} \\ -\underline{v} \\ \underline{0} \\ \underline{0} \end{bmatrix} - \frac{\partial z}{\partial \underline{\bar{J}}}\underline{u} + \frac{\partial z}{\partial \underline{\bar{\theta}}}\underline{v}, \tag{27}$$

where X_{H_ε} is the Hamiltonian vector field of H_ε. Moreover from (7) we have

$$\frac{\partial}{\partial \underline{\bar{J}}} z_{\alpha,0}(0) = \begin{bmatrix} 0 \\ 0 \\ \mathrm{Id} \\ 0 \\ 0 \\ 0 \end{bmatrix}, \quad \frac{\partial}{\partial \underline{\bar{\theta}}} z_{\alpha,0}(0) = \begin{bmatrix} 0 \\ 0 \\ 0 \\ \mathrm{Id} \\ 0 \\ 0 \end{bmatrix},$$

where Id is the $N \times N$ identity matrix. Therefore, for ε small enough,

$$\frac{\partial z}{\partial \underline{\bar{J}}}\underline{u} = \begin{bmatrix} 0 \\ 0 \\ \underline{u} \\ 0 \\ \underline{0} \\ \underline{0} \end{bmatrix} + \mathcal{O}(\varepsilon|\underline{u}|), \quad \frac{\partial z}{\partial \underline{\bar{\theta}}}\underline{v} = \begin{bmatrix} 0 \\ 0 \\ 0 \\ \underline{v} \\ \underline{0} \\ \underline{0} \end{bmatrix} + \mathcal{O}(\varepsilon|\underline{v}|). \tag{28}$$

In view of (25), (26) and (28), it is proved that $z_{\alpha(t),\varepsilon}(0)$ is an approximate solution of H_ε to order $\mathcal{O}(\varepsilon|\underline{u}|, \varepsilon|\underline{v}|)$, meaning that

$$\frac{\mathrm{d}}{\mathrm{d}t} z_{\alpha_\varepsilon(t),\varepsilon} - X_{H_\varepsilon}(z_{\alpha_\varepsilon(t),\varepsilon}) = \mathcal{O}(\varepsilon|\underline{u}|, \varepsilon|\underline{v}|). \tag{29}$$

If $(\underline{\bar{J}}_\varepsilon, \underline{\bar{\theta}}_\varepsilon)$ is an equilibrium point of (24),

$$\alpha_\varepsilon(t) = (a, f_\varepsilon(a, \underline{\bar{J}}_\varepsilon, \underline{\bar{\theta}}_\varepsilon)\, t + \varphi_{0\varepsilon}, \underline{\bar{J}}_\varepsilon, \underline{\bar{\theta}}_\varepsilon)$$

is a periodic orbit of (22) with any $\varphi_{0\varepsilon}$, and

$$\underline{u}_\varepsilon(\alpha_\varepsilon(t)) = \underline{v}_\varepsilon(\alpha_\varepsilon(t)) = 0.$$

This means that $z_{\alpha_\varepsilon(t),\varepsilon}(0)$ is an exact periodic orbit of H_ε. □

5. Linear Stability of Multibreathers

In this section we consider the relation between the linear stability of an equilibrium $(\underline{\bar{J}}_\varepsilon, \underline{\bar{\theta}}_\varepsilon)$ of $H^{\text{eff}}_{a,\varepsilon}$ and of the corresponding periodic orbit γ_ε defined by $\gamma_\varepsilon(t) := z_{\alpha_\varepsilon(t),\varepsilon}(0)$. We obtain the linear stability type in the case that the equilibrium has distinct eigenvalues which move with non-zero speed with ε.

The linear stability of $(\underline{\bar{J}}_\varepsilon, \underline{\bar{\theta}}_\varepsilon)$ under $H^{\text{eff}}_{a,\varepsilon}$ is determined by the eigenvalues $\mu_{\varepsilon,1}, \ldots, \mu_{\varepsilon,2N}$ of

$$B_\varepsilon = \begin{pmatrix} -D_{\underline{\bar{\theta}}} D_{\underline{\bar{J}}} H^{\text{eff}}_{a,\varepsilon}(\underline{\bar{J}}_\varepsilon, \underline{\bar{\theta}}_\varepsilon) & -D^2_{\underline{\bar{\theta}}} H^{\text{eff}}_{a,\varepsilon}(\underline{\bar{J}}_\varepsilon, \underline{\bar{\theta}}_\varepsilon) \\ D^2_{\underline{\bar{J}}} H^{\text{eff}}_{a,\varepsilon}(\underline{\bar{J}}_\varepsilon, \underline{\bar{\theta}}_\varepsilon) & D_{\underline{\bar{J}}} D_{\underline{\bar{\theta}}} H^{\text{eff}}_{a,\varepsilon}(\underline{\bar{J}}_\varepsilon, \underline{\bar{\theta}}_\varepsilon) \end{pmatrix}. \tag{30}$$

Since $(\underline{\bar{J}}_\varepsilon, \underline{\bar{\theta}}_\varepsilon)$ is an equilibrium point of a Hamiltonian system, if μ is an eigenvalue of B_ε, then $-\mu$, $\bar{\mu}$, and $-\bar{\mu}$ are also eigenvalues. In the limit $\varepsilon \to 0$, one can see that (24) becomes degenerate and $\lim_{\varepsilon\to 0} \mu_{\varepsilon,i} = 0$ for all i.

On the other hand, the linear stability of $\gamma_\varepsilon(t)$ is determined by its Floquet multipliers. At $\varepsilon = 0$, they consist of a $(2N+2)$-fold degenerate eigenvalue $+1$ and other eigenvalues on the unit circle which are due to elliptic equilibria of the units in S''. For ε small, under certain 'signature' conditions on the elliptic equilibria in S'' as in [10], all eigenvalues except $+1$ remain on the unit circle, but $+1$ turns into $2N+2$ eigenvalues near $+1$ of which $+1$ remains at least double. In this case the linear stability is determined by the eigenvalues around $+1$. In fact, the $2N$ eigenvalues near $+1$ not constrained to be $+1$ are in one-to-one correspondence with $\mu_{\varepsilon,1}, \ldots, \mu_{\varepsilon,2N}$ in the non-degenerate case. To prove this fact we need a general result on perturbations of eigenvalue problems in Banach spaces.

Let L_ε be a continuous one-parameter family of bounded linear operators on a Banach space X with a norm $||\cdot||$ for $0 \le \varepsilon \le \varepsilon_0$. Suppose L_ε has a two dimensional invariant subspace $\mathbb{E}_\varepsilon$ with eigenvalue $+1$. Let the spectrum of L_ε be separated into two parts Σ'_ε, Σ''_ε in such a way that a simple closed curve σ encloses an open set containing Σ'_ε in its interior and Σ''_ε in its exterior, and Σ'_ε consists of a finite number of eigenvalues including $+1$. Then one defines the spectral projection associated with Σ'_ε by

$$P_\varepsilon = -\frac{1}{2\pi i} \int_\sigma R_\varepsilon(\zeta)\, \mathrm{d}\zeta,$$

where $R_\varepsilon(\zeta)$ is the resolvent of L_ε

$$R_\varepsilon(\zeta) = (L_\varepsilon - \zeta)^{-1}.$$

P_ε gives a decomposition $X = U_\varepsilon \oplus V_\varepsilon$, where $U_\varepsilon = \operatorname{Range} P_\varepsilon$ and $V_\varepsilon = \operatorname{Range}(1 - P_\varepsilon)$, such that the spectra of $L_\varepsilon|_{U_\varepsilon}$, $L_\varepsilon|_{V_\varepsilon}$ coincide with Σ'_ε, Σ''_ε respectively [26]. Let $n+2$ be the dimension of U_ε. We will use the following lemma.

LEMMA 2. *Suppose that there are non-zero* $\mu_{\varepsilon,i} \in \mathbb{C}$, $\xi_{\varepsilon,i} \in X$, $i = 1, \ldots, n$, *such that for* $0 < \varepsilon \le \varepsilon_0$,

$$L_\varepsilon\, \xi_{\varepsilon,i} - \mathrm{e}^{\mu_{\varepsilon,i}}\, \xi_{\varepsilon,i} = \varepsilon\, \mu_{\varepsilon,i}\, \eta_{\varepsilon,i}, \tag{31}$$

$$\lim_{\varepsilon\to 0} \mu_{\varepsilon,i} = 0, \quad \lim_{\varepsilon\to 0} ||\eta_{\varepsilon,i}|| < \infty \quad \textit{for} \quad 1 \le i \le n,$$

$\xi_{\varepsilon,i}$ *satisfies*

$$\lim_{\varepsilon\to 0} ||P_\varepsilon\, \xi_{\varepsilon,i}|| \neq 0 \quad \textit{for} \quad 1 \le i \le n, \tag{32}$$

and

$$U_0 = \lim_{\varepsilon\to 0} \mathbb{E}_\varepsilon \oplus span\{\lim_{\varepsilon\to 0} P_\varepsilon\, \xi_{\varepsilon,i}\}_{1\le i\le n}.$$

Then there are constants $b > 0$ *and* $\varepsilon_1 > 0$ *such that for* $\varepsilon \in [0, \varepsilon_1]$, *the set*

$$\mathcal{D}_{\varepsilon,i}(b) = \{\lambda \in \mathbb{C} \,:\, |\lambda - e^{\mu_{\varepsilon,i}}| \le \varepsilon|\mu_{\varepsilon,i}|b\}$$

has an eigenvalue of L_ε *belonging to* Σ'_ε *for all* $1 \le i \le n$.

Proof. Let $\{\lambda_{\varepsilon,0} = 1, \lambda_{\varepsilon,1}, \ldots, \lambda_{\varepsilon,s}\}$ be the whole set of eigenvalues in Σ'_ε and $P_{\varepsilon,i}$, $D_{\varepsilon,i}$ be the eigenprojection and the eigennilpotent associated to $\lambda_{\varepsilon,i}$ respectively. Then the resolvent $R_\varepsilon(\mu)$ has the following Laurent expansion [26]:

$$R_\varepsilon(\mu) = -\sum_{j=0}^{s}\left[\frac{P_{\varepsilon,j}}{\mu - \lambda_{\varepsilon,j}} + \sum_{k=1}^{m_j-1}\frac{D^k_{\varepsilon,j}}{(\mu - \lambda_{\varepsilon,j})^{k+1}}\right] + R_\varepsilon(\mu)\,(1 - P_\varepsilon), \tag{33}$$

where m_j is the algebraic multiplicity of $\lambda_{\varepsilon,j}$. We note that $1 - P_\varepsilon$ is the projection onto V_ε and $R_\varepsilon(\mu)\,(1 - P_\varepsilon)$ is uniformly bounded for μ inside σ and $0 \le \varepsilon \le \varepsilon_0$.

First we show that under the conditions of the lemma, $\lambda_{\varepsilon,1}, \ldots, \lambda_{\varepsilon,s}$ are simple. Equation (31) can be written as

$$\xi_{\varepsilon,i} = R_\varepsilon(e^{\mu_{\varepsilon,i}})\,(\varepsilon\,\mu_{\varepsilon,i}\,\eta_{\varepsilon,i}), \quad \text{for} \quad 1 \le i \le n. \tag{34}$$

If $\lambda_{\varepsilon,i}$ is not simple for some $1 \le i \le n$, then for some j, k,

$$\lim_{\varepsilon\to 0} P_{\varepsilon,i}\, \xi_{\varepsilon,j} \neq 0, \quad \lim_{\varepsilon\to 0} P_{\varepsilon,i}\, \xi_{\varepsilon,k} \neq 0.$$

Taking the limit $\varepsilon \to 0$ in (33) and (34),

$$\lim_{\varepsilon\to 0} P_{\varepsilon,i}\, \xi_{\varepsilon,j} = \lim_{\varepsilon\to 0} \frac{D^{m_i-1}_{\varepsilon,i}(\varepsilon\mu_{\varepsilon,j}\eta_{\varepsilon,j})}{(e^{\mu_{\varepsilon,j}} - \lambda_{\varepsilon,i})^{m_i}}.$$

A similar equation holds for k. However, the range of $D^{m_i-1}_{\varepsilon,i}$ is the one-dimensional subspace spanned by the eigenvector of $\lambda_{\varepsilon,j}$. This contradicts the linear independence of $\lim_{\varepsilon\to 0} P_{\varepsilon,i}\, \xi_{\varepsilon,j}$ and $\lim_{\varepsilon\to 0} P_{\varepsilon,i}\, \xi_{\varepsilon,k}$.

Thus, $R_\varepsilon(\mu)$ can be written as:

$$R_\varepsilon(\mu) = -\sum_{j=1}^{n}\frac{P_{\varepsilon,j}}{\mu - \lambda_{\varepsilon,j}} - \frac{P_{\varepsilon,0}}{\mu - 1} - \frac{D_{\varepsilon,0}}{(\mu - 1)^2} + R_\varepsilon(\mu)\,(1 - P_\varepsilon).$$

Now we fix j and consider the equation

$$\lim_{\varepsilon\to 0} P_{\varepsilon,i}\, \xi_{\varepsilon,j} = \lim_{\varepsilon\to 0} P_{\varepsilon,i}\, R_\varepsilon(e^{\mu_{\varepsilon,j}})\,(\varepsilon\mu_{\varepsilon,j}\eta_{\varepsilon,j}) \quad \text{for} \quad 1 \le i \le n.$$

By the hypothesis, there is some i, $1 \le i \le n$ for which $\lim_{\varepsilon\to 0} P_{\varepsilon,i}\, \xi_{\varepsilon,j} \neq 0$. Then

$$\lim_{\varepsilon\to 0} P_{\varepsilon,i}\, \xi_{\varepsilon,j} = -\lim_{\varepsilon\to 0} \frac{P_{\varepsilon,i}(\varepsilon\mu_{\varepsilon,j}\eta_{\varepsilon,j})}{e^{\mu_{\varepsilon,j}} - \lambda_{\varepsilon,i}} = -\lim_{\varepsilon\to 0} \frac{\varepsilon\mu_{\varepsilon,j}}{e^{\mu_{\varepsilon,j}} - \lambda_{\varepsilon,i}} P_{\varepsilon,i}\,\eta_{\varepsilon,j}$$

implies that there is a constant $b^{(j)}$, $\varepsilon^{(j)}$ such that $\mathcal{D}_{\varepsilon^{(j)}}(b^{(j)})$ contain $\lambda_{\varepsilon,i}$ for $\varepsilon \in [0, \varepsilon^{(j)}]$. Choosing $b = \min b^{(j)}$ and $\varepsilon_1 = \min \varepsilon^{(j)}$, the proof is completed. □

The following theorem gives a condition for one-to-one correspondence between the eigenvalues of B_ε in (30) and the Floquet multipliers of γ_ε near 1.

THEOREM 3. *Let $\mu_{\varepsilon,1}, \ldots, \mu_{\varepsilon,2N}$ be eigenvalues of B_ε and $\xi_{\varepsilon,1}, \ldots, \xi_{\varepsilon,2N}$ be corresponding unit eigenvectors. Let T_ε be the period of γ_ε. Suppose that $\mu_{\varepsilon,1}, \ldots, \mu_{\varepsilon,2N}$ are simple eigenvalues and $\lim_{\varepsilon\to 0}\{\xi_{\varepsilon,1}, \ldots, \xi_{\varepsilon,2N}\}$ form a basis of $\mathbb{R}^{2N}$. Then there exist constants $b > 0$ and ε_1 such that for $\varepsilon \in (0, \varepsilon_1]$, there are disjoint balls of radius $\varepsilon\,|\mu_{\varepsilon,i}|\,b$ centred on $\exp(\mu_{\varepsilon,i} T_\varepsilon)$ containing a Floquet multiplier of γ_ε for each i.*

Proof. Let us write (22) as

$$\dot{\alpha} = g_\varepsilon(\alpha), \quad \alpha \in E \times \mathbb{T}^1 \times \mathbb{R}^{2N} \tag{35}$$

and the Hamiltonian equation of H_ε as

$$\dot{z} = X_{H_\varepsilon}(z), \quad z \in \Gamma. \tag{36}$$

The first variational equation around the periodic orbit α_ε defined by $\alpha_\varepsilon(t) := (a, f_\varepsilon t + \varphi_0, \bar{\underline{J}}_\varepsilon, \bar{\underline{\theta}}_\varepsilon)$ is given by

$$\delta\dot{\alpha} = Dg_\varepsilon(\alpha_\varepsilon(t))\,\delta\alpha = \begin{pmatrix} 0 & 0 & 0 \\ C_\varepsilon & 0 & D_\varepsilon \\ E_\varepsilon & 0 & B_\varepsilon \end{pmatrix} \delta\alpha, \tag{37}$$

where

$$C_\varepsilon = \frac{\partial f}{\partial a}(a, \bar{\underline{J}}_\varepsilon, \bar{\underline{\theta}}_\varepsilon), \quad D_\varepsilon = \begin{pmatrix} \dfrac{\partial f}{\partial \underline{J}}(a, \bar{\underline{J}}_\varepsilon, \bar{\underline{\theta}}_\varepsilon) & \dfrac{\partial f}{\partial \underline{\theta}}(a, \bar{\underline{J}}_\varepsilon, \bar{\underline{\theta}}_\varepsilon) \end{pmatrix},$$

$$E_\varepsilon = \begin{pmatrix} -\frac{\partial u}{\partial a}(a, \bar{\underline{J}}_\varepsilon, \bar{\underline{\theta}}_\varepsilon) \\ \frac{\partial v}{\partial a}(a, \bar{\underline{J}}_\varepsilon, \bar{\underline{\theta}}_\varepsilon) \end{pmatrix}.$$

By standard Floquet theory [27] it is readily checked that the Floquet multipliers of α_ε are given by $\exp(\mu_1 T_\varepsilon), \ldots, \exp(\mu_{2N} T_\varepsilon)$ with corresponding eigenvectors

$$\tilde{\xi}_{\varepsilon,1} = \begin{pmatrix} 0 \\ 0 \\ \xi_{\varepsilon,1} \end{pmatrix}, \quad \ldots, \quad \tilde{\xi}_{\varepsilon,2N} = \begin{pmatrix} 0 \\ 0 \\ \xi_{\varepsilon,2N} \end{pmatrix}$$

and double eigenvalue +1.

Now let us consider the relation between the Floquet multipliers of α_ε and γ_ε. By Theorem 1 the map $\Phi_\varepsilon : \alpha \to z_{\alpha,\varepsilon}(0)$ gives approximate solutions to (36) and transforms periodic orbits of (35) into periodic orbits of (36). In other words, Φ_ε satisfies

$$\frac{\mathrm{d}}{\mathrm{d}t}\Phi_\varepsilon(\alpha(t)) = X_{H_\varepsilon}(\Phi_\varepsilon(\alpha(t))) + K_\varepsilon(\alpha(t)),$$

where K_ε is the error term given in the proof of Theorem 1, which vanishes for $\alpha_\varepsilon(t)$. Differentiating the left hand side with respect to t, we obtain

$$D\Phi_\varepsilon(\alpha)\,g_\varepsilon(\alpha) = X_{H_\varepsilon}(\Phi_\varepsilon(\alpha)) + K_\varepsilon(\alpha).$$

Differentiating both sides with respect to α,

$$D^2\Phi_\varepsilon(\alpha)(g_\varepsilon(\alpha),\,\cdot\,) + D\Phi_\varepsilon(\alpha)\,Dg_\varepsilon(\alpha) = DX_{H_\varepsilon}(\Phi_\varepsilon(\alpha))\,D\Phi_\varepsilon(\alpha) + DK_\varepsilon(\alpha). \tag{38}$$

Using (37) and (38)

$$\begin{aligned}\frac{\mathrm{d}}{\mathrm{d}t}(D\Phi_\varepsilon(\alpha_\varepsilon(t))\,\tilde{\xi}_{\varepsilon,i}) &= D^2\Phi_\varepsilon(\alpha_\varepsilon(t))(g_\varepsilon(\alpha_\varepsilon(t)),\tilde{\xi}_{\varepsilon,i})\\ &= DX_{H_\varepsilon}(\gamma_\varepsilon(t))\,D\Phi_\varepsilon(\alpha_\varepsilon(t))\,\tilde{\xi}_{\varepsilon,i} - \mu_{\varepsilon,i}D\Phi_\varepsilon(\alpha_\varepsilon(t))\tilde{\xi}_{\varepsilon,i}\\ &\quad + DK_\varepsilon(\alpha_\varepsilon(t))\,\tilde{\xi}_{\varepsilon,i}.\end{aligned} \tag{39}$$

On the other hand, from (27) and (28)

$$\begin{aligned}K_\varepsilon(\alpha) &= \begin{bmatrix}0\\0\\\underline{u}(\alpha)\\-\underline{v}(\alpha)\\\underline{0}\\\underline{0}\end{bmatrix} - \frac{\partial z_{\alpha,\varepsilon}}{\partial\underline{J}}\,\underline{u} + \frac{\partial z_{\alpha,\varepsilon}}{\partial\underline{\theta}}\,\underline{v}\\ &= \left(D\Phi_\varepsilon(\alpha) - \begin{bmatrix}0&0&0&0\\0&0&0&0\\0&0&\mathrm{Id}&0\\0&0&0&\mathrm{Id}\\0&0&0&0\\0&0&0&0\end{bmatrix}\right)\begin{pmatrix}0\\0\\-\underline{u}\\\underline{v}\end{pmatrix} := \varepsilon\,N_\varepsilon(\alpha)\begin{pmatrix}0\\0\\-\underline{u}\\\underline{v}\end{pmatrix},\end{aligned}$$

and $N_\varepsilon(\alpha)$ is a uniformly bounded linear operator for $\varepsilon\in[0,\varepsilon_0]$. Using the fact $\underline{u}(\alpha_\varepsilon(t)) = \underline{v}(\alpha_\varepsilon(t)) = 0$, it follows that

$$DK_\varepsilon(\alpha_\varepsilon(t)) = \varepsilon N_\varepsilon(\alpha)\begin{pmatrix}0&0&0\\0&0&0\\E_\varepsilon&0&B_\varepsilon\end{pmatrix}. \tag{40}$$

In view of (39) and (40), observe that

$$\begin{aligned}\frac{\mathrm{d}}{\mathrm{d}t}(\mathrm{e}^{\mu_{\varepsilon,i}t}D\Phi_\varepsilon(\alpha_\varepsilon(t))\tilde{\xi}_{\varepsilon,i}) &= DX_{H_\varepsilon}(\gamma_\varepsilon(t))\,\mathrm{e}^{\mu_{\varepsilon,i}t}\,D\Phi_\varepsilon(\alpha_\varepsilon(t))\tilde{\xi}_{\varepsilon,i}\\ &\quad + \varepsilon\,\mu_{\varepsilon,i}\,\mathrm{e}^{\mu_{\varepsilon,i}t}N_\varepsilon(\alpha_\varepsilon(t))\,\tilde{\xi}_{\varepsilon,i}.\end{aligned} \tag{41}$$

Using the variation of constants formula and integrating (41) from 0 to T_ε we obtain

$$\begin{aligned}\mathrm{e}^{\mu_{\varepsilon,i}T_\varepsilon}D\Phi_\varepsilon(\alpha_\varepsilon(0))\,\tilde{\xi}_{\varepsilon,i} &= M_\varepsilon\,D\Phi_\varepsilon(\alpha_\varepsilon(0))\tilde{\xi}_{\varepsilon,i}\\ &\quad + \varepsilon\mu_{\varepsilon,i}\int_0^{T_\varepsilon} M_\varepsilon(T_\varepsilon,s)\,N_\varepsilon(\alpha_\varepsilon(s))\,\mathrm{e}^{\mu_{\varepsilon,i}s}\,\mathrm{d}s\,\tilde{\xi}_{\varepsilon,i},\end{aligned}$$

where $M_\varepsilon(t,s)$ is the fundamental map solution of the first variational equation around $\gamma_\varepsilon(t)$ and M_ε is its monodromy map. We note that the last integral is uniformly bounded in the operator norm from $\mathbb{R}^{2N+2}$ to Γ for $\varepsilon \in [0, \varepsilon_0]$.

The non-resonance condition (8) tells us that the spectrum of M_0 is composed of the isolated eigenvalue 1 and other eigenvalues on the unit circle. Since the spectrum is continuous with respect to ε, for ε small enough, we can choose some closed curve σ enclosing 1 containing all eigenvalues but 1 in its exterior. It is easily checked that if we set $L_\varepsilon := M_\varepsilon$, then $\mu_{\varepsilon,i} T_\varepsilon$, $D\Phi_\varepsilon(\alpha_\varepsilon(0))\tilde{\xi}_{\varepsilon,i}$, and

$$\eta_{\varepsilon,i} := -\frac{1}{T_\varepsilon} \int_0^{T_\varepsilon} M_\varepsilon(T_\varepsilon, s)\, N_\varepsilon(\alpha_\varepsilon(s))\, \mathrm{e}^{\mu_{\varepsilon,i} s}\, \mathrm{d}s\, \tilde{\xi}_{\varepsilon,i}$$

for $1 \le i \le 2N$, satisfy the conditions in Lemma 2. Applying Lemma 2 we obtain constants $b > 0$, $\varepsilon_1 > 0$, and $2N$ balls centered on $\exp(\mu_{\varepsilon,i} T_\varepsilon)$ with radius $\varepsilon|\mu_{\varepsilon,i}|b$ for $\varepsilon \in (0, \varepsilon_1]$. Each ball contains a Floquet multiplier of γ_ε and they are disjoint for ε_1 small enough. □

By the above theorem one can see that if the Floquet multipliers of γ_ε due to the elliptic equilibria of the units in S'' remain on the unit circle for ε small, $(\bar{\underline{J}}_\varepsilon, \bar{\underline{\theta}}_\varepsilon)$ and $\gamma_\varepsilon(t)$ have the same stability type. According to symplectic signature theory (see, e.g., [10]) the Floquet multipliers can leave the unit circle only by a collision of eigenvalues having different symplectic signature or a collision with ± 1. The following theorem provides a condition on symplectic signatures of those Floquet multipliers to establish the coincidence of linear stability.

THEOREM 4. *Suppose that the conditions in Theorem 3 are fulfilled. Let ω_s be the frequency of small oscillations about (p_s^e, q_s^e) for $s \in S''$, signed according to the sign of $D^2 H_s$. Let*

$$\Sigma = \{\mathrm{e}^{i\omega_s T} : s \in S''\}$$

and $\bar{\Sigma}$ its complex conjugate and denote the breather period by T. Suppose there is a constant $C_1 > 0$ such that

$$\mathrm{d}(\Sigma, \bar{\Sigma}) > C_1.$$

Then there is a constant $\varepsilon_2 > 0$ such that the linear stability types of $(\bar{\underline{J}}_\varepsilon, \bar{\underline{\theta}}_\varepsilon)$ and of γ_ε coincide with each other for $\varepsilon \in [0, \varepsilon_2)$.

Proof. As proved in [10], the hypotheses imply that there exists $\varepsilon_2 > 0$ such that the symplectic signatures of eigenvalues continuing Σ and $\bar{\Sigma}$ are unchanged for $\varepsilon \in [0, \varepsilon_2]$, so that the eigenvalues continuing Σ and $\bar{\Sigma}$ remain on the unit circle and their Jordan blocks are diagonal. Therefore, the linear stability type of γ_ε is completely determined by the $2N$ eigenvalues near $+1$. By Theorem 3 they are contained in $2N$ disjoint balls of radius $\varepsilon|\mu_{\varepsilon,i}|b$ centered on $\exp(\mu_{\varepsilon,i} T_\varepsilon)$. If one of $\mu_{\varepsilon,i}$ has non-zero real part the ball centered on $\exp(\mu_{\varepsilon,i} T_\varepsilon)$ has no intersection with the unit ball for ε small enough. The corresponding Floquet multiplier can not be on the unit circle. If all $\mu_{\varepsilon,i}$ have zero real part, that is $(\bar{\underline{J}}_\varepsilon, \bar{\underline{\theta}}_\varepsilon)$ is linearly stable, all $2N$ balls have intersections with the unit circle and contain only one Floquet multiplier each. If μ is an eigenvalue of a symplectic map, $1/\bar{\mu}$ is also an eigenvalue. Thus, those Floquet multipliers inside the balls are on the unit circle. □

6. Examples and Extensions

6.1. Two-Site Generalised Multibreathers

The simplest example to treat is a 1D chain of anharmonic oscillators $H_s = p_s^2/2m_s + V_s(q_s)$, $s \in \mathbb{Z}$, with a nearest neighbour coupling $\varepsilon W(q_{s+1}, q_s)$, and just two neighbouring oscillators excited at $\varepsilon = 0$, say $s = 0, 1$, with equal period T. We will derive 'pendulum dynamics' for the one relative phase $\bar{\theta}$.

Taking action-angle variables (I_s, ψ_s) for $s = 0, 1$, we choose to define the global and relative phase variables ϕ and θ respectively, and associated momenta A, J, by

$$\psi_1 = \phi + \theta, \tag{42}$$

$$\psi_0 = \phi, \tag{43}$$

$$A = I_0 + I_1, \tag{44}$$

$$J = I_1. \tag{45}$$

Thus we obtain unperturbed Hamiltonian of the form $H = H_0(I_0) + H_1(I_1) = H_0(A - J) + H_1(J)$ for the two oscillators. Let $J^{\mathrm{res}}(A)$ be the value of J at which $\partial H/\partial J = 0$ (assumed to exist and depend smoothly on A). Applying our procedure (or the procedure described in [11]), we obtain effective Hamiltonian to first order in ε and in $j = J - J^{\mathrm{res}}(A)$,

$$H^{\mathrm{eff}}_{\varepsilon,a}(j, \theta) = C(a) + \frac{1}{2}\beta j^2 + \varepsilon\langle W(q_1(J^{\mathrm{res}}(a), \theta + \phi), q_0(A - J, \phi))\rangle, \tag{46}$$

where the average is taken over ϕ and

$$C(a) = H_0(a - J^{\mathrm{res}}(a)) + H_1(J^{\mathrm{res}}(a)), \tag{47}$$

$$\beta = H_1''(a - J^{\mathrm{res}}(a)) + H_0''(J^{\mathrm{res}}(a)). \tag{48}$$

In the case of two hard oscillators (i.e. H_i'' positive), then $\beta > 0$, so the effective dynamics oscillates about minima of the effective potential $\varepsilon\langle W\rangle(\theta)$ and librates for energy above its maximum. For soft oscillators (H_i'' negative), the signs are reversed. In the mixed case, the sign of β depends on which oscillator has the stronger anharmonicity.

6.2. Stability Criterion for Multibreathers

The above stability criterion for 2-site multibreathers generalises to the case of more than two excited oscillators, giving a variant (because our effective potential is at constant action rather than constant period) and extension of a result of [5] (which was already anticipated in [3, 4]).

To first order in ε, if all oscillators are hard then the local minima of the effective potential are linearly stable and critical points of positive index are linearly unstable; if all oscillators are soft then the maxima of the effective potential are linearly stable and all critical points of lower index are linearly unstable.

In the mixed case with some hard and some soft oscillators, there is no easy statement. One has to analyse the linear stability of the critical points of the effective Hamiltonian. Many different sorts of dynamics are possible. For example, if there is one heavy near-isochronous oscillator, one light soft oscillator and one light hard oscillator, one can obtain an effective Hamiltonian like

$$H^{\text{eff}}(\underline{J}, \underline{\theta}) = \frac{1}{2}(J_1^2 - J_2^2) + \sin(\theta_1 + \theta_2).$$

Then

$$\ddot{\theta}_1 + \ddot{\theta}_2 = 0,$$

$$\ddot{\theta}_1 - \ddot{\theta}_2 = -2\cos(\theta_1 + \theta_2),$$

so there are solutions with $\theta_1 + \theta_2 = c$, constant, and for them $\theta_1 - \theta_2$ accelerates at rate $-2\cos c$. Alternatively, $\theta_1 + \theta_2 = \alpha t$, some $\alpha \neq 0$, and then $\theta_1 - \theta_2 = (2/\alpha^2)\cos\alpha t + At + B$ for arbitrary constants A, B.

6.3. Travelling Breathers

We believe that our approach may provide an explanation of numerical observations of travelling breathers, e.g. [28–30, 32]. Our approach is based on deriving approximate effective dynamics for loops by elimination of dependence on a global phase, as done throughout this paper.

Consider a translation invariant lattice (e.g. one-dimensional chain) and suppose there exists a 1-site breather solution and a 2-site breather with approximately the same energy and area as the 1-site breather. This is the regime studied by [32], for example, because using $A = \int \mathrm{d}E/\omega$, one can transform figure 4.3 of [32] to Figure 1 here.

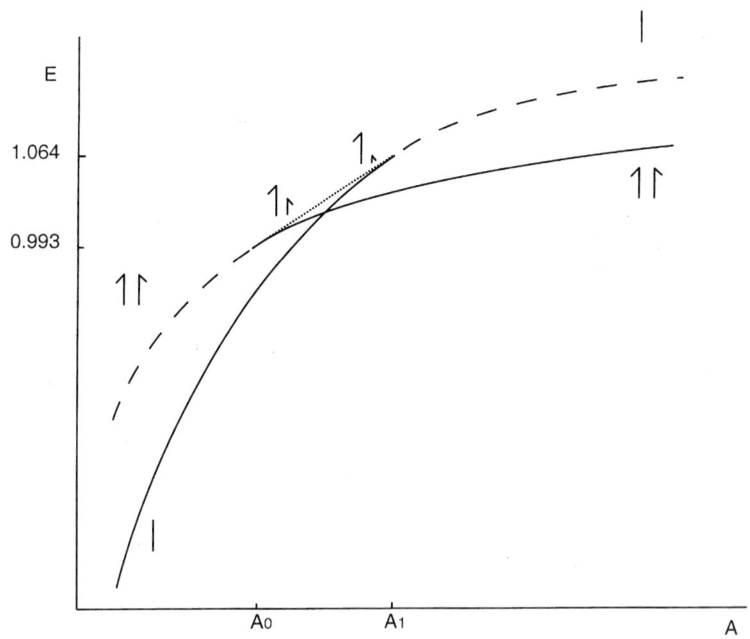

Figure 1. Transformation of figure 4.3 of [32] to E *versus* A, using $A = \int \mathrm{d}E/\omega$. The symbols ↑, etc., indicate the amplitudes and phases of oscillation of the relevant breathers. Solid lines are linearly stable, dashed and dotted are unstable, the dotted line corresponding to a pair of asymmetric breathers.

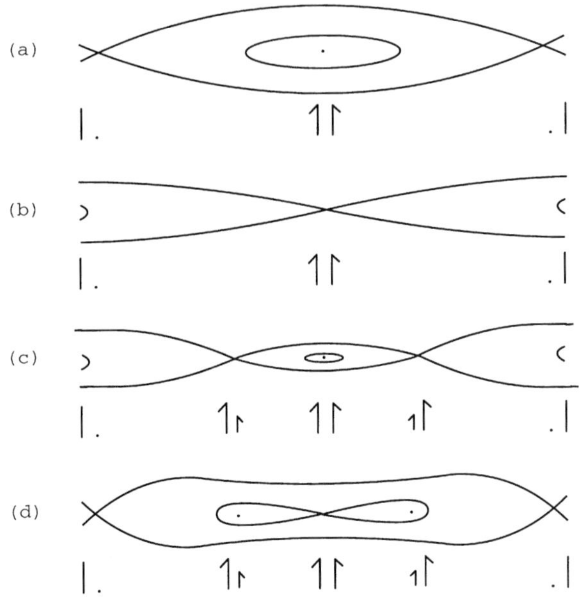

Figure 2. Four possible phase portraits for the relative phase and momentum for the interaction between a 1-site and a 2-site breather. The symbols indicate the amplitudes and phases of oscillation of the two most important sites at the equilibrium points of this effective dynamics.

Note that it also has a second 2-site breather which is in antiphase but it has significantly higher energy than the indicated in-phase 2-site breather and plays no role in what we will describe. In addition, it has a pair of asymmetric 2-site breathers for A in an interval (A_0, A_1), which are not essential to our argument but modify the resulting phase portraits. The oscillators are soft, so the kinetic energy term in H^{eff} is negative.

We can translate the two chosen breathers on the lattice to obtain other equivalent ones. It will be convenient, however, to identify configurations which differ by a lattice translation and hence study the interaction of just these two breathers. Suppose that they are part of an effective Hamiltonian dynamics that looks like one of the cases of Figure 2.

This would require theoretical justification, but the results of [32] strongly suggest case (c) for $A \in (A_0, A_1)$, (b) for $A < A_0$ and (a) for $A > A_1$. Then choosing orbits of H^{eff} just a little outside the separatrix yields travelling breathers.

This approximation is good as long as the dynamics it yields is slow compared with the time scale of the underlying oscillations. This restriction requires the dynamics to remain in a thin zone of $\bar{J}$ where $D_{\bar{J}} H^{\text{eff}}$ is small, and this in turn requires that the island widths be not too large. We believe this is the reason that case (c) and nearby cases of (a) and (b) are good for travelling breathers. When the energy difference between the 1-site and 2-site breathers is too large then the corresponding island is wide, so the dynamics of H^{eff} is a poor approximation, but when they are of roughly equal energy (for the same area) then the approximate dynamics is good.

Note that our interpretation of travelling breathers gives substance to the idea of a 'movability separatrix' proposed in [33]. It also clarifies the applicability of the concept of Peierls–Nabarro (PN) barrier to moving breathers, as follows. Let us define the relative kinetic energy $K_a^{\text{eff}} = H^{\text{eff}}(a) - V_a^{\text{eff}}$, and suppose it has definite sign. As one of us has argued since 1993, the appropriate definition of PN barrier for breathers is the energy difference between certain breathers with the same area. More precisely, the PN barrier for breathers of area a is the minimum change in V_a^{eff} to move a breather by one lattice site, because this is the minimum relative kinetic energy that one has to supply to put the most stable breather onto a travelling trajectory of H_a^{eff}.

Note that in [32], the relative kinetic energy is negative (soft oscillators), so the most stable breather is the one with the highest V^{eff}. Thus, denoting the 1-site breather by 1, the 2-site in-phase symmetric breather by 2 and the asymmetric 2-site breathers by $2a$, the PN barriers for the three cases of [32] are

(a) $V^{\text{eff}}(2) - V^{\text{eff}}(1)$,
(b) $V^{\text{eff}}(1) - V^{\text{eff}}(2)$,
(c) $\max(V^{\text{eff}}(2), V^{\text{eff}}(1)) - V^{\text{eff}}(2a)$.

If the relative kinetic energy is indefinite (this requires at least three excited states, so it is not immediately relevant to the above scenario) then it would be hard to define a useful concept of PN barrier.

6.4. Complete Energy Exchange

Our ideas can be extended to situations in which the amplitude of oscillation on one excited site can go to zero. This could be relevant to ideas of Aubry [31] on energy transfer in biomolecules like myosin.

One might think that we already permitted the amplitude of oscillation on one excited oscillator to go to zero in the previous subsection, but the picture there was that the coupling was strong enough to move us into a regime where one 2-site breather had the same energy as a 1-site breather and so the amplitude of oscillation on the neighbouring site to the 1-site breather is not that small.

The problem with the amplitude of oscillation of an excited oscillator going to zero is that this is a coordinate singularity in the action-angle variables on which our analysis was based. It is not difficult, however, to get round this singularity. If one has $N+1$ oscillators of roughly equal frequency ω (and non-negative energies) for which any of the amplitudes is allowed to go through zero then one simply replaces the space $\mathbb{T}^N \times \mathbb{R}^N$ of relative phases and momenta by $\mathbb{C}P^N$, complex projective N-space. It is the space of 1-complex dimensional subspaces of $\mathbb{C}^{N+1}$. To see this, think of each oscillator phase space as $\mathbb{C}$ rather than $\mathbb{R}^2$. Constant action gives a sphere $\mathbb{S}^{2N+1}$ in the phase space $\mathbb{C}^{N+1}$ of $N+1$ oscillators, and then we eliminate a global phase using the reference dynamics $\dot{z} = i\omega z$, to obtain

$$\{z \in \mathbb{C}^{N+1} : |z| = 1\}/(z \sim e^{i\phi} z) = \mathbb{C}P^N.$$

It has a natural symplectic structure.

To derive an effective Hamiltonian on $\mathbb{C}P^N$, instead of the constraints $\underline{\bar{J}}, \underline{\bar{\theta}}$, we introduce constraint

$$\underline{w} = \int e^{-2\pi i s} \gamma(s)\, ds \in \mathbb{C}^{N+1}$$

and see that $H^{\text{eff}}(\underline{w})$ is independent of global phase rotation. Thus it preserves area A. For each $a \in \mathbb{R}$, the set $\{A = a\}$ is homeomorphic to a $(2N+1)$-sphere in $\mathbb{C}^{N+1}$ thus H_a^{eff} is a Hamiltonian on $\mathbb{C}P^N$.

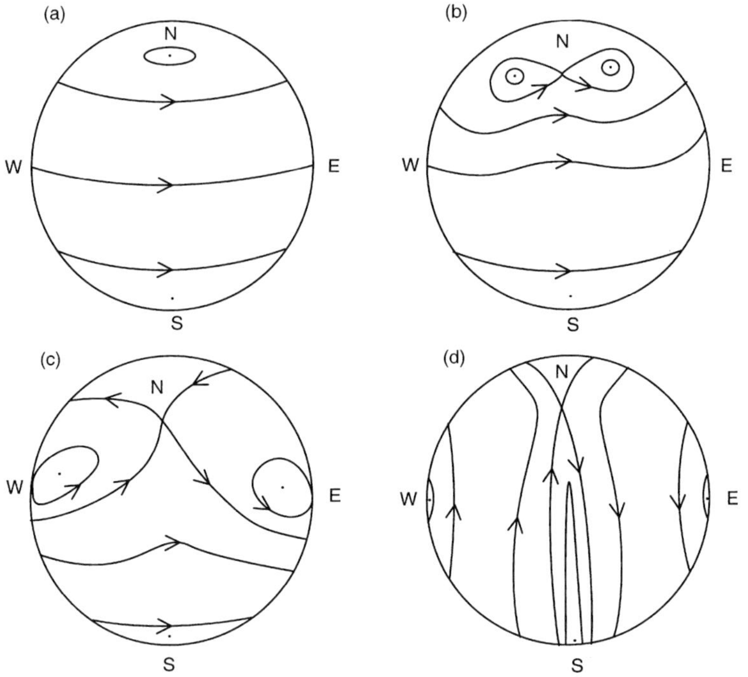

Figure 3. Four phase portraits for weakly coupled pendula, for the following ranges of values of the action a: (a) small, (b) between a_0 and a_1, (c) a little above a_1, (d) large.

For example, the standard physics experiment of two weakly coupled near-identical pendula has relative phase space $\mathbb{C}P^1$, homeomorphic to $\mathbb{S}^2$. Let us draw it so that the 'West' pole corresponds to the right hand pendulum at rest, the 'East' pole to the lefthand pendulum at rest, the North pole to equal amplitude in antiphase and the South pole to equal amplitude in phase. Then coupling of the form $\varepsilon(q_1 - q_0)^2$ yields relative phase portraits like Figure 3, familiar to chemists in the transition from normal to local modes in molecules with two relevant degrees of freedom, and to mechanical engineers in the interaction of nearly resonant undamped nonlinear oscillations.

We see solutions of H^{eff} which completely exchange energy between the two oscillators provided a is not too large, viz. near the equator. For a large, however, there are two 1-site breathers, born at some value a_0, and the separatrix around them crosses the East and West poles as a increases through a higher value a_1, leading to a fraction of the energy remaining for ever on the initially excited pendulum. We could say there is an 'exchange separatrix'.

The connection with the rest of this paper can be seen in the picture for large a. If one restricts attention to a band around the Greenwich meridian and the date line, one obtains a pendulum portrait for 2-site generalised multibreathers.

The approximation by H^{eff} is best if the frequencies of the two excited oscillators remain nearly resonant for all points on the $\mathbb{C}P^1$ for some a: then H_a^{eff} is nearly constant on it and so the relative phase dynamics is very slow. This can be achieved by coupling a soft and hard oscillator with nearly equal but opposite anharmonicities $\mathrm{d}\omega/\mathrm{d}I$. This 'isochronous' regime is very interesting, as already pointed out by Chirikov [12]. Furthermore, it is the only regime where complete energy exchange between two anharmonic localised oscillations can be obtained for arbitrarily weak coupling, else libration around a breather occurs, as in

Figure 3c. Aubry proposes [31] that it may play a role in energy transfer in biomolecules. This may be true but we wish to point out that many phase portraits are possible in this case, as it is a small perturbation of the zero flow, the only constraint being that it is Hamiltonian.

Note that the Liusternik–Schnirelman category theory gives at least $N+1$ critical points of a smooth function on $\mathbb{C}P^N$ (and assuming non-degeneracy does not give any more), so we have at least $N+1$ exactly periodic orbits. The lower bound is achieved in the familiar case of linear systems with distinct normal mode frequencies, but what is interesting here is that one can not reduce the number further.

7. Conclusions

For networks of oscillators, in parts of phase space where some number $N+1$ of oscillators are moving in nearly rational frequency ratio and the remainder are approximately at rest, we have derived an effective Hamiltonian $H_a^{\mathrm{eff}}(\underline{J}, \underline{\theta})$ which describes approximately the evolution of the relative phases $\underline{\theta}$ and momenta $\underline{J}$ of the excited oscillators, a being an adiabatic invariant corresponding to the area of the instantaneous approximate periodic orbit.

Our derivation has the special feature that critical points $(\underline{\bar{J}}, \underline{\bar{\theta}})$ of H_a^{eff} correspond to exact periodic orbits, with mean values of $(\underline{J}, \underline{\theta})$ given by $(\underline{\bar{J}}, \underline{\bar{\theta}})$ and period near that of the unperturbed case. Under an anharmonicity assumption, topological principles allow us to deduce existence of at least $N+1$ such periodic orbits and generically at least 2^N.

The linear stability of these periodic orbits is closely related to that of the critical point of H_a^{eff} under the associated Hamiltonian dynamics. We prove that the linear stability type of the critical point determines that of the periodic orbit in the case that the eigenvalues of the former move at distinct non-zero speeds with the coupling ε.

We believe that the construction of H^{eff} can be slightly modified to make an exact correspondence between the linearised dynamics of the two problems, but have not yet completely worked out the details. The idea is that the remainder term in Equation (29) can be reduced, by near-identity coordinate changes and/or modification of the constraints, to $\mathcal{O}(\varepsilon\tau^2)$ where $\tau = \max(|u|, |v|)$ plus an $\mathcal{O}(\varepsilon\tau)$ multiple of the Hamiltonian vector field, because the remainder has zero mean with respect to φ_0 in the A, $\underline{J}$ and $\underline{\theta}$ directions. In fact, we believe that one could iterate the procedure to reduce the remainder to $\mathcal{O}(\varepsilon\tau^n)$ plus an $\mathcal{O}(\varepsilon\tau)$ multiple of the Hamiltonian vector field, for any $n \in \mathbb{N}$, and in the analytic case by optimising over n one should be able to achieve $\mathcal{O}(\varepsilon e^{-c/\tau^\alpha})$ for some c, α. Note that we must expect to retain a $\mathcal{O}(\varepsilon\tau)$ multiple of the Hamiltonian vector field because generically a geometric phase change ('Hannay angle') results from adiabatic movement round non-trivial loops. Strategies to achieve such results were developed by Nekhoroshev and Neishtadt; for a recent application of the method to a closely related problem, see [34].

We gave an extension to the case where the amplitudes of some excited oscillators are allowed to go to zero, and a tentative explanation of observations of travelling breathers as approximate solutions in certain regimes.

We presented the analysis in the context of coupled one-degree-of-freedom oscillators, but the ideas extend easily to coupled oscillators with more than one degree of freedom: they apply to any choices of normally non-degenerate anharmonic 1-parameter family of periodic orbits for each excited oscillator.

Acknowledgements

This work was partially supported by EC TMR networks on 'Nonlinear Dynamics and Statistical Physics of Extended Systems' and 'Stability and Universality in Classical Mechanics' and the final changes were made under EC network contract HPRN-CT-1999-00163. T. A. was supported by a Chevening Scholarship of the British Council. J.-A. S. was financially supported by Trinity College, Cambridge. They are acknowledged gratefully. A preliminary version of this work was presented at a conference in Toulouse in honour of B. V. Chirikov's 70th birthday in July 1998.

References

1. Sievers, A. J. and Takeno, S., 'Intrinsic localized modes in anharmonic crystals', *Physical Review Letters* **61**, 1988, 970–973.
2. Takeno, S., Kisoda, K., and Sievers, A. J., 'Intrinsic localised vibrational modes in anharmonic crystals', *Progress in Theoretical Physics Supplement* **94**, 1988, 242–269.
3. MacKay, R. S. and Aubry, S., 'Proof of existence of breathers for time-reversible or Hamiltonian networks of weakly coupled oscillators', *Nonlinearity* **7**, 1994, 1623–1643.
4. MacKay, R. S., 'Dynamics of networks: Features which persist from the uncoupled limit', in *Stochastic and Spatial Structures of Dynamical Systems*, S. J. van Strien and S. M. Verduyn Lunel (eds.), North Holland, Amsterdam, 1996, pp. 81–104.
5. Aubry, S., 'Breathers in nonlinear lattices: Existence, linear stability and quantization', *Physica D* **103**, 1997, 201–250.
6. Bambusi, D., 'Exponential stability of breathers in Hamiltonian networks of weakly coupled oscillators', *Nonlinearity* **9**, 1996, 433–457.
7. Sepulchre, J.-A. and MacKay, R. S., 'Localized oscillations in conservative or dissipative networks of weakly coupled autonomous oscillators', *Nonlinearity* **10**, 1997, 679–713.
8. Sepulchre, J.-A. and MacKay, R. S., 'Discrete breathers in disordered media', *Physica D* **113**, 1998, 342–345.
9. Livi, R., Spicci, M., and MacKay, R. S., 'Breathers on a diatomic FPU chain', *Nonlinearity* **10**, 1997, 1421–1434.
10. MacKay, R. S. and Sepulchre, J.-A., 'Stability of discrete breathers', *Physica D* **119**, 1998, 148–162.
11. Ahn, T., 'Multisite oscillations in networks of weakly coupled autonomous oscillators', *Nonlinearity* **11**, 1998, 965–989.
12. Chirikov, B. V., 'A universal instability of many-dimensional oscillator systems', *Physics Reports* **52**, 1979, 264–379.
13. Poincaré, H., *Les méthodes nouvelles de la mécanique céleste*, tome I, Gauthier Villars, Paris, 1892.
14. Laskar, J., 'La stabilité du systéme solaire', in *Chaos et déterminisme*, A. Dahan Dalmedico, J. L. Chabert, and K. Chemla (eds.), Seuil, Paris, 1992, pp. 170–211.
15. Whitham, G. B., *Linear and Nonlinear Waves*, Wiley, New York, 1974.
16. Caputo, J. G., Flytzanis, N., and Sorensen, M. P., 'The ring laser configuration studied by collective coordinates', *Journal of the Optical Society of America B* **12**, 1995, 139–145.
17. Stuart, D., 'Dynamics of Abelian Higgs vortices in the near Bogomolny regime', *Communications in Mathematical Physics* **159**, 1994, 51–91.
18. Moser, J., 'Regularization of Kepler's problem and the averaging method on a manifold', *Communications on Pure and Applied Mathematics* **23**, 1970, 609–636.
19. Weinstein, A., 'Bifurcations and Hamilton's principle', *Mathematische Zeitschrift* **159**, 1978, 235–248.
20. Baesens, J., Guckenheimer, J., Kim, S., and MacKay, R. S., 'Three coupled oscillators: Mode-locking, global bifurcations and toroidal chaos', *Physica D* **49**, 1991, 387–475.
21. Johansson, M. and Aubry, S., 'Existence and stability of quasiperiodic breathers in the discrete Schrödinger equation', *Nonlinearity* **10**, 1997, 1151–1178.
22. Schwartz, J. T., *Nonlinear Functional Analysis*, Gordon and Breach, 1969.
23. Morse, M., 'The calculus of variation in the large', *American Mathematical Society Colloquium Publications* **18**, 1934.

24. Angenent, S., 'A variational interpretation of Melnikov's function and exponentionally small separatrix splitting', in *Symplectic Geometry*, D. Salamon (ed.), Cambridge University Press, Cambridge, 1993, pp. 5–35.
25. Ambrosetti, A., Coti Zelati, V., and Ekeland, I., 'Symmetry breaking in Hamiltonian systems', *Journal of Differential Equations* **67**, 1987, 165–184.
26. Kato, T., *Perturbation Theory for Linear Operators*, Springer-Verlag, Berlin, 1966.
27. Iooss, G. and Joseph, D. D., *Elementary Stability and Bifurcation Theory*, Springer-Verlag, New York, 1980.
28. Dauxois, T. and Peyrard, M., 'Energy localization in nonlinear lattices', *Physical Review Letters* **70**, 1993, 3935–3938.
29. Tamga, J. M., Remoissenet, M., and Pouget, J., 'Breathing solitary waves in a sine-Gordon two-dimensional lattice', *Physical Review Letters* **75**, 1995, 357–361.
30. Aubry, S. and Cretegny, T., 'Mobility and reactivity of discrete breathers', *Physica D* **119**, 1998, 34–46.
31. Aubry, S., 'Discrete breathers in nonlinear lattices: New perspectives in physics and biological physics', unpublished.
32. Cretegny, T., 'Dynamique collective et localisation de l'énergie dans les réseaux non-linéaires', Doctoral Thesis, ENS Lyon, 1998.
33. Flach, S. and Willis, C. R., 'Movability of localized excitations in nonlinear discrete systems', *Physical Review Letters* **72**, 1994, 1777–1780.
34. Bambusi, D. and Nekhoroshev, N. N., 'A property of exponential stability in nonlinear wave equations near the fundamental linear mode', *Physica D* **122**, 1998, 73–104.

Nonlinear Dynamics **25:** 183–205, 2001.

Nonlinear Modal Analysis of Structural Systems Using Multi-Mode Invariant Manifolds

ERIC PESHECK, NICOLAS BOIVIN, and CHRISTOPHE PIERRE*
Department of Mechanical Engineering and Applied Mechanics, University of Michigan, Ann Arbor, MI 48109-2125, U.S.A.

STEVEN W. SHAW
Department of Mechanical Engineering, Michigan State University, East Lansing, MI 48824-1226, U.S.A.

(Received: 8 February 1999; accepted: 10 August 1999)

Abstract. In this paper, an invariant manifold approach is introduced for the generation of reduced-order models for nonlinear vibrations of multi-degrees-of-freedom systems. In particular, the invariant manifold approach for defining and constructing nonlinear normal modes of vibration is extended to the case of multi-mode manifolds. The dynamic models obtained from this technique capture the essential coupling between modes of interest, while avoiding coupling from other modes. Such an approach is useful for modeling complex system responses, and is essential when internal resonances exist between modes. The basic theory and a general, constructive methodology for the method are presented. It is then applied to two example problems, one analytical and the other finite-element based. Numerical simulation results are obtained for the full model and various types of reduced-order models, including the usual projection onto a set of linear modes, and the invariant manifold approach developed herein. The results show that the method is capable of accurately representing the nonlinear system dynamics with relatively few degrees of freedom over a range of vibration amplitudes.

Keywords: Normal modes, modal analysis, nonlinear vibration, invariant manifolds.

1. Introduction

The concept of normal modes of motion is well developed for linear oscillatory systems, due to the special features of the linear differential equations governing their dynamics. These features allow for the definition of normal modes in terms of eigenvectors (or eigenfunctions) and the expression of an arbitrary system response as a superposition of modal responses [1]. In particular, given the invariance of the normal modes, truncation procedures have been developed to allow for the reduction of the number of modeled (i.e., simulated) modes, and yet for the elimination, in many cases, of contamination from the non-modeled modes.

Many themes present in linear systems may be extended to nonlinear systems. For example, much work has been done on the existence and stability of normal modes of motion for two-degrees-of-freedom, conservative systems [2, 3], as well as continuous systems [4, 5]. However, new methodologies have been developed [6–8] that generalize these definitions to a very wide class of systems which include non-conservative, gyroscopic, and infinite-dimensional systems. Essentially, these works define normal modes in terms of motions which occur on low – typically two – dimensional invariant manifolds in the system's phase space. Such motion must be inherently like that of a lower dimensional system, as is desired for a

* Author for correspondence. Phone: (734) 936-0401, fax: (734) 647-3170, E-mail: pierre@engin.umich.edu.

normal mode. A constructive technique for generating such manifolds in terms of asymptotic series, without having to solve the equations of motion, is provided by a simple generalization of the method used in constructing approximate center manifolds in bifurcation theory [9]. Using this approach, it is possible to determine the manifolds which represent the normal modes for weakly nonlinear systems. The equations of motion restricted to these manifolds then provide the dynamics of the associated normal modes. The planes tangent to the manifolds at the equilibrium point are the planes on which the usual modal dynamics of the linearized system take place, i.e., they are the familiar eigenspaces. By definition, these nonlinear normal mode manifolds are invariant, such that any motion starting exactly in one nonlinear normal mode will be comprised only of that one for all time and will not generate any motion in the other nonlinear normal modes. On the contrary, a standard linear modal analysis of the system's dynamics – obtained by projection of the equations of motion onto the linear modes – necessarily results in the exchange of energy, or contamination, between linear modes, due to the nonlinear coupling terms between the obtained equations. As was demonstrated previously [10], this may yield inaccurate results if one includes only a few linear modes, or expensive solutions if one includes many of them. Many of the principles inherent to this approach have been modified and/or extended to achieve various ends [11–16], but the basic ideas are outlined in Section 2.

Just as the primary use for normal modes of motion of linear systems is the modal analysis associated with them, the concept of nonlinear normal modes of vibration suggests the definition of a proper 'nonlinear modal analysis' in order to obtain the response of a system in terms of nonlinear modal coordinates. Moreover, one ought to be able to perform model reductions using the nonlinear modal coordinates – as is done for linear systems – which requires the development of efficient truncation procedures, the ultimate goal being to use fewer nonlinear modes than linear ones to perform equally accurate modal analyses of nonlinear systems.

Other works have been done which investigate coupled normal mode motions. Many of these have concentrated on the phenomenon of internal resonance, and have hence made certain restrictions on the candidate systems, and the dynamics therein [2, 17]. Another approach [18], similar in spirit to that which follows, uses an action-angle transformation to create 'multi-modes', whose manifolds are then determined numerically. The present paper seeks to extend the invariant manifold-based nonlinear normal mode method, such that practical analysis of larger systems, including those with internal resonances, may be conducted.

Given the definition of the nonlinear normal modes in terms of two-dimensional invariant manifolds, it is clear that (1) they will not interact during a pure modal motion, and (2) they are bound to interact during more general motions. As ignoring these interactions has been found to yield reduced-order models of insufficient accuracy [19], it is essential for any nonlinear extension of modal analysis to account for them in a rigorous manner.

Consequently, a new formulation has been developed to ensure the invariance of the set of modeled nonlinear modes with respect to the non-modeled ones. This formulation, described in Section 3, builds on ideas introduced in [20–22] and generalizes the individually invariant nonlinear normal mode manifolds to multi-mode invariant manifolds. A multi-mode manifold is of dimension $2M$ when M nonlinear modes are modeled, and includes the influence of all of the M individual nonlinear manifolds defined previously. The interactions between the various modeled nonlinear modes are accounted for at the first stage of the definition process, thus eliminating the need for later work to recover them. The generation of a multi-mode invariant manifold follows closely that of an individually invariant manifold, and approximations for weakly nonlinear systems can be constructed easily using this method. As with the individual nonlinear modes, the modes constituting a multi-mode manifold do not interact with the non-modeled ones for motions occurring on that manifold, thus eliminating contamination from the non-modeled modes.

Closed form solutions have been generated for a subset of nonlinear structural systems with arbitrary quadratic and cubic coupling between their linear modes. This allows for the automated generation of third-order accurate multi-mode manifolds which govern the dynamics of a chosen modal subset. These high-dimensional manifolds may then be used to produce reduced equations of motion for a wide variety of structural systems.

Reduced-order models have been obtained for two continuous systems with discrete nonlinearities, one analytic, and the other finite-element based. These results illustrate the benefits of the formulation compared to classical linear modal analyses of nonlinear systems (i.e., projections of equations of motion onto the linear modes). The dynamics recovered by the multi-mode manifold methodology are generally more accurate than those obtained by a linear modal analysis using the same number of linear modes. It should be noted that strong nonlinearities may require higher-order expansions than those employed herein in order to depict the associated manifold accurately. However, for nonlinearities of sufficiently modest degree, this model reduction method may yield significant computational savings.

2. Individual Nonlinear Normal Modes

The equations of motion for the structural systems considered are assumed to be of the form

$$\begin{cases} \dot{x} = y_i, \\ \dot{y}_i = f_i(x_1, \ldots, x_N, y_1, \ldots, y_N), \end{cases} \qquad i = 1, \ldots, N \tag{1}$$

or

$$\mathbf{z} = \mathbf{A}(\mathbf{z}) \tag{2}$$

with $\mathbf{z}^T = [x_1, y_1, \ldots, x_N, y_N]^T$ and $\mathbf{A}(\mathbf{z})^T = [y_1, f_1, \ldots, y_N, f_N]^T$ (where ^{T} denotes a transpose), where any necessary discretization has been achieved previously, for instance using the modes of the linearized system. In the case of a discretized continuous system, $\mathbf{x}^T = [x_1, \ldots, x_N]^T$ and $\mathbf{y}^T = [y_1, \ldots, y_N]^T$ would represent some assumed modal coordinates and velocities, while for discrete systems they would represent generalized coordinates (displacements or rotations) and the corresponding generalized velocities. Furthermore, $\mathbf{f}^T = [f_1, \ldots, f_N]^T$ represents some nonlinear, autonomous forcing on the system.

For a nonlinear, autonomous, oscillatory system such as that defined above, a normal mode of motion is a motion which takes place on a two-dimensional invariant manifold in the system's phase-space. This manifold passes through a stable equilibrium point of the system $(\mathbf{x}, \mathbf{y}) = (0, 0)$ and it is tangent to an eigenspace of the system linearized about that equilibrium [6]. This invariant manifold and the dynamics on it can be described by a pair of independent coordinates, which can be chosen to be a single displacement-velocity pair. However, this procedure must be modified for certain degenerate cases [6, 7], and internal resonances [20, 2]. For the kth nonlinear normal mode, it is a natural choice to define $u_k = x_k$

and $v_k = y_k$, so that all displacements and velocities can be related to (u_k, v_k) only – thus enforcing the two-dimensionality and the invariance of the motion – as

$$\begin{cases} x_i = X_i(u_k, v_k), \\ \qquad\qquad\qquad\qquad i = 1, \ldots, N, \quad i \neq k, \\ y_i = Y_i(u_k, v_k). \end{cases} \tag{3}$$

Substitution into the equations of motion yields a set of constraint equations which describe the geometry of the nonlinear invariant manifold, as

$$\begin{cases} \dfrac{\partial X_i}{\partial u_k} v_k + \dfrac{\partial X_i}{\partial v_k} f_k = Y_i, \\ \qquad\qquad\qquad\qquad i = 1, \ldots, N, \quad i \neq k, \\ \dfrac{\partial Y_i}{\partial u_k} v_k + \dfrac{\partial Y_i}{\partial v_k} f_k = f_i, \end{cases} \tag{4}$$

where use has been made of the kth pair of equations of motion, i.e., $\dot{u}_k = v_k$ and $\dot{v}_k = f_k$. Notice no assumption has yet been made on u_k and v_k, and therefore Equation (4) describes the kth nonlinear normal mode in a non-local sense. Thus, if one can find the exact solution of Equation (4), this solution will describe the exact shape of the manifold. However, solving Equation (4) is, in general, not possible.

For weakly nonlinear systems, an approximate local solution can be computed by assuming a polynomial expansion of X_i and Y_i with respect to u_k and v_k up to the desired order as

$$\begin{cases} X_i = a^k_{1,i} u_k + a^k_{2,i} v_k + a^k_{3,i} u_k^2 + a^k_{4,i} u_k v_k + a^k_{5,i} v_k^2 \\ \qquad + a^k_{6,i} u_k^3 + a^k_{7,i} u_k^2 v_k + a^k_{8,i} u_k v_k^2 + a^k_{9,i} v_k^3 + \cdots, \\ Y_i = b^k_{1,i} u_k + b^k_{2,i} v_k + b^k_{3,i} u_k^2 + b^k_{4,i} u_k v_k + b^k_{5,i} v_k^2 \\ \qquad + b^k_{6,i} u_k^3 + b^k_{7,i} u_k^2 v_k + b^k_{8,i} u_k v_k^2 + b^k_{9,i} v_k^3 + \cdots. \end{cases} \tag{5}$$

Substituting Equation (5) into Equation (4) and equating coefficients of like powers in u_k and v_k yields a set of linear equations which can be solved, one order at a time, for the $a^k_{j,i}$'s and $b^k_{j,i}$'s. These coefficients govern the nonlinear contributions (at various orders) of the ith linear mode to the kth nonlinear mode.

The reduced differential equation, which governs motions on the manifold, is determined by back substitution of the X_i's and Y_i's into Equation (1) for $i = k$. The dynamics of $u_k(t)$ may then be determined (for example by numerical integration), and the remaining coordinates obtained through the relations in Equation (5). This process requires solving only one nonlinear ordinary differential equation (ODE) to determine the motion on each nonlinear normal mode, as compared to N coupled ODE's involved in a direct integration of Equation (1). It can be shown that, for systems where the lowest nonlinearity is of order Q, the dynamics are precise through order $(N' + Q - 1)$ where N' is the order of approximation of the manifold. For general nonlinearities, this means that the approximation of the dynamics is precise to one order higher than that of the manifold itself, and two orders higher in the case of purely odd nonlinearities. However, additional higher order terms, while not precise, may yield more accurate equations of motion under certain conditions.

2.1. Specialized Single-Mode Solution

In order to develop approximate solutions for Equation (4) which may be applied to a wide variety of nonlinear structural systems, yet remain somewhat tractable, several assumptions are necessary. In this formulation, the following assumptions are made:

- The equations are decoupled to linear order using N linear system modes, and thus expressed in terms of the linear normal coordinates. This form may be easily obtained for large structural systems, for example using modal analysis in conjunction with a finite element representation.
- Damping is not considered. However, small damping may be subsequently added to the reduced equation set.
- The nonlinearities are of second and third order only in displacement.
- The manifold is single-valued and analytic for the amplitudes under consideration.

These assumptions allow the equations of motion to be written in the following, second-order form:

$$\ddot{\mathbf{x}} + [\Omega]\mathbf{x} + \mathbf{f}^{\mathbf{nl}}(\mathbf{x}) = 0, \tag{6}$$

where

$$[\Omega] = \begin{bmatrix} \omega_1^2 & 0 & 0 \\ 0 & \ddots & 0 \\ 0 & 0 & \omega_N^2 \end{bmatrix}$$

and ω_i is the ith natural frequency of the linearized system. The nonlinear forcing, $\mathbf{f}^{\mathbf{nl}}$ may be expressed as:

$$\mathbf{f}^{\mathbf{nl}} = [\alpha]\mathbf{x}^{2*} + [\beta]\mathbf{x}^{3*},$$

where $\mathbf{x}^{2*}$ contains all unique second-order combinations, and $\mathbf{x}^{3*}$ contains all unique third-order combinations within the vector $\mathbf{x}$. Consequently, the non-square matrices $[\alpha]$ and $[\beta]$ contain the coefficients for every possible quadratic or cubic coupling in the (modal) displacement $\mathbf{x}$. The individual elements in $[\alpha]$ and $[\beta]$ are denoted as: $\alpha^i_{j,l}$, $\beta^i_{j,l,m}$, where i indicates the row in $[\alpha]$ or $[\beta]$; and j, l, m indicate the coupled degrees of freedom. In this notation, the order of the subscripts is extraneous, that is, $\alpha^3_{1,2} = \alpha^3_{2,1}$, and both refer to the same element in $[\alpha]$. Note that in order to obtain these elements, one must transform the nonlinear terms from physical to modal coordinates, a somewhat tedious procedure.

Now, f_i from Equation (4) may be written as:

$$f_i = -\omega_i^2 X_i - \sum_{j=1}^{N}\sum_{l=j}^{N} \alpha^i_{j,l} X_j X_l - \sum_{j=1}^{N}\sum_{l=j}^{N}\sum_{m=l}^{N} \beta^i_{j,l,m} X_j X_l X_m. \tag{7}$$

We seek a local solution for the X_i's and Y_i's in terms of u_k and v_k in the form of the expansion above, Equation (5).

Due to the choice of modal coordinates, the first-order solution is trivial:

$$\begin{aligned} a^k_{1,i} &= \delta_{ik}, \\ a^k_{2,i} &= 0, \\ b^k_{1,i} &= 0, \\ b^k_{2,i} &= \delta_{ik}. \end{aligned} \tag{8}$$

Here, δ_{ik} is the Kronecker delta. The higher-order coefficients may be determined by substituting Equations (7) and (5) into Equation (4) and collecting like powers in u_k and v_k to produce a linear set of equations, a task greatly simplified by symbolic processors such as Mathematica™. Using this formulation, the second-order coefficients are:

$$\begin{aligned} a^k_{3,i} &= \frac{\alpha^i_{k,k}(2\omega_k^2 - \omega_i^2)}{\omega_i^2(\omega_i^2 - 4\omega_k^2)}, \\ a^k_{5,i} &= \frac{2\alpha^i_{k,k}}{\omega_i^2(\omega_i^2 - 4\omega_k^2)}, \\ b^k_{4,i} &= \frac{-2\alpha^i_{k,k}}{\omega_i^2 - 4\omega_k^2}. \end{aligned} \tag{9}$$

In addition, $a^k_{4,i} = b^k_{3,i} = b^k_{5,i} = 0$. At third order, $a^k_{7,i} = a^k_{9,i} = b^k_{6,i} = b^k_{8,i} = 0$, and,

$$\begin{aligned} a^k_{6,i} &= \frac{-C^{1,k}_{1,i} - \beta^i_{k,k,k} + \alpha^k_{k,k} b^k_{4,i}}{\omega_i^2} - \frac{\omega_k^2[C^{1,k}_{2,i} - C^{1,k}_{3,i}(3\omega_k^2 - \omega_i^2)]}{\omega_i^2 D^{1,k}_i}, \\ a^k_{8,i} &= \frac{C^{1,k}_{3,i}}{2\omega_i^2\omega_k^2} - \frac{(3\omega_k^2 - \omega_i^2)[C^{1,k}_{2,i} - C^{1,k}_{3,i}(3\omega_k^2 - \omega_i^2)]}{2\omega_i^2\omega_k^2 D^{1,k}_i}, \\ b^k_{7,i} &= \frac{-[C^{1,k}_{2,i} - C^{1,k}_{3,i}(3\omega_k^2 - \omega_i^2)]}{D^{1,k}_i}, \\ b^k_{9,i} &= a^k_{8,i}. \end{aligned} \tag{10}$$

The constants $C^{1,k}_{p,i}$, and $D^{1,k}_i$ are defined as:

$$\begin{aligned} C^{1,k}_{1,i} &= \sum_{m=1}^{N} a^k_{3,m}\alpha^i_{k,m}, \\ C^{1,k}_{2,i} &= 2\omega_i^2\omega_k^2 \sum_{m=1}^{N} a^k_{5,m}\alpha^i_{k,m}, \\ C^{1,k}_{3,i} &= -3(C^{1,k}_{1,i} + \beta^i_{k,k,k} - \alpha^k_{k,k} b^k_{4,i}) - 2\omega_i^2\alpha^k_{k,k} a^k_{5,i}, \\ D^{1,k}_i &= (\omega_k^2 - \omega_i^2)(\omega_i^2 - 9\omega_k^2). \end{aligned} \tag{11}$$

Note that the third-order coefficients depend on the second-order results, as is typical in perturbation expansions. Similar results have been found for fourth- and fifth-order coefficients, but the general analytic results are omitted for the sake of brevity.

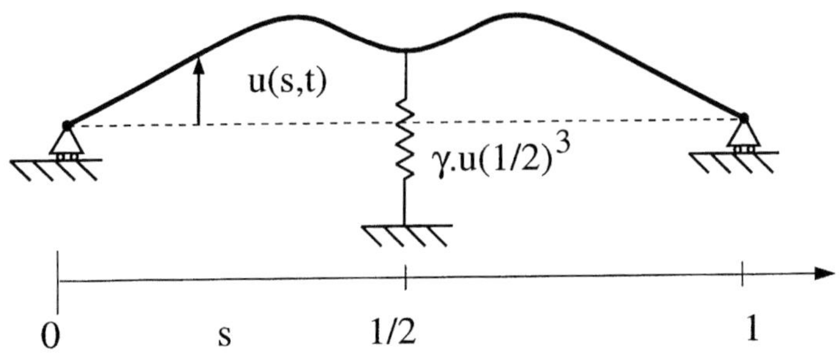

Figure 1. Simply-supported Euler–Bernoulli (linear) beam connected to a purely cubic spring.

Once the above expansion coefficients are calculated, all degrees of freedom are known, albeit approximate, functions of u_k and v_k. These expressions may be substituted into the kth equation of motion to produce a single, second-order, nonlinear oscillator in u_k and v_k. This oscillator governs the local dynamics on the kth invariant manifold.

2.2. Analytic Example: A Simply Supported Euler–Bernoulli Beam Constrained by a Nonlinear Spring

The above methodology is applied to a homogeneous, uniform, simply supported Euler–Bernoulli beam with a nonlinear cubic spring attached at the center – see Figure 1. While the beam itself is assumed to deform in the linear range, the spring is chosen as purely cubic so that the linearized system's normal modes are those of the simply supported beam alone (i.e., pure sine waves). With this choice, the influence of the various linear modes on the nonlinear ones can be visualized easily. Notice that since the spring is located at a node of the antisymmetric (even numbered) modes, they remain unaffected (linear). Therefore, only the symmetric (odd numbered) normal modes are influenced by the nonlinear spring and, furthermore, they feature only contributions from the other symmetric linear modes.

The beam has the following equation for transverse motion, in non-dimensional form:

$$\ddot{u} + \lambda u_{,ssss} + \mu u^3 \delta\left(s - \frac{1}{2}\right) = 0, \quad s \in]0, 1[, \tag{12}$$

where $\lambda = EI/m$, $\mu = \gamma/m$, E is Young's modulus, I is the second moment of area of a cross section, m is the mass per unit length, γ is the nonlinear stiffness of the spring, s represents the spatial coordinate along the beam, $u(s, t)$ is the transverse deflection of the beam, $(\cdot)_{,s}$ denotes a derivative with respect to s, an overdot represents a derivative with respect to time, and $\delta(\cdot)$ is the Dirac function. The associated boundary conditions are $u(0) = u(1) = 0$, and $u_{,ss}(0) = u_{,ss}(1) = 0$. For this example, we take $\lambda = 1$, and $\mu = 10^4$ when numerical results are shown.

The beam deflection, $u(s, t)$, is first discretized using the natural modes of the linearized system, $\phi_j(s) = \sin(j\pi s)$,

$$u(s, t) \backsimeq \sum_{j=1}^{N} x_j(t)\phi_j(s), \tag{13}$$

where N is the number of terms in the expansion, i.e., the number of modes that would be retained for a linear modal analysis of the nonlinear system. Projection of the equation of motion onto the ith linear mode yields

$$\ddot{x}_i + \lambda(i\pi)^4 x_i + 2\mu \left[\sum_{j=1}^{N} x_j \sin\left(j\frac{\pi}{2}\right) \right]^3 \sin\left(i\frac{\pi}{2}\right) = 0, \quad i = 1, \ldots, N, \tag{14}$$

which can be written in first-order form as

$$\begin{cases} \dot{x}_i = y_i, \\ \dot{y}_i = f_i(x_1, \ldots, x_N, y_1, \ldots, y_N), \end{cases} \quad i = 1, \ldots, N, \tag{15}$$

where the inter-modal coupling effects are captured by

$$f_i = -\lambda(i\pi)^4 x_i - 2\mu \left[\sum_{j=1}^{N} x_j \sin(j\frac{\pi}{2}) \right]^3 \sin\left(i\frac{\pi}{2}\right).$$

The set of differential equations, Equation (15), is what is simulated for a typical linear modal analysis. Alternatively, the procedure described earlier in this section can be applied to Equation (15) to determine the third- or higher-order approximation of the nonlinear normal modes of the system. The kth nonlinear mode is given by, to third order [10]:

$$x_k = u_k, \quad \dot{x}_k = v_k$$

and for $i = 1, \ldots, N, i \neq k$:

$$\begin{cases} x_i = \dot{x}_i = 0, & (i \text{ even}), \\ x_i = a_{6,i}^k u_k^3 + a_{8,i}^k u_k v_k^2 + \cdots, & (i \text{ odd}), \\ \dot{x}_i = b_{7,i}^k u_k^2 v_k + b_{9,i}^k v_k^3 + \cdots, & (i \text{ odd}), \end{cases} \tag{16}$$

where, if k is even, $a_{6,i}^k = a_{8,i}^k = 0$, and if k is odd, for $i = 1, \ldots, N, i \neq k$,

$$\begin{cases} a_{6,i}^k = 2\mu(-1)^{(k+i)/2} \dfrac{(i^4 - 7k^4)}{\lambda\pi^4(i^4 - k^4)(i^4 - 9k^4)}, \\ a_{8,i}^k = -12\mu(-1)^{(k+i)/2} \dfrac{1}{\lambda^2\pi^8(i^4 - k^4)(i^4 - 9k^4)}, \end{cases} \tag{17}$$

and, for all k with $i = 1, \ldots, N, i \neq k, i$ odd:

$$\begin{cases} b_{7,i}^k = -2\lambda(\pi k)^4 a_{8,i}^k + 3a_{6,i}^k, \\ b_{9,i}^k = a_{8,i}^k. \end{cases} \tag{18}$$

Though different in form, equivalent expressions may be obtained using Equations (9)–(11).

Next, the deflection and velocity of the beam in the kth nonlinear mode, $u^k(s,t)$ and $v^k(s,t)$, can be expressed in terms of the kth nonlinear modal coordinate, $u_k(t)$, and the associated modal velocity, $v_k(t)$, as

$$u^k(s,t) = u_k \sin(k\pi s) + \sum_{\substack{i \text{ odd} \\ i \neq k}} \left[a_{6,i}^k u_k^3 + a_{8,i}^k u_k v_k^2\right] \sin(i\pi s) + \cdots, \tag{19}$$

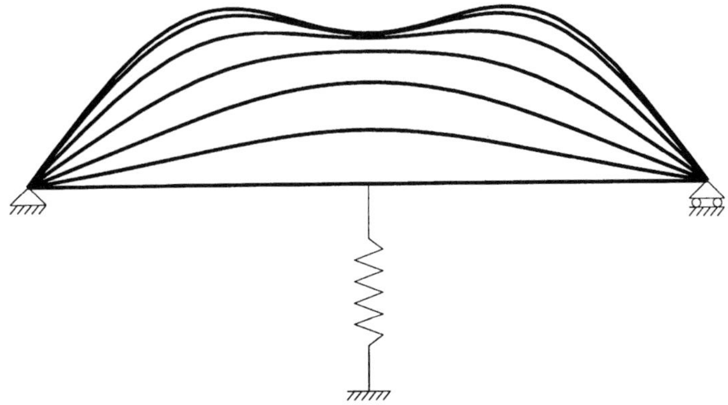

Figure 2. Beam displacement through a quarter-period of motion in the first nonlinear normal mode, sampled at seven equal time intervals. $u_1(0) = 0$, $v_1(0) = 6.93$.

$$v^k(s,t) = v_k \sin(k\pi s) + \sum_{\substack{i \text{ odd} \\ i \neq k}} \left[b^k_{7,i} u_k^2 v_k + b^k_{9,i} v_k^3\right] \sin(i\pi s) + \cdots . \tag{20}$$

Note that the nonlinear normal modes indicate synchronous motions for this system. That is, when $v_k = 0$, the entire velocity field, $v^k(s,t)$, must also vanish. Similarly, since the nonlinearities are odd (that is, symmetric about the equilibrium), all points along the beam pass through zero simultaneously – specifically, when $u_k = 0$, $u^k(s,t) = 0$ over the entire beam. (For an example of a synchronous motion with even-order nonlinearities present, in which case the latter observation does not hold, see [8].) Here, $u^k(s,t)$ is the deflection of the point located at coordinate s at time t when the system undergoes a motion in the kth nonlinear normal mode. It should not to be confused with $u_k(t)$, which is the nonlinear modal coordinate and is not meant to represent the motion of any particular point.

The dynamics of the nonlinear modes are governed by

$$\begin{aligned} \ddot{u}_k &+ \lambda(k\pi)^4 u_k + 2\mu \sin^2(k\pi/2) u_k^3 \\ &+ 6\mu u_k^3 \left(\sum_{\substack{i \text{ odd} \\ i \neq k}} [a^k_{6,i} u_k^2 + a^k_{8,i} v_k^2] \sin(i\pi/2) \right) \sin(k\pi/2) = 0, \end{aligned} \tag{21}$$

for $k = 1, \ldots, N$. Note that the dynamics of the N nonlinear modal oscillators are individually decoupled from one another, and that a third-order manifold accurately produces these equations of motion through fifth order.

Figure 2 shows the position of the beam at seven equidistant instants of time for a motion purely in the first nonlinear mode, as generated by Equations (19–21). Note that as the beam undergoes its oscillation, the shape does not remain constant, and that the spring influence increases with displacement amplitude. These geometric effects, and their mathematical repercussions cannot be captured through a single linear mode truncation of the system. Simply adding linear modes does increase model accuracy. However, without the constraints provided by the manifold, the response loses periodicity.

Figures 3 and 4 display simulation results obtained using this procedure, along with results obtained with classical linear modal analyses of the nonlinear system performed with various

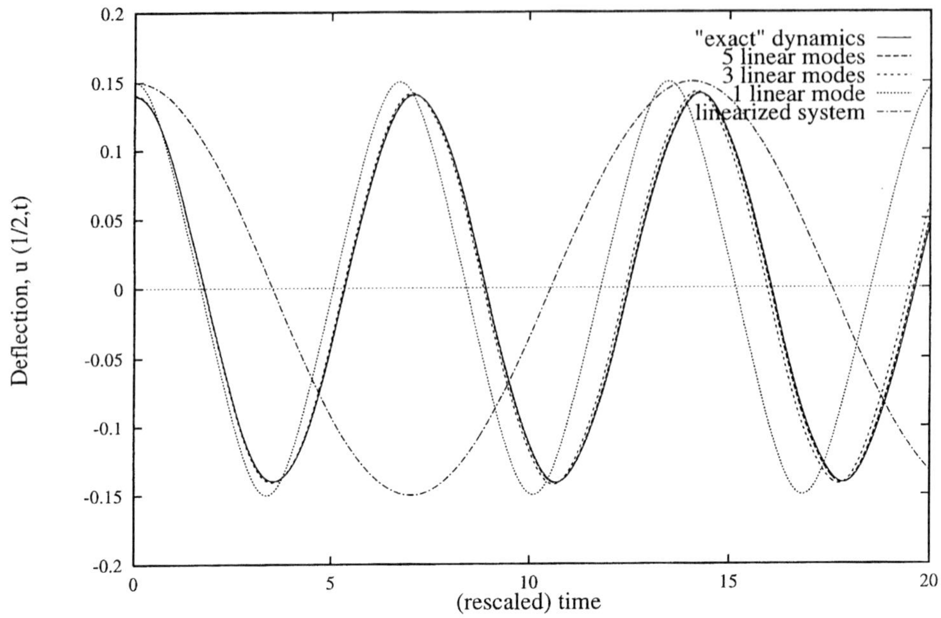

Figure 3. Deflection of the middle-point of the beam obtained by various linear modal analysis simulations initiated on the first nonlinear normal mode manifold. Initial conditions: $u_1(0) = 0.15$, $v_1(0) = 0$.

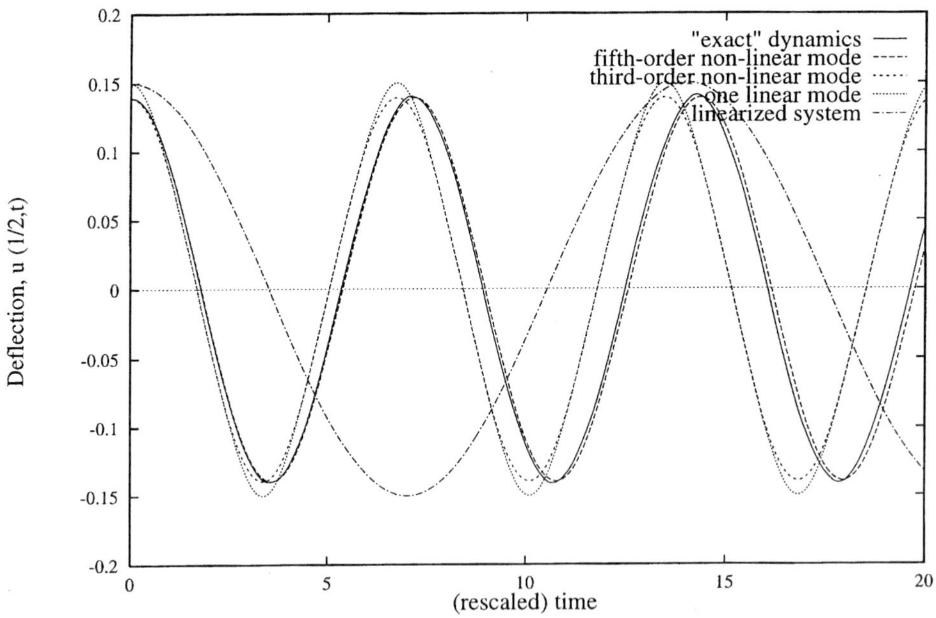

Figure 4. Deflection of the middle-point of the beam obtained by various simulations initiated on the first nonlinear normal mode manifold. Initial conditions: $u_1(0) = 0.15$, $v_1(0) = 0$.

numbers of modeled linear modes. In both figures, the 'exact' solution was determined using a linear modal analysis with 25 linear modes; that is, by simulating the 25-degrees-of-freedom system obtained by projecting the equations of motion onto the lower 25 linear modes, using initial conditions which lie on the third-order approximation of the first nonlinear normal mode manifold in the 50-dimensional phase space. In this particular case it is seen that at least three to five linear modes are necessary to achieve an accuracy comparable to that obtained with a single fifth-order nonlinear mode as obtained above. The nonlinear mode results are obtained by simulation of a single second-order differential equation (Equation (21)), which includes the influence of the other linear modes. In the case where one uses a single linear mode through the entire analysis, the influence of the other linear modes is completely missing, whereas it is naturally embedded in the nonlinear normal mode. See Figure 4 which compares several simulations, all utilizing only one ODE, to the 'exact' solution. It should be noted that the initial displacements shown in these figures include contributions from chosen mode (u_1), as well as the non-modeled modes. Consequently, the initial displacement will vary as additional modes are included, i.e., as the manifold is projected into a larger configuration space.

3. Multi-Mode Invariant Manifolds

The potential of nonlinear normal modes is evident from the previous section. However, it is important to note that, by definition, they are only individually invariant. Therefore, though they do not interact when the system undergoes a motion in any one of the modal manifolds, nothing prevents them from interacting during an arbitrary motion. This immediately reminds one of the contamination problems encountered during the linear modal analysis of nonlinear systems – an issue which was precisely at the origin of the definition of individual nonlinear normal modes.

In order to properly remove the contamination due to nonlinear modes which are not included in the reduced-order model, a new, more general, formulation is necessary, such that the individually invariant nonlinear normal modes are but a special case. The underlying idea is to generate high-dimensional invariant manifolds, referred to as multi-mode manifolds, essentially in the same manner as the individually invariant manifolds were produced in Section 2. These multi-mode manifolds, when comprising the influence of M nonlinear modes, are of dimension $2M$ in the phase space for the oscillatory systems typically of interest in structural dynamics. These multi-mode manifolds are still not completely invariant, as two different multi-mode manifolds would interact during a general motion, just as the individual nonlinear normal modes did. However, for motions on a given multi-mode manifold, invariance is ensured between itself and the rest of the (non-modeled) nonlinear modes.

Consequently, for a system for which M nonlinear modes are to be modeled and for which the remaining ones are to be merely ignored, the multi-mode manifold should comprise all M modes, so that (1) the interactions between those M modes can be accounted for, and (2) the interactions with the non-modeled modes can be completely removed. If a mode is non-modeled despite an internal resonance with a modeled one, the mathematical process of generating the multi-mode manifold will become singular, thereby detecting the anomaly.

The procedure to determine multi-mode invariant manifolds follows closely the one presented in Section 2. If S_M denotes the subset of indices corresponding to the modeled modes, and $\mathbf{u}_M$ and $\mathbf{v}_M$ represent the vectors of the corresponding nonlinear modal coordinates and velocities, then the various linear modal coordinates are expressed as functions of the modeled modes as

$$\begin{cases} x_k = u_k, \\ \qquad\qquad \text{for} \quad k \in S_M \\ y_k = v_k, \end{cases} \tag{22}$$

$$\begin{cases} x_j = X_j(\mathbf{u}_M, \mathbf{v}_M), \\ \qquad\qquad\qquad \text{for} \quad j \notin S_M, \\ y_j = Y_j(\mathbf{u}_M, \mathbf{v}_M). \end{cases} \tag{23}$$

Taking the time-derivatives of Equation (23) yields:

$$\begin{cases} Y_j = \dot{X}_j = \displaystyle\sum_{k \in S_M} \left(\frac{\partial X_j}{\partial u_k} v_k + \frac{\partial X_j}{\partial v_k} f_k \right), \\ \qquad\qquad\qquad\qquad\qquad\qquad \text{for} \quad j \notin S_M, \\ f_j = \dot{Y}_j = \displaystyle\sum_{k \in S_M} \left(\frac{\partial Y_j}{\partial u_k} v_k + \frac{\partial Y_j}{\partial v_k} f_k \right). \end{cases} \tag{24}$$

As before, exact analytical solutions to these equations are not readily available, and in most cases (namely, for weakly nonlinear systems) approximations will be sought in a series expansion form. One has, to third order:

$$\begin{aligned} X_j(\mathbf{u}_M, \mathbf{v}_M) &= \sum_{k \in S_M} (a_{1,j}^k u_k + a_{2,j}^k v_k) + \sum_{k \in S_M} \sum_{l \in S_M} (a_{3,j}^{k,l} u_k u_l + a_{4,j}^{k,l} u_k v_l + a_{5,j}^{k,l} v_k v_l) \\ &\quad + \sum_{k \in S_M} \sum_{l \in S_M} \sum_{q \in S_M} (a_{6,j}^{k,l,q} u_k u_l u_q + a_{7,j}^{k,l,q} u_k u_l v_q \\ &\quad + a_{8,j}^{k,l,q} u_k v_l v_q + a_{9,j}^{k,l,q} v_k v_l v_q) + \cdots, \end{aligned} \tag{25}$$

$$\begin{aligned} Y_j(\mathbf{u}_M, \mathbf{v}_M) &= \sum_{k \in S_M} (b_{1,j}^k u_k + b_{2,j}^k v_k) + \sum_{k \in S_M} \sum_{l \in S_M} (b_{3,j}^{k,l} u_k u_l + b_{4,j}^{k,l} u_k v_l + b_{5,j}^{k,l} v_k v_l) \\ &\quad + \sum_{k \in S_M} \sum_{l \in S_M} \sum_{q \in S_M} (b_{6,j}^{k,l,q} u_k u_l u_q + b_{7,j}^{k,l,q} u_k u_l v_q \\ &\quad + b_{8,j}^{k,l,q} u_k v_l v_q + b_{9,j}^{k,l,q} v_k v_l v_q) + \cdots. \end{aligned} \tag{26}$$

Note that this decomposition is not unique, and that the number of coefficients of order p when M nonlinear modes are modeled is, for each X_j and Y_j:

$$C_p^{2M-1+P} = \frac{(2M-1+p)!}{(2M-1)!p!}$$

which increases very rapidly with both p and M. Substituting Equations (25) and (26) into the jth pair of manifold equations, Equation (24), and equating like powers in $\mathbf{u}_m$ and $\mathbf{v}_m$, one obtains the coefficients, one order at a time.

3.1. Specialized Multi-Mode Solution

As with the single-mode case, it is necessary to restrict the eligible systems in order that the solutions remain marginally tractable. Consequently, the same restrictions as those introduced in Section 2.1 are applied. In addition, in order to ensure a unique solution, it is assumed that the non-unique coefficients of Equations (25) and (26) are equal. That is, because

$$u_k u_l v_q = u_l u_k v_q \neq u_k u_q v_l,$$

we enforce

$$a_{7,j}^{k,l,q} = a_{7,j}^{l,k,q},$$

but allow

$$a_{7,j}^{k,l,q} \neq a_{7,j}^{k,q,l}.$$

Also, as in the single mode case, these are local results, and their fidelity depends on the amplitude of motion as well as the proximity to internal resonances between modeled and non-modeled modes.

Although each order may be solved sequentially in its entirety, resulting in large matrix equations for each complete set of second- and third-order coefficients, the process may be simplified by dividing it into several subproblems according to the number of distinct modes contained in the coefficient superscripts. That is, one first solves for terms like $a_{6,i}^{k,k,k}$, then for terms like $a_{6,i}^{k,k,l}$, where $k \neq l$, and finally for those like $a_{6,i}^{k,l,q}$ where k, l and q are all distinct. These are referred to as the one-mode, two-mode, and three-mode coefficients, respectively. The one-mode problem is almost identical to that of the single mode manifold, and the same solutions may be used. However, the numerical answers are not identical, as the inclusion of more modes eliminates some of the expansion coefficients (e.g., $a_{3,j}^{k,k} = 0$ if $j \in S_m$), and the values of the constants $C_{p,i}^{1,k}$, as defined in Equation (11), are consequently altered. That is, certain constraints are eliminated when an additional mode is included in S_m, and this change is reflected in the reduced equations of motion.

The two-mode problem may be solved first for the second-order expansion coefficients, and then for the appropriate third-order coefficients. The problems are both linear, but due to the size and complexity of the explicit solutions, they are presented here in matrix form only. In addition, the coefficients which are zero have been eliminated from the matrix formulation. It is found that in the one-, two- and three-mode problems the expansion coefficients $a_4, a_7, a_9, b_3, b_5, b_6, b_8$ are all zero. This is due to the fact that we are considering conservative, non-gyroscopic systems, which eliminates the possibility of such terms appearing. For the two-mode problem, the second-order coefficients may be determined by solving

$$\begin{bmatrix} -2 & 2\omega_k^2 & 1 & 0 \\ -2 & 2\omega_l^2 & 0 & 1 \\ -2\omega_i^2 & 0 & \omega_l^2 & \omega_k^2 \\ 0 & 2\omega_i^2 & 1 & 1 \end{bmatrix} \begin{bmatrix} a_{3,i}^{k,l} \\ a_{5,i}^{k,l} \\ b_{4,i}^{k,l} \\ b_{4,i}^{l,k} \end{bmatrix} = \begin{bmatrix} 0 \\ 0 \\ \alpha_{k,l}^i \\ 0 \end{bmatrix}, \tag{27}$$

while the third-order coefficients are obtained by solving

$$\begin{bmatrix} -3 & 2\omega_k^2 & 0 & 1 & 0 & 0 \\ -6 & 2\omega_l^2 & 2\omega_k^2 & 0 & 2 & 0 \\ 0 & -2 & -1 & 0 & 0 & 3 \\ -3\omega_i^2 & 0 & 0 & \omega_l^2 & 2\omega_k^2 & 0 \\ 0 & -2\omega_i^2 & 0 & -2 & -2 & 6\omega_k^2 \\ 0 & 0 & -\omega_i^2 & 0 & -2 & 3\omega_l^2 \end{bmatrix} \begin{bmatrix} a_{6,i}^{k,k,l} \\ a_{8,i}^{k,k,l} \\ a_{8,i}^{l,k,k} \\ b_{7,i}^{k,k,l} \\ b_{7,i}^{l,k,k} \\ b_{9,i}^{k,k,l} \end{bmatrix} = \begin{bmatrix} -C_{1,i}^{3,k,l} \\ -C_{2,i}^{3,k,l} \\ 0 \\ C_{7,i}^{3,k,l} \\ C_{5,i}^{3,k,l} \\ C_{6,i}^{3,k,l} \end{bmatrix}. \tag{28}$$

For the three-mode problem, the third-order coefficients are the solution of:

$$\begin{bmatrix} -6 & 2\omega_k^2 & 2\omega_l^2 & 0 & 2 & 0 & 0 & 0 \\ -6 & 0 & 2\omega_m^2 & 2\omega_k^2 & 0 & 2 & 0 & 0 \\ -6 & 2\omega_m^2 & 0 & 2\omega_l^2 & 0 & 0 & 2 & 0 \\ 0 & -1 & -1 & -1 & 0 & 0 & 0 & 3 \\ -3\omega_i^2 & 0 & 0 & 0 & \omega_m^2 & \omega_l^2 & \omega_k^2 & 0 \\ 0 & -2\omega_i^2 & 0 & 0 & -2 & 0 & -2 & 6\omega_l^2 \\ 0 & 0 & -2\omega_i^2 & 0 & -2 & -2 & 0 & 6\omega_k^2 \\ 0 & 0 & 0 & -2\omega_i^2 & 0 & -2 & -2 & 6\omega_m^2 \end{bmatrix} \begin{bmatrix} a_{6,i}^{k,l,m} \\ a_{8,i}^{l,m,k} \\ a_{8,i}^{k,l,m} \\ a_{8,i}^{m,l,k} \\ b_{7,i}^{k,l,m} \\ b_{7,i}^{k,m,l} \\ b_{7,i}^{l,m,k} \\ b_{9,i}^{k,l,m} \end{bmatrix} = \begin{bmatrix} C_{8,i}^{3,k,l,m} \\ C_{8,i}^{3,k,m,l} \\ C_{8,i}^{3,l,m,k} \\ 0 \\ C_{9,i}^{3,k,l,m} \\ C_{10,i}^{3,l,m,k} \\ C_{10,i}^{3,k,l,m} \\ C_{10,i}^{3,m,l,k} \end{bmatrix}. \tag{29}$$

In Equations (27–29), each equation depends on a specific $i \notin S_M$ and $k, l, m \in S_M$. The constants $C_{p,i}^{3,k,l}$ depend on the first- and second-order results, and are shown explicitly in the appendix. These matrix expressions were initially formulated by hand, and then checked using the symbolic processor Mathematica^{TM}. Due to the relatively sparse nature of these matrices, they may be symbolically inverted to yield analytic expressions for the expansion coefficients which depend only on $[\alpha]$, $[\beta]$, the ω's and the lower-order expansion coefficients. Though these expressions are rather large and unwieldy, they may easily be implemented in a computer code and used to calculate systematically all of the expansion coefficients. It should be noted that the symbolic approach used here allows k, l, m in Equations (27–29) to represent any three modeled modes, and the corresponding (symbolic) solutions will be valid for any three modeled modes. Alternatively, numerical values may be used in Equations (27–29), and the inversion may be repeated for each combination of modeled modes. If several modes are chosen, this may entail considerable work each time a solution is sought. The symbolic approach avoids this shortcoming by producing a single, explicit, general purpose solution. However, due to its complexity, this approach may no longer be practical if additional nonlinear terms or expansion coefficients are included in the formulation (for example, those arising from gyroscopic terms). However, even if additional effects are added, making symbolic inversion impractical, the sub-problems above are still more desirable than the explicit, simultaneous solution of all coefficients at a given order.

While internal resonances between modeled modes are automatically accounted for, internal resonances between modeled and non-modeled modes will cause singularities in the analytic solutions. There are many possible internal resonances. Through third order, they are:

$$\begin{aligned} &\omega_i = \omega_k, && \omega_i = 3\omega_k, \\ &\omega_i = 2\omega_k, && \omega_i = |2\omega_k \pm \omega_l|, \\ &\omega_i = |\omega_k \pm \omega_l|, && \omega_i = |\omega_k \pm \omega_l \pm \omega_m|, \end{aligned} \tag{30}$$

where $(i, k, l, m) \in S_M$.

Once the expansion coefficients are known, the non-modeled degrees of freedom can be expressed as approximate functions of the modeled coordinates. These functions may be substituted back into the equations of motion corresponding to the modeled modes, thus eliminating the non-modeled modes from the original set of second-order differential equations. This reduction process may be automated, producing a set of equations with polynomial nonlinearities in the modeled coordinates. As in the single-mode case, a third-order expansion of the non-modeled coordinates produces equations which are precise through fourth order if $[\alpha]$ is non-zero, and fifth order if $[\alpha] = 0$. In addition, it has been found that, even when $[\alpha]$ is non-zero, model fidelity is considerably improved when the fifth-order terms are retained. Empirical evidence suggests that, while not precise, these additional terms contain essential information.

As alluded to above, a computer program has been developed for the multi-mode case which – given $[\alpha]$, $[\beta]$, and the system's linear natural frequencies – can reduce a system of N equations which are formulated in the linear modal coordinates, down to a system of M equations of motion, in M chosen nonlinear modal coordinates i.e., a reduced-order model. This program generates the complete, third-order, M-mode manifold via the analytic solutions to Equations (27–29), enabling the M equations of motion to be expressed to fifth order. This means it is now feasible to systematically (and automatically) reduce many nonlinear structural systems to include only the modes of interest without eliminating all effects due to non-modeled modes. For example, a two mode manifold may be produced to examine an internal resonance between two modes, yielding equations which not only account for all interactions between the two chosen modes, but include relevant effects due to all other modes. Of course, these additional effects are embedded in the fourth- and fifth-order terms, of which there may be a great many. Consequently, some examination is necessary to determine when 'reducing' N third-order equations to M fifth-order equations will result in significant time savings; this matter is considered next.

3.2. Some Computational Considerations

Depending on the ratio of M/N, the simplification obtained from the nonlinear mode approach may be significant. One indication of the complexity of a given system is the total number of nonlinear terms in the model. For example, this is a measure of both the number of computations necessary at a given time step in a numerical simulation. In the original system, there are N equations, with

$$\frac{1}{6}(N^3 + 6N^2 + 5N)$$

possible nonlinear terms in each equation (this and the following results are obtained by counting the terms at each order). By comparison, in the reduced system there are M equations with

$$\frac{M}{30}(4M^4 + 30M^3 + 85M^2 + 105M + 46)$$

terms each, when all generated fifth-order terms are retained (in order to be conservative). At the highest order, this difference reduces to an $M \times (4/5)M^5$ *versus* an $N \times N^3$ set of nonlinear terms. Consequently, there is not significant reduction in the number of terms if

$$M \geq \approx N^{2/3}$$

in the limiting case of the fifth-order result. It can be shown that for $N > 4$, reduction will always occur if $M \leq \sqrt{N}$. This is not a distinct boundary, but it may be used as a general guide when applying this technique. These criteria are indicative of the most nonlinear cases. In simpler cases, when the $[\alpha]$ and $[\beta]$ matrices are sparse, the resulting reduced set may contain many terms which are also zero or of negligible magnitude. Due to these effects, an absolute description of the efficiency of this technique is not possible, and each case must be evaluated individually. For example, if $N = 20$, and $[\alpha]$ and $[\beta]$ are full, there will be $20 \times 1750 = 35000$ original nonlinear terms. If this is reduced to $M = 4$ equations, there will be, at most, $4 \times 636 = 2544$ nonlinear terms in the reduced set.

In many structural systems, it is primarily the low frequency modes which are of interest. However, when nonlinearities are present, it is often necessary to simulate the high frequency modes as well in order to achieve the desired accuracy in the modes of interest. This requires that the time steps of any integration scheme be small enough to reproduce the high frequency dynamics accurately. When these systems are reduced to their low frequency modes using a multi-mode manifold, the effects of the high frequency modes are captured in the reduced equations, and it is no longer necessary to simulate them explicitly, hence allowing for a time step governed by the low frequency modes. When this time savings is compounded with that due to the size reduction of the equations, the necessary computer time may be reduced by several orders of magnitude. In addition, it should be noted that the size of the reduced equations does not grow with N. Consequently, it is possible to produce models which are reduced from many more modes than would have previously been practical. Also, as N is increased, the convergence of the reduced equations (with M held constant) may be studied to guarantee their accuracy.

3.3. Analytic Example: A Simply Supported Euler–Bernoulli Beam Constrained by a Nonlinear Spring

In the particular case of the system depicted in Figure 1, a two-mode invariant manifold is computed with the aid of symbolic manipulation. Here, the first-order terms vanish except for those corresponding directly to the linear modes (since the linear modes are used to discretize the system), and all second-order terms are zero (since there are no quadratic nonlinearities). If $S_m = \{k, l\}$, the non-zero a's are as follows:

$$a_{6,j}^{k,k,k} = \frac{2(-j^4 + 7k^4)\mu \sin^3(k\pi/2)\sin(j\pi/2)}{\lambda\pi^4(j^8 - 10j^4k^4 + 9k^8)}, \tag{31}$$

$$\begin{aligned} a_{6,j}^{k,k,l} &= 6\mu \sin^2(k\pi/2)\sin(l\pi/2)\sin(j\pi/2) \\ &\times (j^8 - 6j^4l^4 - 2j^4k^4 - 2k^4l^4 + l^8 + 8k^8) \Big/ \big[(\lambda\pi^4(l^4 - j^4) \\ &\times (-8k^4l^4 - 8j^4k^4 + 16k^8 + j^8 - 2j^4l^4 + l^8)\big], \end{aligned} \tag{32}$$

$$a_{8,j}^{k,k,k} = \frac{12\mu \sin^3(k\pi/2)\sin(j\pi/2)}{\lambda^2\pi^8(j^8 - 10j^4k^4 + 9k^8)}, \tag{33}$$

$$a_{8,j}^{k,l,l} = \frac{-12\mu(3k^4 + j^4 - 4l^4)\sin^2(l\pi/2)\sin(j\pi/2)\sin(k\pi/2)}{\lambda^2\pi^8(k^4 - j^4)(-8k^4l^4 - 8j^4l^4 + 16l^8 + j^8 - 2j^4k^4 + k^8)}, \tag{34}$$

$$a_{8,j}^{k,k,l} = \frac{24\mu \sin^2(k\pi/2)\sin(l\pi/2)\sin(j\pi/2)}{\lambda^2\pi^8(-8k^4l^4 - 8j^4k^4 + 16k^8 + j^8 - 2j^4l^4 + l^8)}, \tag{35}$$

with similar relations when k and l are switched, from which one obtains, $b_{6,j}^{p,q,r} = b_{8,j}^{p,q,r} = 0$ for all $\{p, q, r\} \in S_m$, and

$$b_{7,j}^{k,k,k} = \frac{6(-j^4 + 3k^4)\mu \sin^3(k\pi/2)\sin(j\pi/2)}{\lambda\pi^4(j^8 - 10j^4k^4 + 9k^8)}, \tag{36}$$

$$\begin{aligned} b_{7,j}^{k,k,l} &= 6\mu \sin^2(k\pi/2)\sin(l\pi/2)\sin(j\pi/2) \\ &\times (j^8 - 2j^4k^4 - 2j^4l^4 - 6k^4l^4 + l^8 + 8k^8) \Big/ \left[(\lambda\pi^4(l^4 - j^4)\right. \\ &\times \left.(-8k^4l^4 - 8j^4k^4 + 16k^8 + j^8 - 2j^4l^4 + l^8)\right], \end{aligned} \tag{37}$$

$$b_{7,j}^{k,l,k} = \frac{12\mu \sin^2(k\pi/2)\sin(l\pi/2)\sin(j\pi/2)}{\lambda\pi^4(-8k^4l^4 - 8j^4k^4 + 16k^8 + j^8 - 2j^4l^4 + l^8)}, \tag{38}$$

$$b_{9,j}^{k,k,k} = a_{8,j}^{k,k,k} = \frac{12\mu \sin^3(k\pi/2)\sin(j\pi/2)}{\lambda^2\pi^8(j^8 - 10j^4k^4 + 9k^8)}, \tag{39}$$

$$b_{9,j}^{k,k,l} = \frac{\mu \sin^2(k\pi/2)\sin(l\pi/2)\sin(j\pi/2)(-l^4 - 3j^4 + 4k^4)}{\lambda^2\pi^8(l^4 - j^4)(-8k^4l^4 - 8j^4k^4 + 16k^8 + j^8 - 2j^4l^4 + l^8)}, \tag{40}$$

with, again, similar relations when k and l are switched.

It can be noted by inspection of Equations (31), (33) and (18) that

$$a_{6,j}^{k,k,k} = a_{6,j}^{k} \text{ as obtained for the } k\text{th nonlinear normal mode,} \tag{41}$$

and

$$a_{8,j}^{k,k,k} = a_{8,j}^{k} \text{ as obtained for the } k\text{th nonlinear normal mode.} \tag{42}$$

Although it was noted earlier that the multi-mode manifold formulation should change these coefficients, these changes are due to the presence of quadratic nonlinearities which are not included in this example. Hence, for this system, the one-mode coefficients are the same for both the single- and multi-mode manifolds. Also, as a two-mode manifold was sought, it was not necessary to solve the three-mode subproblem in order to obtain a general solution.

Note that for this example system, as was the case for single-mode manifolds, the dynamics are obtained at order $N'+2$ when the order of approximation of the multi-mode manifold is N'. In general the dynamics are precise through order $N' + Q - 1$ when the lowest nonlinearity is of order Q. In the present case, the dynamics are obtained at fifth order by the coupled equations

$$\begin{cases} \dot{u}_k = v_k, \\ \dot{v}_k = -\lambda(k\pi)^4 u_k - 2\mu \sin(k\pi/2)\left[\sum_{l\in S_M} u_l \sin(l\pi/2)\right]^3 \\ \qquad -6\mu \sin(k\pi/2)\left[\sum_{l\in S_M} u_l \sin(l\pi/2)\right]^2 \left[\sum_{j\notin S_M} X_j^{(3)} \sin(j\pi/2)\right] + \cdots, \end{cases} \tag{43}$$

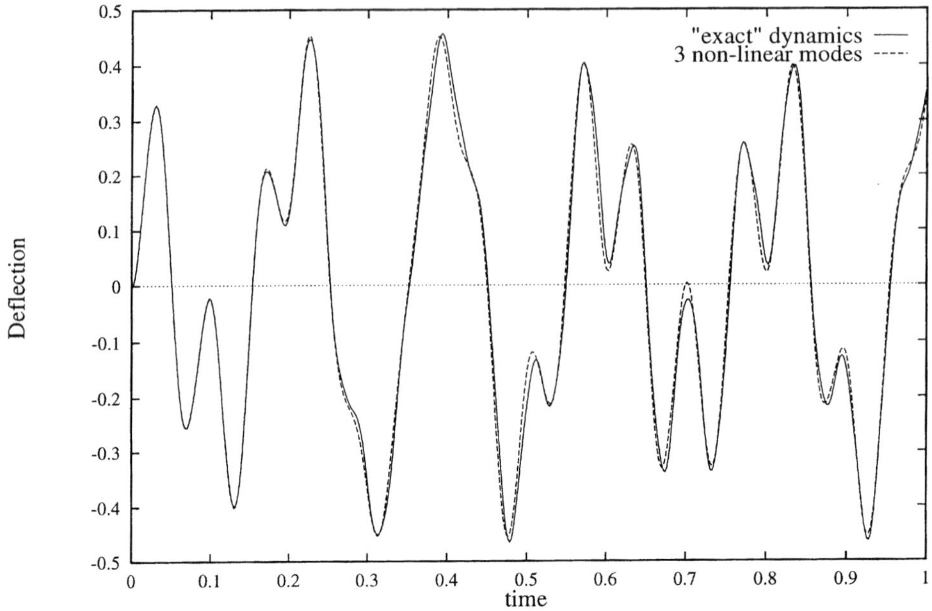

Figure 5. Deflection of the middle-point of the beam obtained by a third-order accurate three-mode invariant manifold. Initial conditions: $u_1(t=0) = u_3(t=0) = 0.2$, $u_2(t=0) = 0.1$, $v_1(t=0) = v_2(t=0) = v_3(t=0) = 0$.

for $k \in S_m$, where $X_j^{(3)}$ represents the cubic part of X_j. Note that, in contrast with the single-mode manifold formulation, the dynamics of the various modeled nonlinear modes are coupled (u_l terms), so that essential interactions between them are allowed, while the coupling to the non-modeled modes has been removed, although their influence is included through the $X_j^{(3)}$ terms.

Results of simulations performed using either the above multi-mode manifold procedure or a linear modal analysis of the nonlinear system are shown in Figures 5 and 6 for a given set of initial conditions on a three-mode manifold, using the parameters $\lambda = 1$, and $\mu = 5000$. In these examples, the three-mode model is composed of the first three modes and can therefore be obtained directly from the two-mode model involving only the first and third modes, as all the coefficients corresponding to the added even mode vanish. Consistent with the theory, a given number of nonlinear modes embedded in the multi-mode manifold yields better results than the same number of linear modes used in a linear modal analysis procedure, all the more so as the influence of the nonlinearity increases.

3.4. Finite-Element Based Example: A Pinned/Spring Euler–Bernoulli Beam Constrained by a Nonlinear Torsional Spring

The computer program mentioned above was used for the analysis of the system shown in Figure 7. The finite element code PATRAN was used to create a beam model with 200 two-noded beam finite elements. The beam in Figure 7 is pinned at one end, with the other constrained with a linear spring, and a nonlinear torsional spring located at the pinned end. Once the linear modes of this model were known, the nonlinear torsional spring was added (analytically) to the pin location. The characteristics of the nonlinear spring and the linear eigenvectors were then used to construct the $[\alpha]$ and $[\beta]$ matrices. Using these, reduced equations for any chosen

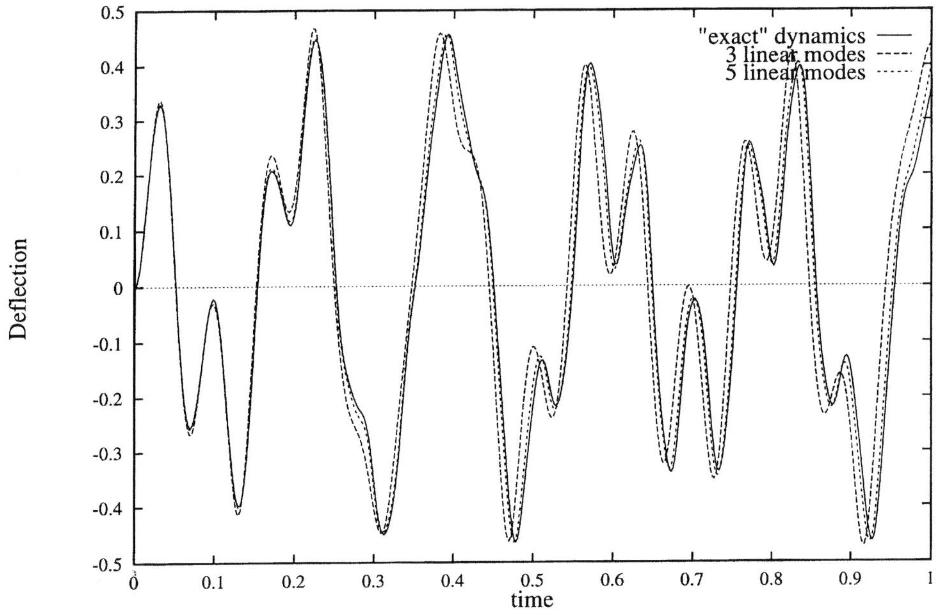

Figure 6. Deflection of the middle-point of the beam obtained by three-mode and five-mode linear modal analysis of the nonlinear system. Initial conditions: $u_1\ (t = 0) = u_3\ (t = 0) = 0.2$, $u_2\ (t = 0) = 0.1$, $v_1\ (t = 0) = v_2\ (t = 0) = v_3\ (t = 0) = 0$.

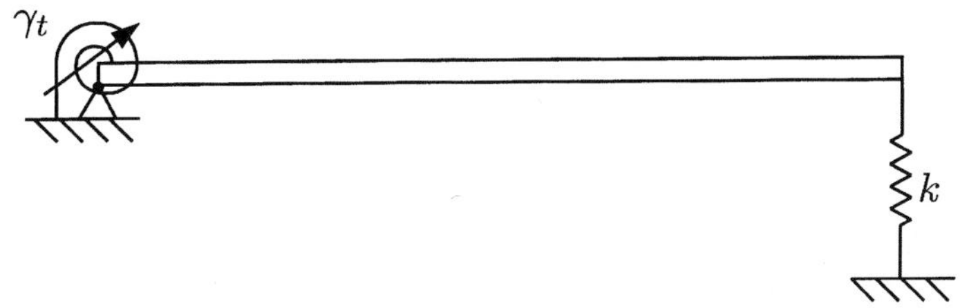

Figure 7. Schematic of finite element model: 200 beam elements were used to construct a beam with: $L = 1$ m, $\rho = 7860$ kg/m^3, $E = 2 \times 10^{11}$ N/m^2, $I = 5 \times 10^{-8}$ m^4. The parameters k and γ_t vary by case.

set of nonlinear normal modes may easily be generated. Once the reduced modal dynamics have been determined (here, through fourth-order Runge–Kutta numerical integration) the manifold coefficients may be used to reconstruct the non-modeled modal displacements, and the corresponding physical displacements. Two results are shown: in the first, the system is not internally resonant and the torsional spring contains both quadratic and cubic nonlinearities, and in the second, the linear spring is tuned such that a 2:1 internal resonance in generated between the second and third linear modes, and the torsional spring is purely quadratic.

The results for the first case, shown in Figure 8, illustrate the displacements at the beam's quarter span, predicted by an 'exact' solution (25 linear modes), a three-mode nonlinear manifold, and a three-linear-mode truncation. The first three modes were used in both three-mode models. For this case it can be seen that the linearly truncated model yields results which slowly diverge, while the other two match quite well. However, these small errors hardly seem to justify the effort necessary to generate the six-dimensional manifold, and the reduced equations it yields. One must keep in mind that the small errors shown here simply indicate

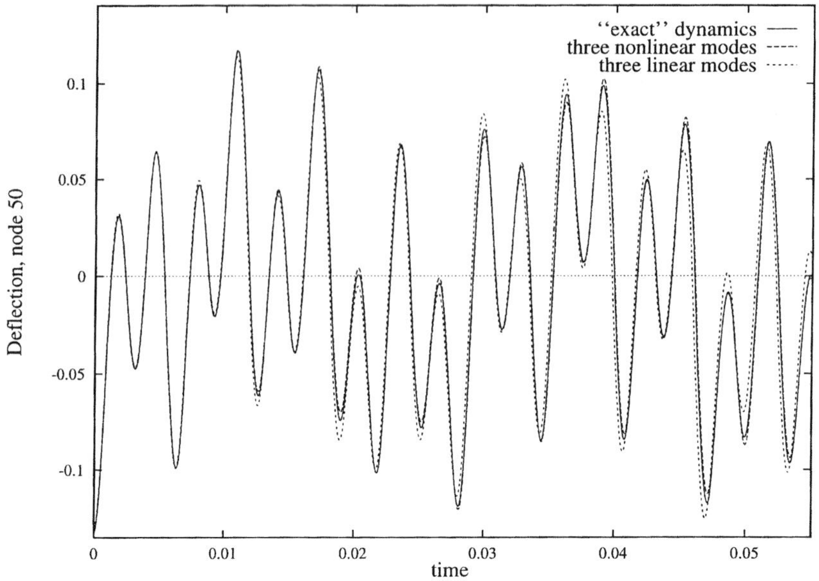

Figure 8. Deflection of node 50 (quarter span of beam) *versus* time using various methods. Here, $k = 10^8$, and the nonlinear spring strength is $\gamma_t = 5000u'(0)^2 + 20000u'(0)^3$, with initial conditions: $u_1(t = 0) = -0.15$, $u_2(t = 0) = 0.12$, $u_3(t = 0) = 0.25$, $v_1(t = 0) = v_2(t = 0) = v_3(t = 0) = 0$.

that for this system, at this amplitude, with these initial conditions, the contributions of the non-modeled modes are not very significant. Changes in any of these parameters may affect the results in unexpected ways.

In the second case, Figure 9, the displacements at half span predicted by the 'exact' solution, the two-mode manifold, and a two-linear-mode truncation are shown. Both two-mode solutions use the second and third (internally resonant) linear modes. Here, a significant error, in the form of a nearly constant offset, is present from the outset. Analysis shows that this offset is due to contributions from the first mode, a phenomenon which deserves some exploration. The two-mode manifold solution explicitly simulates the second and third modes, and then assembles the contributions of the remaining modes, including the first, to obtain the dynamics of the entire system. Hence, the obtained offset is a polynomial function of u_2, u_3, v_2, and v_3. The 'exact' solution is initiated on the manifold, so the initial conditions in the second and third modes, together with the manifold coefficients, specify the initial conditions of all the modes, which are then simulated. For this solution, all 25 modes interact dynamically to maintain the offset. The two-mode truncation simply assumes that all other modes remain quiescent and do not contribute, an approach with no hope of yielding accurate results in this situation. For this case it is apparent that modes two and three are not easily separable from the system. That is, that mode one (at least!) must be accounted for or contamination is ensured. Also, the manifold solution reproduces the 'exact' one exceptionally well, indicating considerable manifold accuracy.

The two examples above illustrate that, while exceptional results are not guaranteed, important (and unexpected) effects, when present, may be captured. Indeed, it is these unpredictable modal interactions which prove the utility of this approach for large scale systems. Due to its finite element origin, and the systematic reduction procedure, the above example could just as

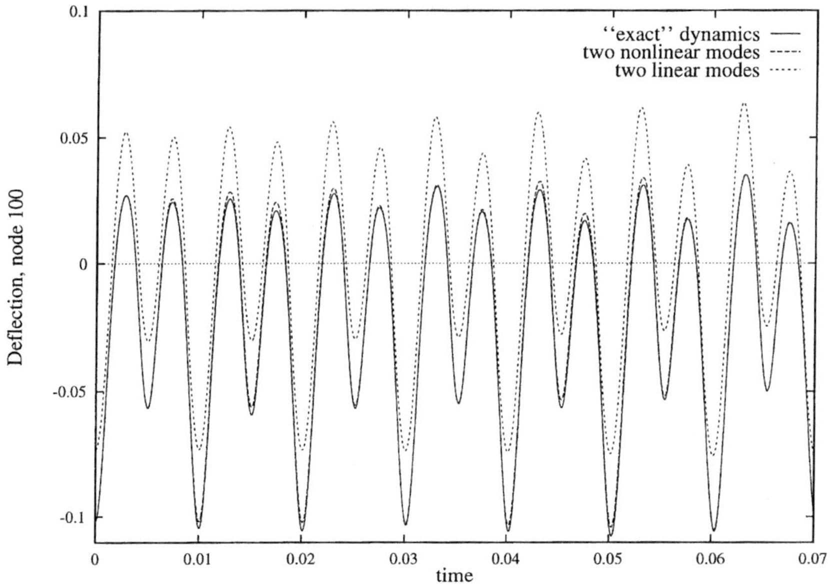

Figure 9. Deflection of node 100 (half span of beam) *versus* time using various methods. Here, $k = 1.185 \times 10^6$, and the nonlinear spring strength is $\gamma_t = 10000u'(0)^2$, with initial conditions: $u_2\ (t = 0) = 0.2$, $u_3\ (t = 0) = -0.35$, $v_2\ (t = 0) = v_3\ (t = 0) = 0$.

easily have been a plate, shell, or general structure, and the reduction procedure would remain unchanged. As such, though the reduction process is analytically complex, its automation enables the generation of rigorous, uncontaminated, reduced-order models of a large variety of nonlinear structural systems.

4. Conclusions

The developments in Sections 2 and 3 suggest that the concept of invariant manifolds has potentially important implications for nonlinear structural dynamics problems. This is a powerful way of defining a nonlinear modal analysis for nonlinear systems, in the sense that proper interaction between the various modeled modes is allowed and accounted for, while contamination with the non-modeled modes is eliminated. This property, which is essential for proper simulation of the dynamics of a system once its most important modes have been selected, is an extension to nonlinear systems from the modal analysis that exists for linear systems.

In addition, when the original equations of motion are given in terms of the linear modal coordinates, each nonlinear mode or multi-mode manifold contains, as its kernel, the corresponding system in which non-modeled modes are entirely ignored. Hence, in regions where the manifold description is accurate, results are guaranteed to be more precise than those produced through simple truncation, as the reduction procedure embeds the influence of additional, non-modeled, modes.

The utility of this approach is enhanced by the fact that the coefficients are generated through linear subproblems, and furthermore by the generalizations which enable additional reduction and analytical solutions. The advent of powerful symbolic manipulation packages has enabled the generation of these solutions for some restricted classes of systems (at least).

Though the solutions outlined herein do not account for all possible nonlinear terms, a considerable subset of structural systems are included. Thus, the generation of rigorous, reduced-order models for a wide variety of nonlinear systems is reduced to a subroutine.

Illustrative results are provided for both an analytically manageable system, and a more complex, finite-element based system. The analytical results depict typical expressions for the manifold expansion coefficients and the corresponding reduced equations of motion, while the finite-element based system emphasizes that practical application is not limited to highly idealized, analytically tractable systems. Numerical integration results are provided for both systems as an indication of the accuracy and utility of the approach.

The methods proposed herein to generate reduced-order models have potentially important implications for many areas, including structural dynamics and control, where accurate low-order models are of interest.

Appendix A: Additional Formulas

The constants from the two and three mode problems are as follows:

$$C_{1,i}^{3,k,l} = 2\sum_{q\in S_M} \alpha_{k,k}^{q} a_{5,i}^{q,l},$$

$$C_{2,i}^{3,k,l} = 2\sum_{q\in S_M} \alpha_{k,l}^{q} a_{5,i}^{q,k},$$

$$C_{7,i}^{3,k,l} = \beta_{k,k,l}^{i} + \sum_{q\notin S_M} (2\alpha_{k,q}^{i} a_{3,q}^{k,l} + \alpha_{l,q}^{i} a_{3,q}^{k,k}) - \sum_{q\in S_M} (\alpha_{k,k}^{q} b_{4,i}^{l,q} + \alpha_{k,l}^{q} b_{4,i}^{k,q}),$$

$$C_{5,i}^{3,k,l} = 2\sum_{q\notin S_M} \alpha_{k,q}^{i} a_{5,q}^{k,l},$$

$$C_{6,i}^{3,k,l} = \sum_{q\notin S_M} \alpha_{l,q}^{i} a_{5,q}^{k,k},$$

$$C_{8,i}^{3,k,l,m} = 2\sum_{q\in S_M} \alpha_{k,l}^{q} a_{5,i}^{q,m},$$

$$\begin{aligned} C_{9,i}^{3,k,l,m} &= \sum_{q\notin S_M} (\alpha_{k,q}^{i} a_{3,q}^{l,m} + \alpha_{m,q}^{i} a_{3,q}^{k,l} + \alpha_{l,q}^{i} a_{3,q}^{m,k}) \\ &\quad + \frac{1}{2}\sum_{q\in S_M} (\alpha_{m,k}^{q} b_{4,i}^{l,q} + \alpha_{l,k}^{q} b_{4,i}^{m,q} + \alpha_{m,l}^{q} b_{4,i}^{k,q}) + \frac{1}{2}\beta_{m,l,k}^{i}, \end{aligned}$$

$$C_{10,i}^{3,k,l,m} = 2\sum_{q\notin S_M} \alpha_{k,q}^{i} a_{5,q}^{l,m}.$$

Acknowledgment

This work has been supported by a grant from the Army Research Office, under Program Manager Dr. Gary L. Anderson.

References

1. Meirovitch, L., *Analytical Methods in Vibrations*, MacMillan, New York, 1967.
2. Nayfeh, A. H., Chin, C., and Nayfeh, S. A., 'On nonlinear normal modes of systems with internal resonance', *ASME, Journal of Vibration and Acoustics* **118**, 1996, 340–345.
3. Rosenberg, R. M., 'On nonlinear vibrations of systems with many degrees of freedom', *Advances in Applied Mechanics* **9**, 1966, 155–242.
4. Aubrecht, J. and Vakakis, A. F., 'Localized and non-localized nonlinear normal modes in a multi-span beam with geometric nonlinearities', *ASME, Journal of Applied Mechanics* **118**, 1996, 533–542.
5. Chakraborty, G., Mallik, A. K., and Hatwal, H., 'Normal modes and near-resonance response of beams with non-linear effects', *Journal of Sound and Vibration* **210**(1), 1998, 19–36.
6. Shaw, S. W. and Pierre, C., 'Normal modes for non-linear vibratory systems', *Journal of Sound and Vibration* **164**(1), 1993, 85–124.
7. Shaw, S. W. and Pierre, C., 'Normal modes of vibration for non-linear continuous systems', *Journal of Sound and Vibration* **169**(3), 1994, 319–347.
8. Shaw, S. W., 'An invariant approach to nonlinear normal modes of vibration', *Journal of Nonlinear Science* **4**, 1994, 419–448.
9. Carr, J., *Applications of Center Manifold Theory*, Springer-Verlag, Berlin, 1981.
10. Boivin, N., Pierre, C., and Shaw, S. W., 'Non-linear normal modes, invariance, and modal dynamics approximations of non-linear systems', *Nonlinear Dynamics* **8**, 1995, 315–346.
11. King, M. E. and Vakakis, A. F., 'An energy based formulation for computing nonlinear normal modes in undamped continuous systems', *ASME, Journal of Vibration and Acoustics* **116**, 1994, 332–340.
12. Chen, S.-L. and Shaw, S. W., 'Normal modes for piecewise linear vibratory systems', *Nonlinear Dynamics* **10**, 1996, 135–164.
13. Slater, J. C., 'A numerical method for determining nonlinear normal modes', *Nonlinear Dynamics* **10**, 1996, 19–30.
14. Burton, T. B. and Hamdan, M. N., 'On the calculation of non-linear normal modes in continuous systems', *Journal of Sound and Vibration* **197**(1), 1996, 117–130.
15. Slater, J. C. and Inman, D. J., 'On the effect of weak non-linearities on linear controllability and observability norms, an invariant manifold approach', *Journal of Sound and Vibration* **199**(3), 1997, 417–429.
16. Qasi, M. I., 'Nonlinear normal modes of a lumped parameter system', *Journal of Sound and Vibration* **205**(2), 1997, 205–211.
17. King, M. E. and Vakakis, A. F., 'An energy based approach to computing resonant nonlinear normal modes', *ASME, Journal of Applied Mechanics* **63**, 1996, 810–819.
18. Leung, A. Y. T. and Ge, T., 'Normal multi-modes of non-linear Euler beams', *Journal of Sound and Vibration* **202**(2), 1997, 145–160.
19. Pellicano, F. and Mastroddi, F., 'Applicability conditions of a non-linear superposition technique', *Journal of Sound and Vibration* **200**(1), 1997, 3–14.
20. Boivin, N., Pierre, C., and Shaw, S. W., 'Non-linear modal analysis of structural systems featuring internal resonances', *Journal of Sound and Vibration* **182**(2), 1995, 336–341.
21. Boivin, N., Pierre, C., and Shaw, S. W., 'Non-linear modal analysis of structural systems using multi-mode invariant manifolds', in *Proceedings of the AIAA Dynamics Specialists Conference*, Hilton Head, SC, 1994, Paper No. 94-1672.
22. Boivin, N., 'Non-linear modal analysis of structural systems using invariant manifolds', Ph.D. Thesis, The University of Michigan, Ann Arbor, MI, 1995.

Nonlinear Dynamics **25:** 207–220, 2001.

Localization in Nonlinear Mistuned Systems with Cyclic Symmetry

WALTER SEXTRO and KARL POPP
Institute of Mechanics, University of Hannover, Appelstr. 11, D-30167 Hannover, Germany

TOMASZ KRZYZYNSKI
Faculty of Mechnical Engineering, Koszalin University of Technology, Raclawicka 15–17, PL-75620 Koszalin, Poland

(Received: 16 October 1998; accepted: 31 August 1999)

Abstract. In forced systems with cyclic symmetry localization can occur due to parameter uncertainties. Often, Monte-Carlo simulations are used to find regions, where the system response is sensitive to parameter uncertainties. These simulations require a large computation time. Therefore, an approximate method to calculate the envelopes of the frequency response functions is developed in this paper. An example of a nonlinear system with cyclic symmetry is a bladed disk assembly with friction dampers. Friction dampers can be installed underneath the blade platforms of turbine blades. Due to dry friction and the relative motion between blades and dampers, energy is dissipated, which results in a reduction of blade vibration amplitudes. By optimizing the mass of the friction dampers, the best damping effects are obtained, which lead to an increase in the reliability of the turbine. In this paper, the calculated response of a mistuned bladed disk assembly with friction dampers is discussed. An approximate method is developed to calculate the envelopes of the corresponding frequency response function for statistically varying eigenfrequencies of the blades. Regions where localization can occur with a high probability, are calculated by this method.

Keywords: Bladed disk assembly, mistuning, sensitivity, localization.

Nomenclature

a_Z	= half width of contact area (m)
c_R	= global tangential contact stiffness (N/m)
$E[\,]$	= mean value of [ ]
F	= force (N)
F_H	= centrifugal force (N)
F_N	= average normal force (N)
i	= complex unit
K	= stiffness coefficient (N/m)
k	= exponent of the Weibull-distribution
ℓ_K	= length of contact area (m)
m_R	= mass of friction damper (kg)
n_M	= number of modes for one blade
n_S	= number of blades
p	= parameter
p_H	= Hertzian contact pressure (N/m^2)
q_N	= distributed normal load (N/m)
r_C	= distance from rotor axis to center of gravity of the friction damper (m)
R	= radius (m)
s	= measure for sensitivity or strength for localization (–)

S	= sensitivity
u	= relative displacement (m)
w	= amplitude (m)
w_0	= parameter of the Weibull-distribution (m)
x, y, z	= coordinates (m)
$\mathbf{A}$	= system matrix
$\mathbf{B}$	= matrix
$\mathbf{F}$	= generalized force vector
$\mathbf{K}$	= stiffness matrix
$\mathbf{q}$	= vector of modal coordinates
$\mathbf{T}$	= modal matrix of the bladed disk
$\mathbf{u}$	= vector of generalized relative displacements
$\mathbf{w}$	= vector of generalized displacements
β	= blade root angle (rad)
η, ζ	= coordinates fixed to the platform (m)
$\Gamma()$	= gamma function
μ	= friction coefficient (–)
γ	= platform angle (rad)
σ	= standard deviation
Ω	= angular velocity (rad/s)
ω	= excitation angular frequency (rad/s)
ω_0	= angular eigenfrequency (rad/s)
$\hat{()}$	= complex
$()^i$	= imaginary part
$()^r$	= real part
$()^T$	= transpose of ()
$()^*$	= dimensionless
$()_E$	= related to points of excitation
$()_m$	= mean value
$()_{\text{res}}$	= resultant
$()_S$	= related to the S-coordinate system
$()_w$	= related to the amplitude w
$()_Z$	= related to the Z-coordinate system
$()_{\eta,\zeta}$	= in η, ζ-direction
$()_0$	= related to the point 0

1. Introduction

Anderson [1] investigated mistuned periodic systems with chain structure. He was the first in finding localization phenomena. Since then many papers have been published on this topic, i.e. [2, 3, 10, 13]. In [13], Pierre gives a review on localization phenomena. The so-called localization factor indicates the exponential decay of the amplitudes of a system with chain structure. Localization phenomena have been also investigated in coupled bladed disk assemblies, see [8, 9, 20, 21]. To be able to use the localization factor for a cyclic structure, the system is uncoupled at one connection. Assuming an infinite number of blades, the localization factor can be calculated. However, if the system is coupled with the disk, this approach is not applicable. In [20, 21], Wei and Pierre investigated the periodic structure of a bladed disk with shrouds and developed a measure for the strength of localization for a simplified model with chain structure of the system. The Gaussian distribution of the system response holds only for very small parameter uncertainties, see [16]. In [11] the Weibull-distribution of the system response of a coupled bladed disk was verified by Monte-Carlo simulation.

In the present paper an approximate method to calculate the envelopes of the frequency response functions for mistuned systems with cyclic symmetry is developed. In principle, this method is based on a sensitivity analysis and the Weibull-distribution of the vibration amplitudes. From here, a measure for the strength of localization for mistuned systems with cyclic symmetry is developed. Furthermore, it is shown, that the approximate method holds for nonlinear systems as well. The proposed method is described using a bladed disk assembly coupled by means of friction dampers.

2. System Description

Friction dampers are widely used to reduce dynamic stresses in turbine blades. In terms of efficiency, such damping devices should be placed outside the hot gas path underneath the blade platforms, between adjacent blades, see Figure 1. The major problem is still the calculation and the optimization of the three-dimensional dynamics of bladed disk assemblies with friction dampers. The description of statics and dynamics of geometric complex structure, including the effects of temperature, centrifugal forces and stationary gas forces is possible by means of the Finite Element Method. In case of a tuned bladed disk, cyclic boundary condition can be used for the calculation of the dynamic response. For a mistuned bladed disk, the complete system has to be modeled and solved, which leads to an enormous increase in computation time. To reduce the computation time the so-called substructure method or component mode synthesis, see [5, 6], is applied. This method is based on the modal description of each substructure, combined with a reduction of the degrees of freedom, to describe the dynamics of each component. The three-dimensional dynamical behavior of the disk, as well as the blades are considered and described by the eigenvectors, eigenfrequencies and modal damping. The amplitudes of the excitation forces of the system can be approximated in a common way, using the stationary gas force and introducing the stimulus, which describes the percentage of the excitation force amplitudes relative to the stationary gas force. Furthermore, it is assumed, that the excitation forces are monofrequent, while the excitation angular frequency is a multiple of the angular velocity of the rotor described by the engine order. The blade vibration is described in the Z-coordinate system, see Figure 1, which is fixed to the root of each blade. The y-axis of the S-coordinate system is parallel to the axis of the friction elements. The transformation from the Z- into the S-coordinate system is defined by the foot angle β. The system rotates with the angular velocity Ω. For modeling the line contact between platforms and friction elements, the Hertzian theory can be used. Examples of platform design and friction elements with a Hertzian line contact are shown in Figure 2. For example, in Figure 2a the friction element with radius R_1 is pressed onto the platforms with radius R_2, because of the centrifugal forces. Figures 2b to 2f show further possibilities of design. The advantage of friction damper with Hertzian line contacts is, that impacts between platform and friction element do not occur for a wide range of excitation forces, while for wedge or cottage roof dampers, see [22, 23], impacts can occur. To be able to calculate the friction forces, the relative displacement of the contact line needs to be known. This relative displacement is dependent on the displacement of the platforms and the friction element. Sanliturk [14] and Sextro et al. [15, 16], have developed a two-dimensional point contact model based on dry friction, contact elasticity and elliptical trajectories using the Harmonic Balance

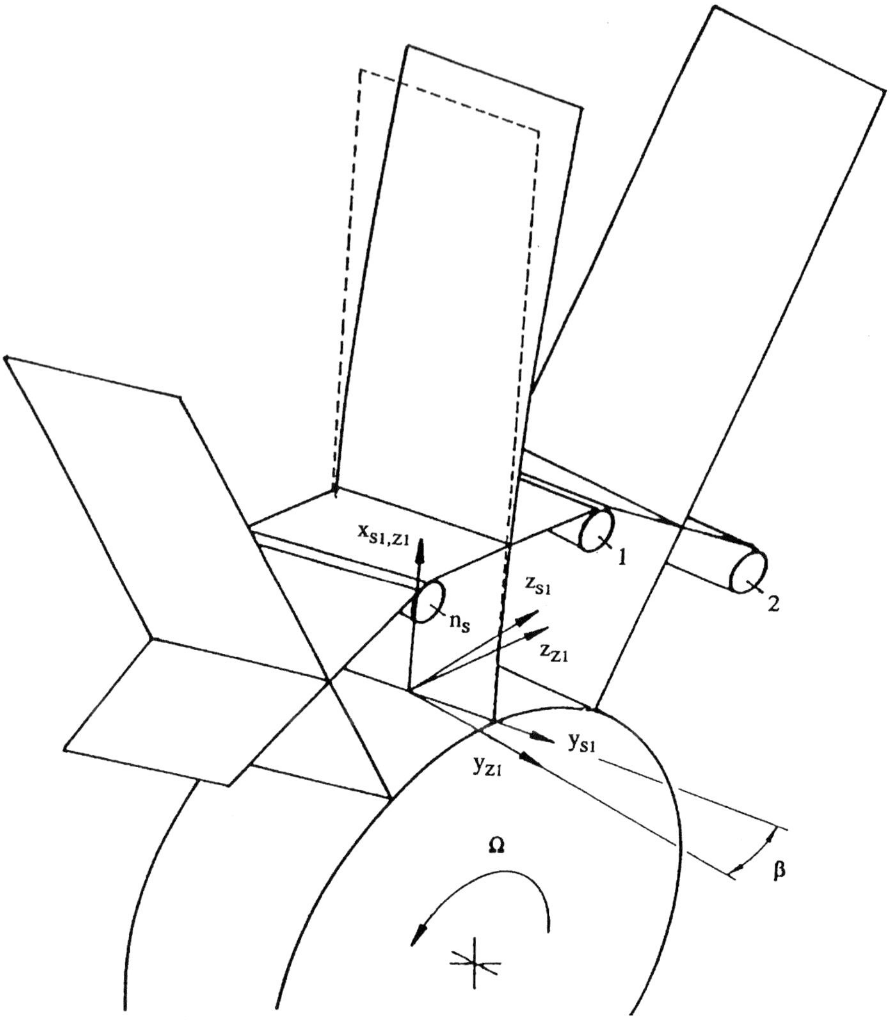

Figure 1. Bladed disk assembly with friction dampers.

Method. This point contact model is applied to a Hertzian line contact including translation and rotation of the contact area. Assuming a Hertzian line contact with the length ℓ_K, see Figure 3, the theory of Hertz can be used for calculating the width $2a_Z$ of the contact area and the contact pressure p_H [7]. For simplicity, the friction dampers are modeled as cylinders with radius R and the platforms as planes, see Figure 2b. The plane motion of the friction damper can be described in the η, ζ-coordinate system, which is connected to the platforms. The origin 0 of the coordinate system is placed in the center of the contact area. The contact area is discretized and the point contact model is used to describe the forces in each area element. Because of simplicity, the complex notation is used to describe the monofrequent motions. In [15] the complex stiffnesses are calculated. In Figure 4 the stiffness (real part) and the damping (imaginary part) for translation are shown, using the following dimensionless contact stiffness and relative displacement in the η-direction of the point 0,

$$\hat{K}_0^* = \frac{\hat{K}_0}{c_R}, \quad u_0^* = \frac{c_R u_0}{\mu F_N}. \tag{1}$$

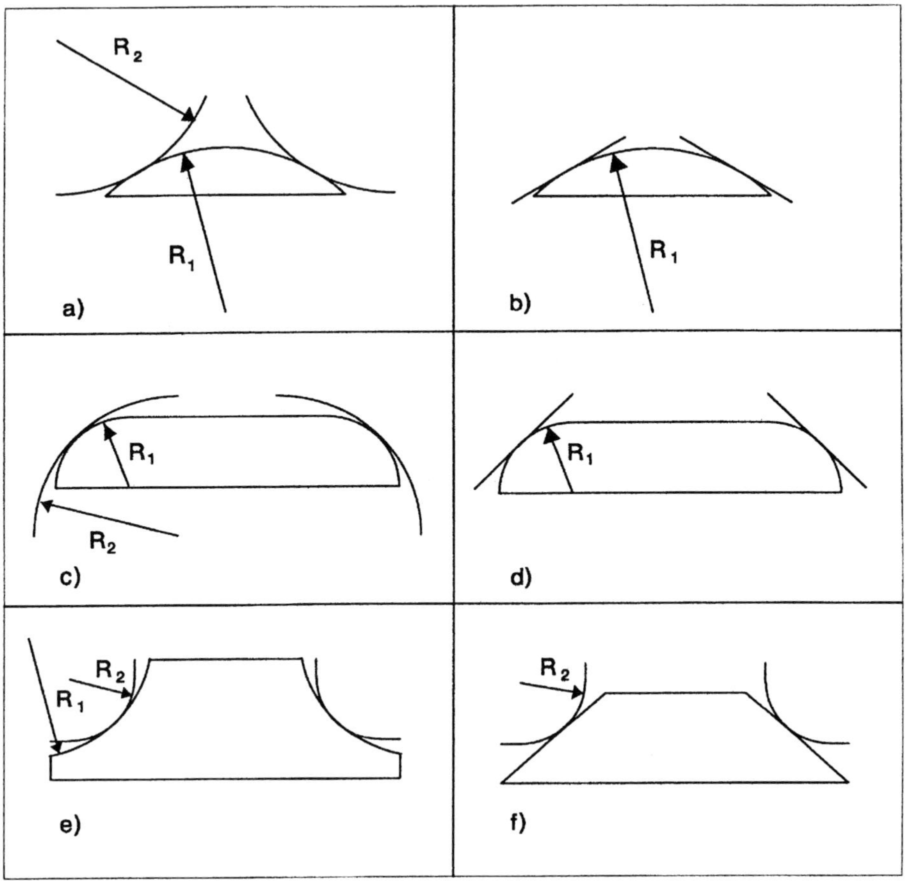

Figure 2. Friction damper with Hertzian line contacts.

The parameters of the contact model are as follows: the global tangential stiffness c_R, the friction coefficient μ and the average normal force F_N,

$$F_N = \frac{F_H}{2\cos\gamma}, \tag{2}$$

which is proportional to the centrifugal force of the friction element

$$F_H = m_R r_C \Omega^2. \tag{3}$$

Knowing the friction forces and moments which act onto the friction element, the reaction forces and moments in the contact area can be calculated from the equilibrium condition of the friction element. This leads to nonlinear coupled equations for modeling the bladed disk assemblies with friction contacts, using the modal description of the elastic components, see [17],

$$\mathbf{A}(\mathbf{u})\mathbf{q} = \mathbf{T}_E^T \mathbf{F}_E, \tag{4}$$

with the system matrix $\mathbf{A}$ of the bladed disk including friction dampers, the vector of modal coordinates $\mathbf{q}$, the modal matrix $\mathbf{T}_E$, the vector $\mathbf{F}_E$ of generalized excitation forces and the relative displacement vector $\mathbf{u}$ calculated by

$$\mathbf{u} = \mathbf{B}\mathbf{q}. \tag{5}$$

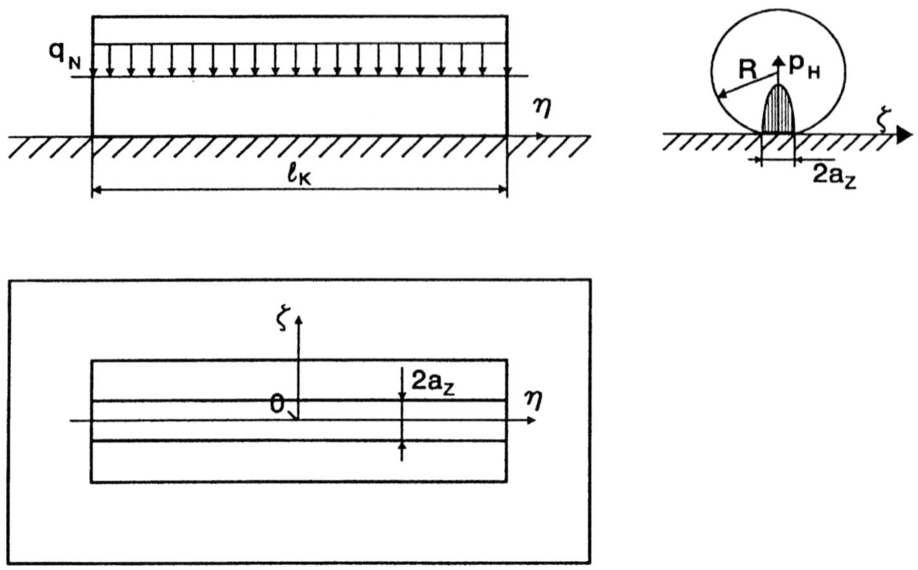

Figure 3. Cylindrical friction damper.

In the case of a tuned system these equations can be reduced to one segment using cyclic boundary conditions. Because of the nonlinearity they must be solved iterative, e.g. by means of the damped Newton–Raphson algorithm. The amplitudes of the system response can be calculated by the modal transformation

$$\mathbf{w} = \mathbf{Tq}, \tag{6}$$

where

$$\mathbf{w}^T = [w^r \;\; w^i]. \tag{7}$$

Then, the maximum amplitude w is given by

$$w = \sqrt{w^{r2} + w^{i2}}. \tag{8}$$

3. Tuned System Response

The disk with $n_S = 25$ blades, which are coupled by means of the friction dampers ($c_R = 86.2 \times 10^6$ N/m, $\mu = 0.33$, $\gamma = 34.0°$, $r_C = 0.25$ m, $\beta = 0°$) is investigated for plane motion. The excitation force at the tip of the blade is 10.0 N and the engine order is one. In Figure 5, the response of the blade tip as a function of the excitation frequency is shown for different masses of the friction elements. For the mass of $m_R = 30.0$ g the maximum amplitude in the z-direction is minimal, see also [4]. The resonance frequencies for different engine orders are shown in Figure 6. The resonance frequencies increase with increasing engine order, because the relative displacements between friction damper and platform and, hence, the contact forces increase. In case of the mistuned system response, these resonance frequencies will occur again. Furthermore, they indicate, where the system behavior is relatively sensitive to parameter uncertainties.

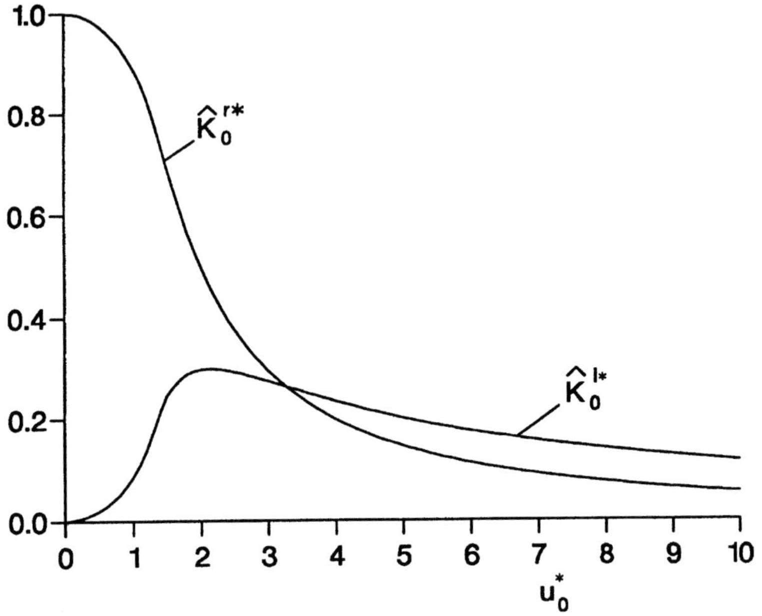

Figure 4. Stiffness and damping *versus* relative amplitude.

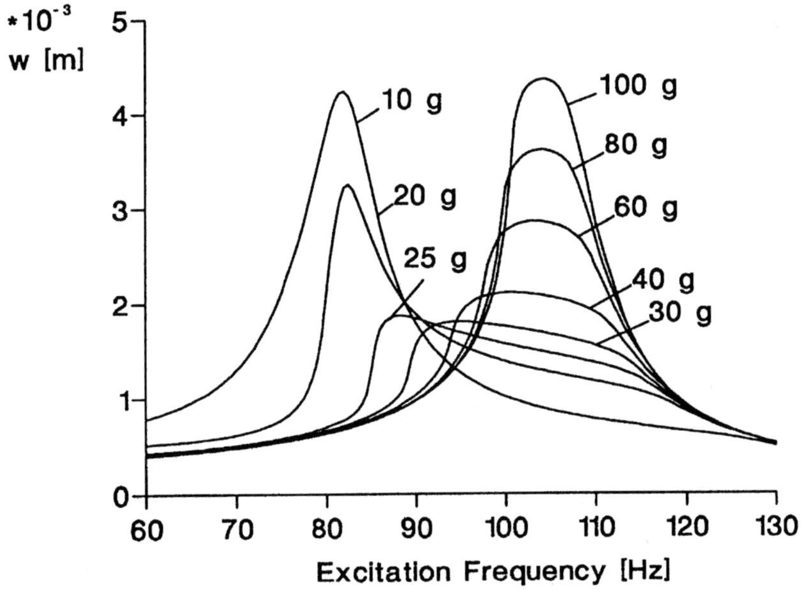

Figure 5. Amplitude in the z-direction *versus* excitation frequency.

4. Approximated Analysis of the Mistuned System

Because of large computation time of Monte-Carlo simulations, an approximate method is developed to calculate the system response. In the following this approximate method will be derived assuming small standard deviations of the eigenfrequencies of the blades. The approximate method is an extension of the theory described in [18, 19]. Expanding Equation (8) into a Taylor series with respect to a parameter p yields

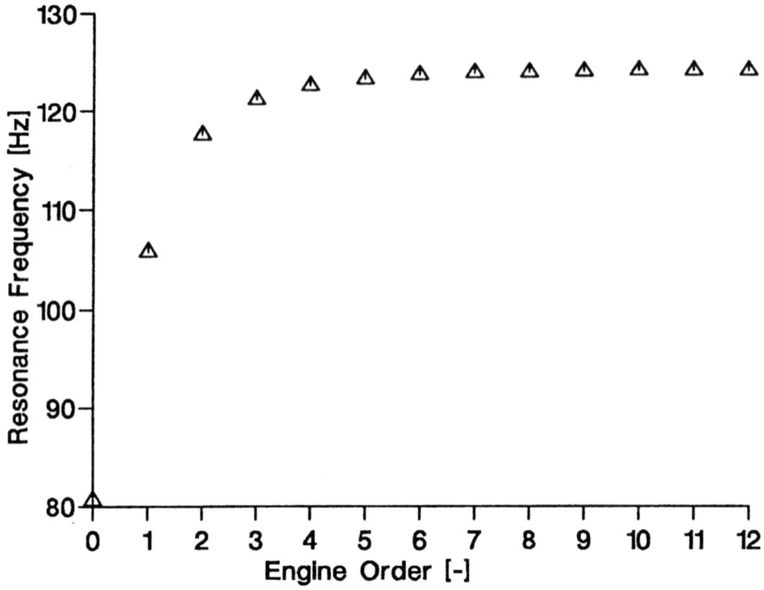

Figure 6. Resonance frequency *versus* engine order.

$$w(p) \approx w(p_m) + \left[\frac{\partial w(p)}{\partial p}\right]_{p=p_m} (p - p_m), \tag{9}$$

where

$$\frac{\partial w(p)}{\partial p} = \frac{1}{w(p)} \mathbf{w}^T(p) \frac{\partial \mathbf{w}(p)}{\partial p}. \tag{10}$$

Differentiating the system equation given by Equation (4) yields

$$\frac{\partial \mathbf{q}(p)}{\partial p} = -\mathbf{A}^{-1}(p) \frac{\partial \mathbf{A}(p)}{\partial p} \mathbf{q}(p). \tag{11}$$

Then the sensitivity S is given by

$$S(p_m) = \left[\frac{\partial w(p)}{\partial p}\right]_{p=p_m} = -\frac{1}{w(p_m)} \mathbf{w}^T(p_m) \mathbf{T} \mathbf{A}^{-1}(p_m) \left[\frac{\partial \mathbf{A}(p)}{\partial p}\right]_{p=p_m} \mathbf{q}(p_m). \tag{12}$$

The mean value and the standard deviation of the parameter p are assumed to be

$$E[p - p_m] = 0, \quad E[(p - p_m)^2] = \sigma_p^2. \tag{13}$$

If the standard deviation of the parameter p is small, then the mean value of the mistuned system response is the response of the tuned system

$$E[w(p)] = w(p_m) = w_m. \tag{14}$$

The standard deviation of the system response can be calculated using the sensitivity S by

$$\sigma_w^2 = S^2(p_m) \sigma_p^2. \tag{15}$$

In the case of a mistuned bladed disk, all sensitivities due to the eigenfrequencies of each blade have to be calculated. The cross correlation of the eigenfrequencies can be neglected. Hence, the resultant standard deviation of the system response is given by

$$\sigma_w^2 = \sum_{j=1}^{n_M} \sum_{i=1}^{n_S} S^2(\omega_{0i,jm}) \sigma_{\omega_{0i,j}}^2 . \tag{16}$$

To be able to calculate the envelopes of a mistuned system, the distribution of the amplitudes has to be known. In [11] it was shown, that the response of a coupled bladed disk follows a Weibull-distribution. The Weibull probability distribution, see [12], is given by

$$P_p(w^*) = 1 - \mathrm{e}^{-(\ln 2) w^{*k}} \tag{17}$$

with the dimensionless amplitude

$$w^* = \frac{w}{w_0}. \tag{18}$$

Both, the exponent k and the parameter w_0 are unknown and are calculated in the following by means of the mean value w_m and the standard deviation σ_w. The corresponding probability density function is given by differentiating Equation (17) to obtain

$$p_p(w^*) = k(\ln 2) w^{*k-1} \, \mathrm{e}^{-(\ln 2) w^{*k}} . \tag{19}$$

The nth moment can be calculated by

$$E[w^{*n}] = (\ln 2)^{-n/k} \Gamma\left(1 + \frac{n}{k}\right). \tag{20}$$

The standard deviation follows from

$$\sigma_{w^*}^2 = E[w^{*2}] - E[w^*]^2. \tag{21}$$

Inserting Equation (20) with $n = 1$ for the first moment and $n = 2$ for the second moment in Equation (21) and solving for the parameter w_0 gives

$$w_0 = \frac{(\ln 2)^{1/k}}{\sqrt{\Gamma\left(1 + \frac{2}{k}\right) - \Gamma^2\left(1 + \frac{1}{k}\right)}} \sigma_w . \tag{22}$$

Using Equation (20) with $n = 1$ for the mean value and inserting Equation (22) results in

$$w_m = \frac{\Gamma\left(1 + \frac{1}{k}\right)}{\sqrt{\Gamma\left(1 + \frac{2}{k}\right) - \Gamma^2\left(1 + \frac{1}{k}\right)}} \sigma_w . \tag{23}$$

Rearranging Equation (23) gives a formula to determine the exponent k of the Weibull-distribution

$$1 + s^2 - \frac{\Gamma\left(1 + \frac{2}{k}\right)}{\Gamma^2\left(1 + \frac{1}{k}\right)} = 0, \tag{24}$$

knowing the measure for sensitivity or strength for localization

$$s = \frac{\sigma_w}{w_m}. \tag{25}$$

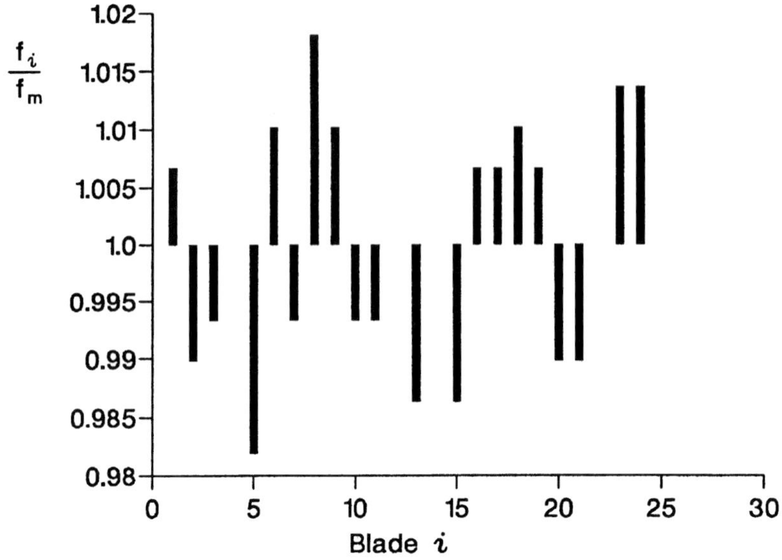

Figure 7. Eigenfrequencies of the blades (1. mode).

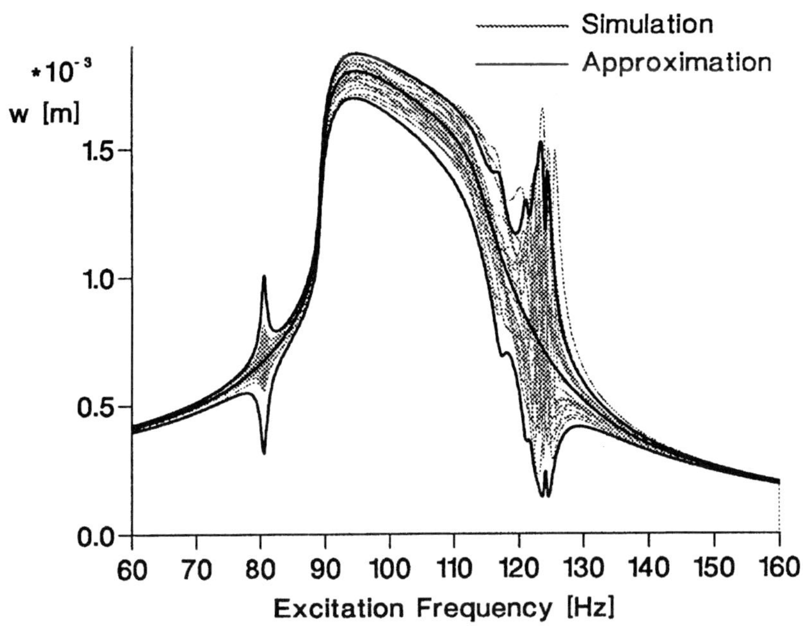

Figure 8. Response of the mistuned bladed disk ($\sigma_f = 1.0\%$).

For example, if this measure s is equal to one, the exponent k is also equal to one. For the Rayleigh distribution ($k = 2$) the measure is $s = 0.523$. The measure can be reduced to the strength for localization developed by Wei and Pierre [20, 21]. There are two possibilities to calculate the exponent k. On the one hand, the strength for localization can be calculated from the mistuned response of the system and on the other hand by the approximate method. This numerical comparison is done in the next section.

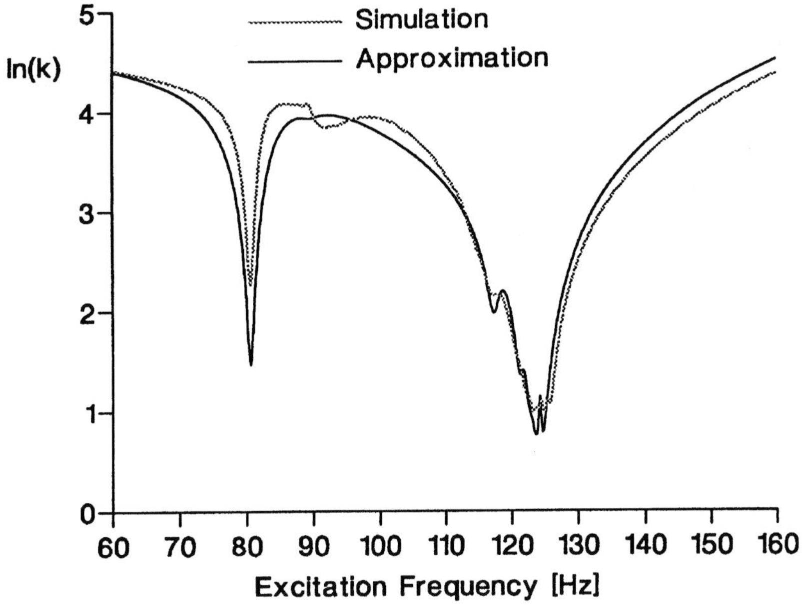

Figure 9. Comparison of the calculated exponents.

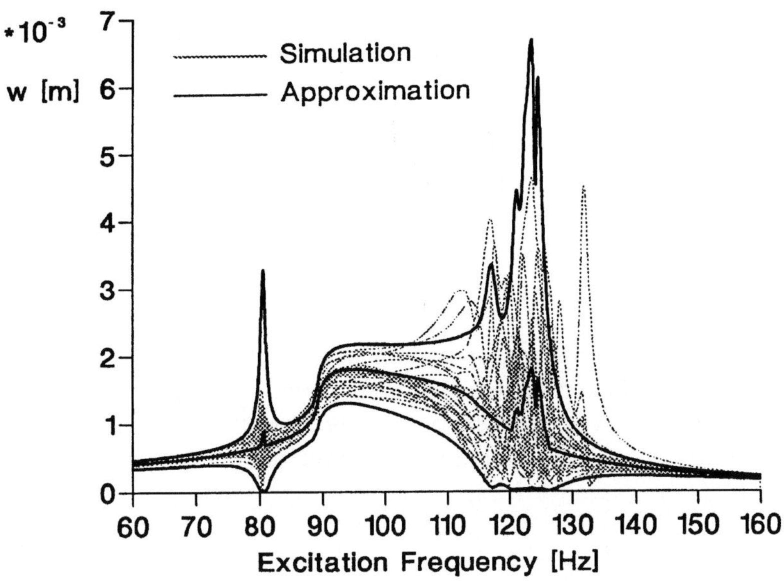

Figure 10. Response of the mistuned bladed disk ($\sigma_f = 5.0\%$).

5. Response of the Mistuned System

The bladed disk described above is now mistuned by the eigenfrequencies of the blades. The mean value of each mode corresponds to the eigenfrequencies of the tuned system and the standard deviation is 1.0% of the mean value for each of the first four modes ($n_M = 4$). For example, the blades are distributed statistically around the disk, with the eigenfrequencies for the first mode shown in Figure 7. The mass of the friction elements is the optimal

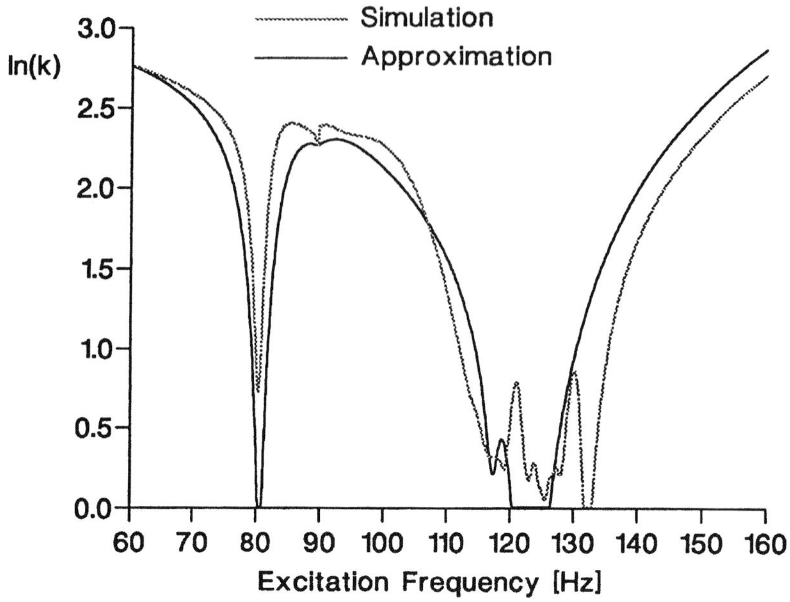

Figure 11. Comparison of the calculated exponents.

mass $m_R = 30.0$ g. The amplitudes in z-direction of the blades of the mistuned system are shown in Figure 8. The side peaks correspond to the resonance frequencies due to the tuned system for different engine orders, see Figure 6. Because of the large computation time of the simulation for one arrangement of the blades around the disk, the approximate method is used to calculate the mean value w_m and the standard deviation σ_w of the amplitudes. By means of this method the envelopes of the amplitudes can be calculated. With a probability of 95% the amplitudes are within these envelopes. Calculating the mean value and the standard deviation of the mistuned systems response, the exponent k can be calculated as well, using Equation (24). In Figure 9 the exponent k is illustrated for the simulated system response and for the approximation. The agreement of both methods is very good. If the standard deviation of the eigenfrequencies for each mode is increased from 1.0 up to 5.0%, see Figure 10, then the amplitudes increase as well. The exponent k reaches the limiting value, see Figure 11, where the probability, that blades vibrate with a large amplitude and localization occurs, is very high. Hence, the measure for localization or the exponent k of the Weibull-distribution can be used to find regions where localization occurs with a high probability. The differences between simulations and the approximation are due to the assumptions in the approximate method. However, the approximate method holds for all arrangements of mistuned blades distributed around the disk. It appears, that one peak is outside of the envelopes calculated by the approximate method at an excitation frequency of 132.2 Hz. The amplitudes of each blade are shown in Figure 12. The blade number with the maximum response is 23. The vibration energy is concentrated in this blade. Hence, localization occurs at this excitation frequency.

6. Conclusion

Different methods have been presented for calculating the envelopes of the frequency response functions for mistuned systems with cyclic symmetry. These methods have been applied to a

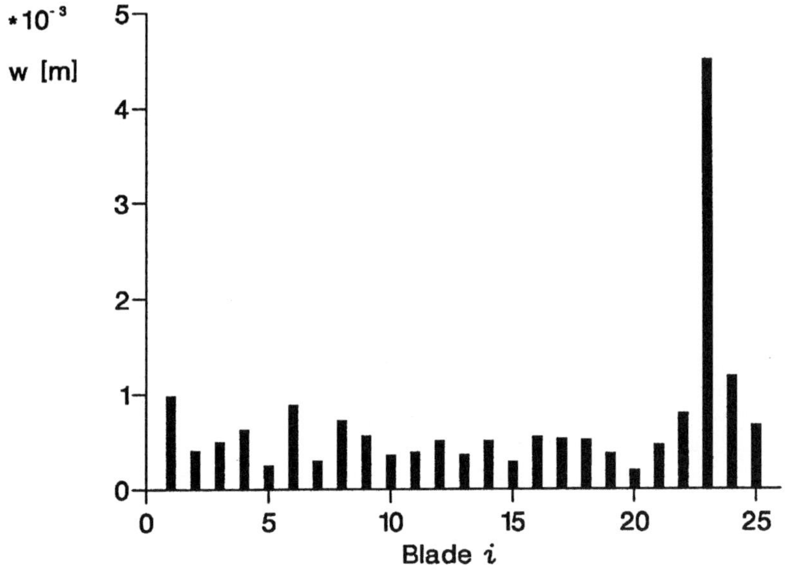

Figure 12. Amplitudes *versus* blade number.

bladed disk assembly with friction dampers. Because of dry friction, the system behavior is nonlinear. The mean value and the standard deviation of the vibration amplitudes can be calculated by an approximate method. Using the sensitivity analysis and the Weibull- distribution, the envelopes of the frequency response function have been determined. A measure for sensitivity or strength for localization has been derived based on a sensitivity analysis and the Weibull-distribution. Regions, where localization occurs with a high probability, can be evaluated by this measure or by the Weibull exponent, because both parameter are related to each other. The comparison between the approximate method and the simulation shows a very good agreement. Therefore, by means of this method the computation time can be reduced dramatically and gives a quick insight into the behavior of mistuned systems with parameter uncertainties and can be also applied to systems, which include the elasticity of the disk [16], and bladed disk assemblies with shrouds [17].

References

1. Anderson, P. W., 'Absence of diffusion in certain random lattices', *Physical Review* **109**(5), 1958, 1492–1505.
2. Ariaratnam, S. T., 'Localization in randomly disordered periodic structures', in *Proceedings of the IUTAM Symposium on Nonlinear Stochastic Dynamic Engineering Systems*, Springer-Verlag, Berlin, 1988, pp. 125–133.
3. Cai, C. W., Cheung, Y. K., and Chan, H. C., 'Mode localization phenomena in nearly periodic systems', *ASME, Journal of Applied Mechanics* **62**, 141–149.
4. Cameron, T. M., Griffin, J. H., Kielb, R. E., and Hoosac, T. M., 'An integrated approach for friction damping design', ASME Bound Volume DE-5, ASME, New York, 1987, pp. 205–212.
5. Hurty, W. C., 'Vibrations of structural systems by component mode synthesis', *ASCE, Journal of Engineering Mechanics Division* **86**, 1960, 51–69.
6. Hurty, W. C., 'Dynamic analysis of structural systems using component modes', *AIAA Journal* **3**(4), 1965, 678–685.
7. Johnson, K. L., *Contact Mechanics*, Cambridge University Press, Cambridge, 1989.

8. Kruse, M. J. and Pierre, C., 'An experimental investigation of vibration localization in bladed disks, Part I: Free response', in *Proceedings of the International Gas Turbine & Aeroengine Congress & Exhibition*, ASME 97-GT-501, Orlando, FL, June 2–5, ASME, New York, 1997.
9. Kruse, M. J. and Pierre, C., 'An experimental investigation of vibration localization in bladed disks, Part II: Forced response', in *Proceedings of the International Gas Turbine & Aeroengine Congress & Exhibition*, ASME 97-GT-502, Orlando, FL, June 2–5, ASME, New York, 1997.
10. Lin, Y. K., 'Dynamics of disordered periodic structures', *ASME, Applied Mechanics Reviews* **49**(2), 1996, 57–64.
11. Mignolet, M. P., Hu, W., and Jadic, I., 'On the forced response of harmonically and partially mistuned bladed disks, Part II: Partial mistuning and applications', in *Proceedings of the 7th International Symposium on Transport Phenomena and Dynamics of Rotating Machinery*, Vol. B, Honolulu, HI, Bird Rock Publishing House, Minden, USA, 1998, pp. 603–613.
12. Newland, D. E., *An Introduction to Random Vibrations and Spectral Analysis*, second edition, Longman Scientific & Technical, New York, 1986.
13. Pierre, C., 'Wave localization in multi-coupled periodic structures: Application to truss beams', *ASME, Applied Mechanics Reviews* **49**(2), 1996, 65–86.
14. Sanliturk, K. Y. and Ewins, D. J., 'Modeling two-dimensional friction contact and its application using harmonic balance method', *Journal of Sound and Vibration* **193**(2), 1996, 511–523.
15. Sextro, W., Popp, K., and Wolter, I., 'Three-dimensional vibration of bladed disk assemblies with friction dampers', in *Proceedings of the 7th International Symposium on Transport Phenomena and Dynamics of Rotating Machinery*, Vol. B, Honolulu, HI, Bird Rock Publishing House, Minden, USA, 1998, pp. 621–630.
16. Sextro, W., Popp, K., and Wolter, I., 'Improved reliability of bladed disks due to friction dampers', in *Proceedings of the International Gas Turbine & Aeroengine Congress & Exhibition*, ASME 97-GT-189, Orlando, FL, June 2–5, ASME, New York, 1997.
17. Sextro, W., 'Forced vibration of elastic structures with friction contact', in *Proceedings of the 1999 ASME Design Engineering Technical Conferences*, DETC99/VIB-8180, Las Vegas, NV, September 12–15, ASME, New York, 1999.
18. Sinha, A., 'Calculating the statistics of forced response of a mistuned bladed disk assembly', *AIAA Journal* **24**, 1986, 1797–1801.
19. Sinha, A. and Chen, S., 'A higher order technique to compute the statistics of forced response of a mistuned bladed disk', *Journal of Sound and Vibration* **130**, 1989, 207–221.
20. Wei, S. T. and Pierre, C., 'Localization phenomena in mistuned assemblies with cyclic symmetry, Part I: Free vibrations', *Journal of Vibration, Acoustic, Stress, and Reliability in Design* **110**(4), 1988, 429–438.
21. Wei, S. T. and Pierre, C., 'Localization phenomena in mistuned assemblies with cyclic symmetry, Part II: Forced vibrations', *Journal of Vibration, Acoustic, Stress, and Reliability in Design* **110**(4), 1988, 439–449.
22. Yang, B. D. and Menq, C. H., 'Characterization of contact kinematics and application to the design of wedge dampers in turbomachinery blading, Part I: Stick-slip contact kinematics', in *Proceedings of the International Gas Turbine & Aeroengine Congress & Exhibition*, ASME 97-GT-19, Orlando, FL, June 2–5, ASME, New York, 1997.
23. Yang, B. D. and Menq, C. H., 'Characterization of contact kinematics and application to the design of wedge dampers in turbomachinery blading, Part II: Prediction of forced response and experimental verification', in *Proceedings of the International Gas Turbine & Aeroengine Congress & Exhibition*, ASME 97-GT-20, Orlando, FL, June 2–5, ASME, New York, 1997.

Nonlinear Dynamics **25:** 221–236, 2001.

Mode Localization Induced by a Nonlinear Control Loop

ROBERT T. M'CLOSKEY
Mechanical and Aerospace Engineering, University of California, Los Angeles, CA 90095, U.S.A.

ALEX VAKAKIS
Mechanical and Industrial Engineering, University of Illinois at Urbana-Champaign, 138 Mech. Eng., 1206 W. Green Street, Bldg. MC244, Urbana, IL 61801, U.S.A.

ROMAN GUTIERREZ
Jet Propulsion Laboratory, 4800 Oak Grove Dr., Pasadena, CA 91109, U.S.A.

(Received: 7 April 1999; accepted: 19 January 2000)

Abstract. We present the analysis of a nonlinear control system that is used to excite and maintain a specified amplitude of oscillation in the Jet Propulsion Laboratory vibratory gyroscope. This experimental application shows that nonlinear localization through active means can be implemented in a practical system when it is desirable to confine the response to a favorable mode. The closed-loop system response predicted by the model shows very close agreement with the experimental results for a significant range of controller parameters. We also experimentally demonstrate that the actively localized motion is eliminated through bifurcation, similar to what was observed in previous passive localization studies applied to extended flexible oscillators.

Keywords: Nonlinear localization, nonlinear control, automatic gain control, vibratory gyroscope.

1. Introduction

We consider active nonlinear localization of energy in an oscillator. Our main application is the micromachined vibratory rate sensor designed by the Jet Propulsion Laboratory (JPL) for future spacecraft missions [1, 2]. This sensor's mechanical structure possesses three dominant elastic modes representing translational motion and rotation of the micro-structure. Proper sensor operation requires that one of the rotational degrees of freedom be excited to a periodic oscillation with a stable, specified amplitude. We approach this problem by designing a nonlinear feedback control loop, termed Automatic Gain Control (AGC), that selectively localizes energy into the desired mode. We demonstrate that by proper design of the AGC, a single-degree-of-freedom (SDOF) oscillator model driven by the control loop accurately describes this active nonlinear localization phenomenon in the physical device. The SDOF case is thoroughly analyzed using standard averaging and phase-plane techniques. Simulations of the averaged equations show excellent agreement with reality, especially given the complexity of the physical device and the fact that the nonlinear controller is implemented on a microprocessor and is governed by discrete-time equations.

Passive nonlinear localization was extensively studied in previous works for discrete [3, 4] and continuous systems [3, 5, 6]. In these works it was shown that weakly coupled nonlinear oscillators can possess strong and weak localized modes corresponding to energy confinement in one or more substructures of the system. In contrast to the well known linear (Anderson) localization, nonlinear localization occurs even in the absence of structural disorder. The generating mechanisms for localization in spatially periodic systems were also studied in previous works. These can either be mode bifurcations, or smooth transitions of non-localized branches of nonlinear modes. To the authors' knowledge, few previous studies exist addressing energy localization in systems with active nonlinear control. In two such works [7, 8], nonlinear motion confinement induced by active nonlinearities was investigated in a system of coupled beams. These works demonstrated that active control can be used to 'alter' the linear dynamics of a structure and induce nonlinear confinement phenomena that are favorable to the design objectives.

The structure of the paper is as follows. In Section 2 we introduce and analyze using the method of averaging, the SDOF oscillator with automatic gain control. The AGC can be physically implemented using rectifiers, low-pass filters, and comparators and this simplicity make it attractive for use in applications such as the JPL sensor. A description of the sensor is given in Section 3.1 and experimentally identified models of the sensor and a simplified SDOF model are presented in Section 3.2. This section also presents experimental results of active localization in the sensor and makes a comparison with theoretical predictions based on the SDOF model. Excellent agreement is observed, validating the SDOF simplification. Limitations of the SDOF model, however, are revealed in Section 3.3 through additional experiments where quasi-periodic response of the closed-loop system is observed after a modest increase in the sensor transfer function gain. Indeed, these experiments show what appears to be a Hopf bifurcation that destabilizes the localized motion and gives rise to nonlocalized multi-mode response. This behavior is not reproducible with the SDOF model and motivates the need for developing multi-degrees-of-freedom models to provide a more thorough understanding of the sensor/AGC dynamics. Section 4 discusses the predictive qualities of the SDOF model and proposes some topics for further investigation.

2. Analysis of an Oscillator with Automatic Gain Control

This section presents the analysis of an AGC loop regulating a single-degree-of-freedom oscillator. The operation of the AGC is intuitive and is explained with the assistance of Figure 1. In this configuration, the velocity of the oscillator is applied to the oscillator input after being multiplied by a gain so in essence the AGC manipulates the damping of the oscillator: if the desired velocity amplitude is less than the *reference amplitude* then the multiplicative gain is chosen to be negative in order to dampen the oscillations; conversely, if the reference amplitude is larger than the current amplitude then the AGC gain is positive and the 'anti-damping' causes the oscillation to grow. This idea forms the basis of a simple automatic gain control, however, implementation in the JPL sensor requires additional compensation as discussed in Section 3.1.

The equations governing the oscillator/AGC system are

$$\begin{aligned} \ddot{x} + c\dot{x} + \Omega^2 x &= K(r - z)\alpha\dot{x}, \\ \dot{z} &= \lambda(|\alpha\dot{x}| - z), \end{aligned} \tag{1}$$

where x is the oscillator degree of freedom, and z is the state of the low-pass filter which estimates the amplitude of a scalar multiple of the oscillator velocity $\alpha\dot{x}$ (the filter has corner frequency λ). The AGC gain is simply the difference between the reference amplitude, r, and filter state, z, multiplied by the gain K.

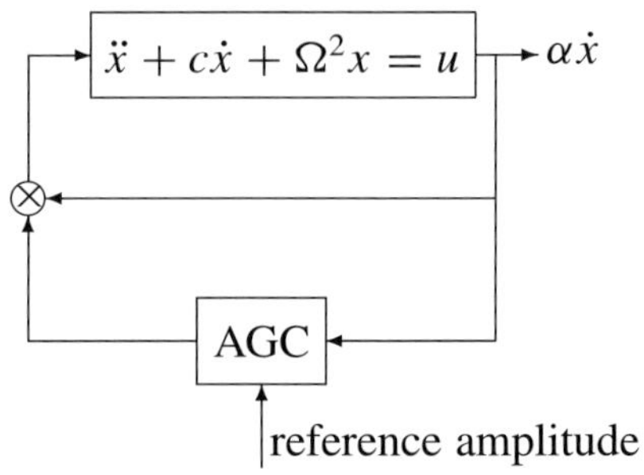

Figure 1. Automatic gain control loop.

Though this model is a SDOF oscillator with essentially nonlinear damping, the control method is applied to the JPL rate sensor that has three modes that dominate the frequency response. By proper design of linear filters in series with the sensor's inputs and outputs, the AGC excites only the selected mode, thereby localizing the energy in this mode. The analysis of the SDOF oscillator model is shown to predict with great accuracy the response of the physical system in most cases.

Exact analysis of Equation (1) is difficult because of the nonlinear amplitude detection. Perturbation analysis using periodic averaging, however, leads to simplified nonlinear differential equations. The separation of time scales between the oscillator dynamics and the low-pass filter and AGC gain may be used as the perturbation parameter. The analysis proceeds in several steps. First, a change of coordinates is performed, then the equations are averaged to remove the fast time scale and, lastly, the approximate system of equations is analyzed.

A coordinate change transforms the oscillator states x and $\dot{x}$ into slowly varying amplitude and phase variables. Define

$$x(t) := a(t)\cos(\Omega t + \phi(t)), \tag{2}$$

where a and ϕ are amplitude and phase functions, respectively. Differentiating Equation (2) with respect to t yields

$$\dot{x} = -a\Omega\sin(\Omega t + \phi) + \dot{a}\cos(\Omega t + \phi) - a\dot{\phi}\sin(\Omega t + \phi). \tag{3}$$

The sum of last two terms in this expression is set to zero and is used in determining the equations for a and ϕ:

$$\dot{a}\cos(\Omega t + \phi) - a\dot{\phi}\sin(\Omega t + \phi) \equiv 0. \tag{4}$$

Differentiate Equation (3) once more (keeping in mind Equation (4)):

$$\ddot{x} = -\dot{a}\Omega\sin(\Omega t + \phi) - a(\Omega + \dot{\phi})\cos(\Omega t + \phi). \tag{5}$$

Substituting Equations (2), (3) and (5) into the oscillator equation yields

$$\big(-\Omega\dot{a} - \Omega a(c - \alpha K(r - z))\big)\sin(\Omega t + \phi) + \big(-\Omega a(\Omega + \dot{\phi}) + \Omega^2 a\big)\cos(\Omega t + \phi) = 0 \tag{6}$$

Differential equations for a and ϕ are obtained by pre-multiplying $[(4)\ \ (6)]^T$ by the non-singular matrix

$$\begin{bmatrix} \cos(\Omega t + \phi) & -\frac{1}{\Omega}\sin(\Omega t + \phi) \\ \frac{1}{\Omega}\sin(\Omega t + \phi) & \frac{1}{\Omega^2}\cos(\Omega t + \phi) \end{bmatrix}.$$

Simplifying the equations after this step and including the equation for z yields the system

$$\begin{aligned}
\dot{a} &= [\alpha K(r-z)-c]a\sin^2(\Omega t+\phi), \\
\dot{\phi} &= [\alpha K(r-z)-c]\sin(\Omega t+\phi)\cos(\Omega t+\phi), \quad a\neq 0, \\
\dot{z} &= \lambda\left(|-\alpha\Omega a\sin(\Omega t+\phi)|-z\right). \qquad (7)
\end{aligned}$$

No approximations are made for these equations. We may apply averaging [9] by changing the time scale to $\tau = \Omega t$ and noting that vector field is 2π-periodic and scaled by a small parameter if the following relation is satisfied

$$\frac{K}{\Omega} \propto \frac{\lambda}{\Omega} \propto \frac{c}{\Omega} \ll 1. \qquad (8)$$

These conditions restrict the AGC gain K, low-pass filter corner frequency λ, and the damping c to be small compared to the oscillator frequency Ω. Since K and λ are parameters that the designer can specify, the analysis really hinges on the requirement that c/Ω be sufficiently small. Fortunately, this requirement is not difficult to satisfy in our application because the damping in sensor is very small. Thus, the quantitative behavior of Equation (7) can be described by the averaged system if Equation (8) is satisfied.

The averaged equations are simply

$$\begin{aligned}
\dot{\bar{a}} &= \frac{1}{2}[\alpha K(r-\bar{z})-c]\bar{a}, \\
\dot{\bar{\phi}} &= 0, \\
\dot{\bar{z}} &= \lambda\left(\frac{2}{\pi}\alpha\Omega\bar{a}-\bar{z}\right). \qquad (9)
\end{aligned}$$

The averaged variables are denoted with the bar. Note that $\bar{\phi}(t)$ is constant and may be set to zero for the analysis.

The equilibria of Equations (9) are

$$\begin{aligned}
\bar{a}_0 &= 0, \quad \bar{a}_1 = \frac{\pi}{2\Omega}\left(r-\frac{c}{K}\right), \\
\bar{z}_0 &= 0, \quad \bar{z}_1 = \frac{1}{\Omega}\left(r-\frac{c}{K}\right).
\end{aligned}$$

The Jacobian linearizations at these equilibria are

$$\begin{aligned}
D_x f(\bar{a}_0,\bar{z}_0) &= \begin{bmatrix} \frac{1}{2}(\alpha K r-c) & 0 \\ \frac{2}{\pi}\alpha\Omega\lambda & -\lambda \end{bmatrix}, \\
D_x f(\bar{a}_1,\bar{z}_1) &= \begin{bmatrix} 0 & -\frac{\pi K}{4\Omega}\left(r-\frac{c}{\alpha K}\right) \\ \frac{2}{\pi}\alpha\Omega\lambda & -\lambda \end{bmatrix}. \qquad (10)
\end{aligned}$$

We must assume $\alpha K r > c$ for perturbations near equilibrium $(\bar{a}_0, \bar{z}_0)$ to grow, i.e., the linearized system has one unstable eigenvalue and the AGC loop is excited if $\bar{a}(0) \neq 0$. This condition merely states that the anti-damping added by the AGC loop must overcome the natural damping present in the system.

The $(\bar{a}_1, \bar{z}_1)$ equilibrium point is stable when $\alpha K r > c$ since the eigenvalues of $D_x f(\bar{a}_1, \bar{z}_1)$ are

$$-\frac{\lambda}{2}\left(1 \pm \sqrt{1 - \frac{2}{\lambda}(\alpha K r - c)}\right).$$

In this case, the response of the linearized equations is critically damped with rate of convergence $\lambda/2$ when the system parameters satisfy $\lambda = 2(\alpha K r - c)$. The eigenvalues are complex conjugates when $\lambda < 2(\alpha K r - c)$ and oscillations in the amplitude are observed. The design parameters in the AGC are λ and K and these may be used to alter the response the system. Increasing K reduces the effect of the natural system damping in the steady state amplitude but produces a more oscillatory response. Note that the response depends on the reference amplitude r and a larger reference amplitude tends to cause oscillation.

A global picture of the $\bar{a}$-$\bar{z}$ dynamics is desirable. In particular it is of interest to determine the existence of limit cycles. A limit cycle indicates periodic solutions of $\bar{a}$ and this implies the existence of quasi-periodic motion of the oscillator. This situation is undesirable since a constant oscillator amplitude is required in applications such as the JPL rate sensor. We now show that limit cycles cannot exist for the averaged equations.

Recall that the sum of indices of equilibrium points enclosed by a limit cycle must be one [10]. This implies that a limit cycle, if it exists, must enclose the equilibrium point $(\bar{a}_1, \bar{z}_1)$. The closed set $Z =: \{(\bar{a}, \bar{z}) : \bar{a} \geq 0, \bar{z} \geq 0\}$ is invariant under the action of Equation (9) because $\dot{z} \geq 0$ when $z = 0$ (the vector field points into Z along the $\bar{a}$ axis, for $\bar{a} > 0$). Thus, any limit cycle that passes through part of Z must actually lie entirely in the interior of Z and enclose $(\bar{a}_1, \bar{z}_1)$. We discount this last possibility by applying Dulac's Criterion [11]. Define

$$\begin{aligned} f(\bar{a}, \bar{z}) &:= \frac{1}{2}\left[\alpha K(r - \bar{z}) - c\right]\bar{a}, \\ g(\bar{a}, \bar{z}) &:= \lambda\left(\frac{2}{\pi}\alpha\Omega\bar{a} - \bar{z}\right), \end{aligned}$$

i.e., $\dot{\bar{a}} = f$ and $\dot{\bar{z}} = g$, and let $\rho = \bar{a}^m$, where $m \in \mathbb{R}$ is chosen below. Dulac's Criterion states that if the sign of $(\partial(\rho f)/\partial\bar{a}) + (\partial(\rho g)/\partial\bar{z})$ is constant in a simply connected region of the phase space then limit cycles do not exist in the region. Computing this quantity yields

$$\begin{aligned} \frac{\partial(\rho f)}{\partial\bar{a}} + \frac{\partial(\rho g)}{\partial\bar{z}} &= \left[\frac{m+1}{2}(\alpha K(r - z) - c) - \lambda\right]\bar{a}^m \\ &= -\frac{m+1}{2}\alpha K\bar{z}\bar{a}^m \quad \text{when} \quad m := \frac{2\lambda}{\alpha K r - c} - 1 \\ &< 0 \quad \text{on the interior of } Z. \end{aligned}$$

Thus, the interior of Z cannot contain a limit cycle. Further arguments are necessary to determine whether $(\bar{a}_1, \bar{z}_1)$ is globally attractive (except when $\bar{a}(0) = 0$) but the fact that limit cycles cannot occur in this system is encouraging and shows that the AGC is robust to initial conditions – under practical circumstances the AGC loop always drives the oscillator to a constant amplitude.

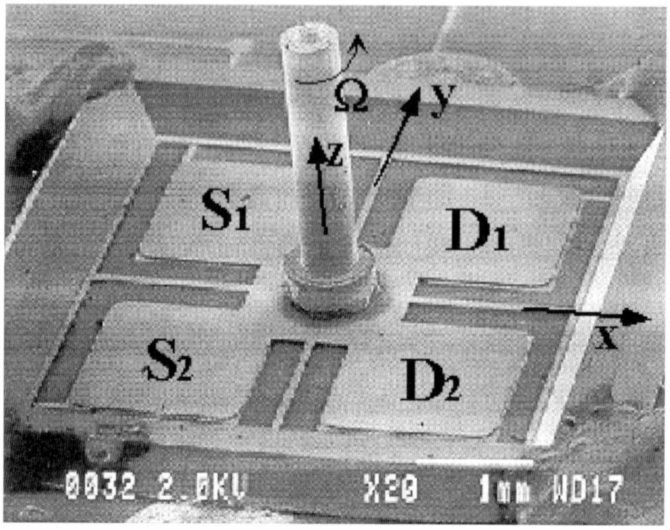

Figure 2. Photo of Jet Propulsion Laboratory microgyro with sense and actuator electrodes labeled (courtesy of T. Tang).

3. AGC Implementation in a Novel Angular Rate Sensor

3.1. The JPL Vibratory Gyroscope

The Jet Propulsion Laboratory micromachined gyroscope is an angular rate sensor whose operation depends on the coupling of energy between two degrees of freedom within the sensor via coriolis terms in a sensor-fixed frame. Vibratory 'gyros' such as the JPL instrument do not employ a spinning mass but rather require a selected degree of freedom in an elastic structure to be driven to a constant amplitude of periodic oscillation. An introduction to the principles of vibratory gyroscopes is covered in [12]. Like most vibratory gyros, the JPL prototype resides in a high vacuum to reduce dissipation and maximize the response of the mechanical structure to the relatively weak electrostatic excitation forces.

The task of driving a very lightly damped degree of freedom to a constant amplitude may be accomplished in many ways. The modal frequencies of the JPL sensor are temperature sensitive, however, so it is desirable to use a method which successfully excites the selected mode without requiring a precise estimate of the modal frequency. This is the primary motivation for using automatic gain control in the JPL sensor since it is essentially a nonlinear velocity feedback scheme that always excites an oscillator at its modal frequency.

The drive and sense electrodes of the JPL microgyro (denoted D_1 and D_2, and S_1 and S_2, respectively, in Figure 2) are suspended above four matching electrodes by silicon springs visible in the figure along the x and y axes. The sensor dynamics are excited by applying potentials between the drive electrodes and their corresponding base-plate electrodes. The potentials create an electrostatic force that pulls the opposing electrodes towards each other, thus rotating and translating the cloverleaf formed by D_1, D_2, S_1 and S_2.

The large central post, which is rigidly attached to the cloverleaf, adds inertia and increases sensor sensitivity. Displacements of the cloverleaf structure are measured using the S_1 and S_2 electrode pairs. The sensor dynamics are dominated by the three rigid body degrees of freedom of the cloverleaf corresponding to axial motion along the z-axis and rotation about the x- and

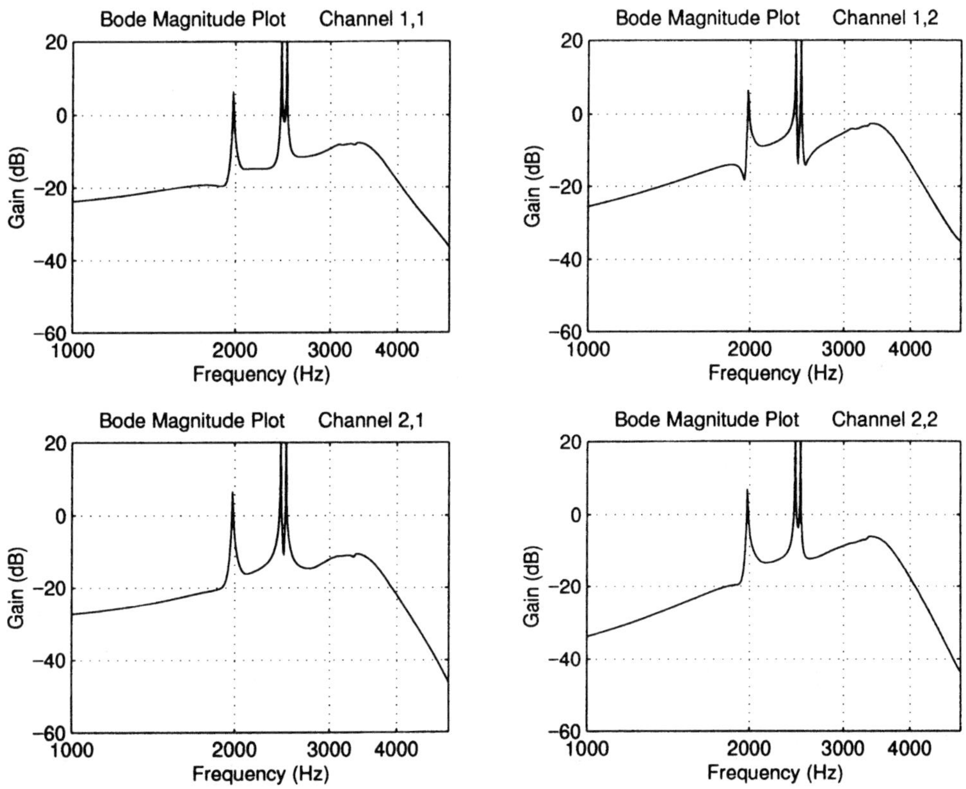

Figure 3. The four frequency responses from D_1 and D_2 to S_1 and S_2.

y-axes. The response amplitude is maximized by driving the sensor at the natural frequencies of these degrees of freedom.

The degree of freedom corresponding to rotation about the y-axis is referred to as the *drive mode*. The *sense mode* corresponds to rotation about the x-axis. A measurement proportional to the sense mode velocity is is provided by taking the difference of the S_1 and S_2 electrode voltages, which are related to the capacitance between the electrodes pairs. Similarly, a measurement proportional to the drive mode velocity is provided by summing the S_1 and S_2 voltage.

3.2. Experimental Results

The frequency response of a linear auto-regressive (ARX) model fit to input-output data taken from a JPL microgyro prototype is shown in Figure 3. Details of the ARX model identification procedure applied to this sensor may be found in [13, 14]. The four magnitude plots represent the frequency responses of the identified model from each drive input to each sense output. It is evident from these plots that there are three lightly damped modes in the sensor in the frequency range shown. The lowest frequency mode near 2000 Hz is the axial mode in which the post-cloverleaf assembly plunges up and down in the z-axis direction. The lower of the two high frequency modes corresponds to the drive mode (rotation about the y-axis in Figure 2) and is denoted $\Omega_{drv} \approx 2431$ Hz. The remaining high frequency mode, denoted $\Omega_{sen} \approx 2492$ Hz, is the sense mode. The rounded peak near 3.5 kHz in the frequency response plots is due to the pre-amplifier used in the capacitive measurement scheme. The sense mode

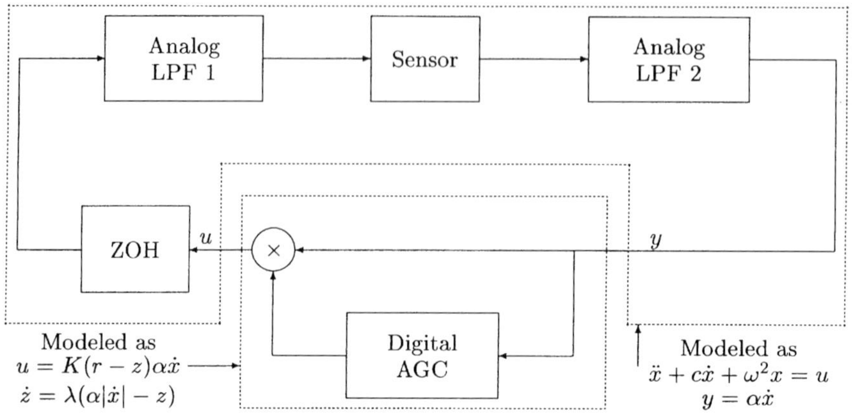

Figure 4. Feedback configuration of sensor, AGC, and analog filters. The zero-order-hold (ZOH) device, which converts the discrete-time signals from the microprocessor into continuous-time signals, is not explicitly modeled.

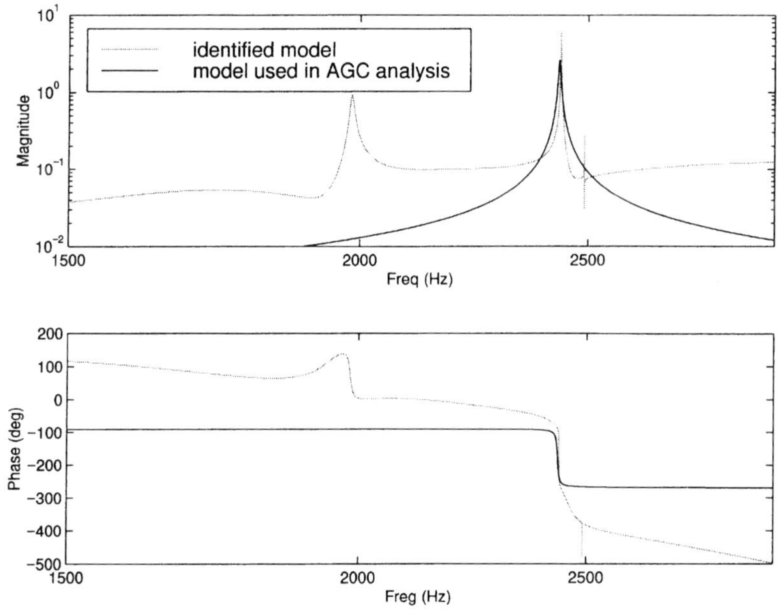

Figure 5. Bode plot of the *drive configuration* including analog filtering.

can be nearly eliminated from the frequency response by driving D_1 and D_2 with the same input and summing the measurements of S_1 and S_2. This single input/single output configuration, which we denote as the *drive configuration*, primarily excites and measures the axial and drive mode, although the sense mode is present in the response to some extent.

The drive configuration is used to test the AGC, however, due to the signal rectification, the controller cannot distinguish between the relative contribution of the three modes so it is necessary to additionally filter the drive and sense signals as shown in Figure 4. These analog filters are designed to provide the controller with the measurement proportional to the drive mode velocity as well as adjust the phase of the remaining modes as described below.

The frequency response of an identified model of the sensor drive configuration with the filters at the input and output is shown in Figure 5 (gray plot). This figure demonstrates that

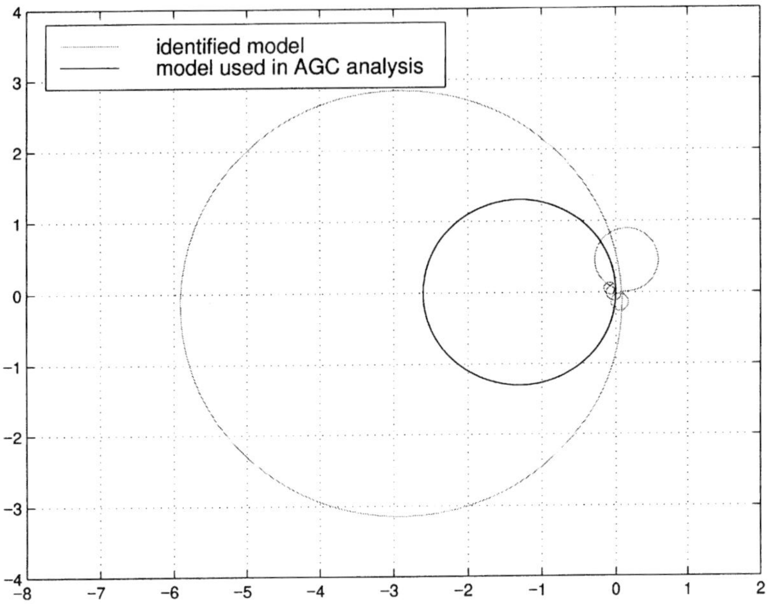

Figure 6. Nyquist plot of the *drive configuration* including analog filtering.

the sense mode has almost been eliminated (compare to the individual channels in Figure 3). The most informative plot for purposes of the AGC design, however, is the Nyquist plot of the drive configuration in Figure 6 (gray plot). The large lobe encircling -1 is the drive mode whose amplitude the AGC is designed to regulate. The small lobe in the upper half-plane is the axial mode while the smallest lobes can be attributed to the sense mode and pre-amplifier dynamics. The most important role of the analog filters now becomes clear: the filters are used to adjust the phase of the drive mode to create a measurement proportional to its modal velocity while also adjusting the phases of the axial and sense modes such that the AGC will not excite these modes.

The AGC loop is implemented with a digital signal processor sampling at 25 kHz. Unfiltered quantized drive signals excite the transient dynamics of the sensor leading to high harmonic content in the sensor response. Thus, another advantage of using filters for the sensor drive electrode signals (denoted LPF1 in Figure 4) is to smooth the quantized signals from the digital-to-analog converters (D/A). The filters at the sensor output, denoted LPF2 in Figure 4, are also used for anti-aliasing, in addition to adjusting the phase of the drive configuration transfer function.

An intuitive explanation of the AGC operation is that it adjusts the magnitude, but not the phase, of the Nyquist diagram in Figure 6. For example, if the drive mode amplitude is smaller than the desired amplitude, the AGC adjusts the loop gain so that the drive mode lobe encircles -1. This has the effect of destabilizing the drive mode by adding anti-damping in the closed-loop system. Conversely, if a smaller drive amplitude is desired, then the AGC changes the sign of the loop gain so that the drive mode lobe no longer encircles -1. This has the effect of adding damping to the drive mode to decrease its amplitude. In steady-state operation the AGC adjusts the loop gain so that the drive mode lobe passes through -1 on the Nyquist plot. The AGC will not excite the other modes unless a large feedback gain is used with a phase

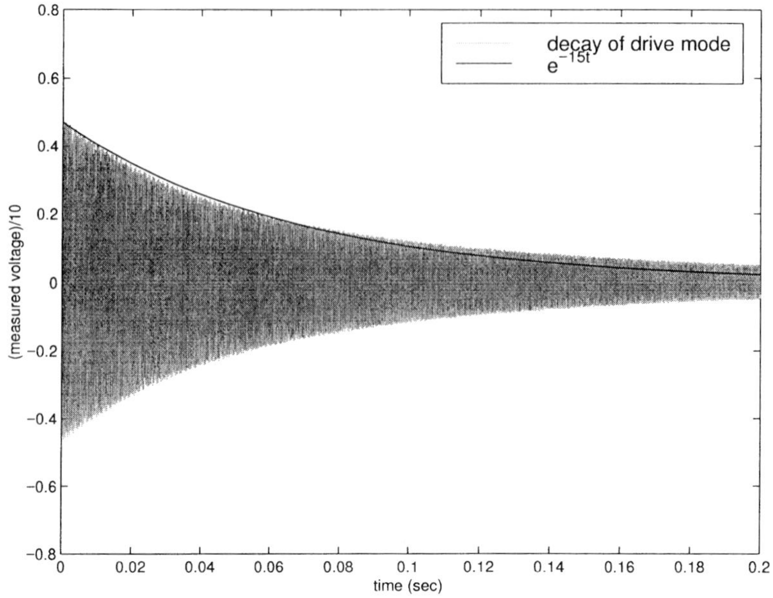

Figure 7. Determining the damping coefficient associated with the drive mode.

change, i.e., the smaller lobes should not encircle -1. The black graphs in Figures 5 and 6 are explained below.

The estimation of Ω_{drv}, c, and α in Equation (1) from experimental data is required to compare this model with the actual device response. In principle these parameters may be extracted from the identified model in Figure 5. This model, however, was obtained from transient response experiments where the input was a chirp signal of short duration. More accurate parameter values may be obtained by using a function synthesizer to excite the drive mode to a constant amplitude and measuring the steady-state gain, decay time when the excitation is removed, etc. Using this method yields $\Omega = 2431$ Hz. The damping is so small in the sensor that the frequency where the transfer function achieves its maximum response may be taken to be the natural frequency in the model. The damping coefficient c is determined from Figure 7. The damping coefficient is twice the magnitude of the exponent used to create the envelope, i.e., $c \approx 2 \cdot 15 = 30$. Figure 7 also illustrates that the damping in the sensor is nonlinear because the exponential envelope does not fit the entire time plot. The gain α is also determined by observing the response amplitude relative to the input amplitude, denoted $H_{\max}$, when driving the sensor with a sinusoid at Ω. The approximate relationship that holds for small damping ratio is $H_{\max} = \alpha/c$. $H_{\max}$ is determined to be 2.6 and so $\alpha = 2.6c = 78$. Thus, the simplified oscillator model used in Equation (1) is

$$\ddot{x} + 30\dot{x} + (2431 \cdot 2 \cdot \pi)^2 = K(r - z) \cdot (78\dot{x}). \tag{11}$$

The Bode and Nyquist plots of this model from the perspective of the AGC are the black graphs in Figures 5 and 6. We see that the ARX model overestimates the peak amplitude by about a factor of two. The ARX model is still indispensable, however, in showing us the phase relationships of the drive mode and the other two modes – facts which are used to argue that the AGC loop regulates the drive mode yet does not excite the axial mode or sense mode.

The experimental closed-loop responses of the sensor are the solid graph traces shown in Figures 8 and 9. These figures illustrate how the AGC loop excites the drive mode to a

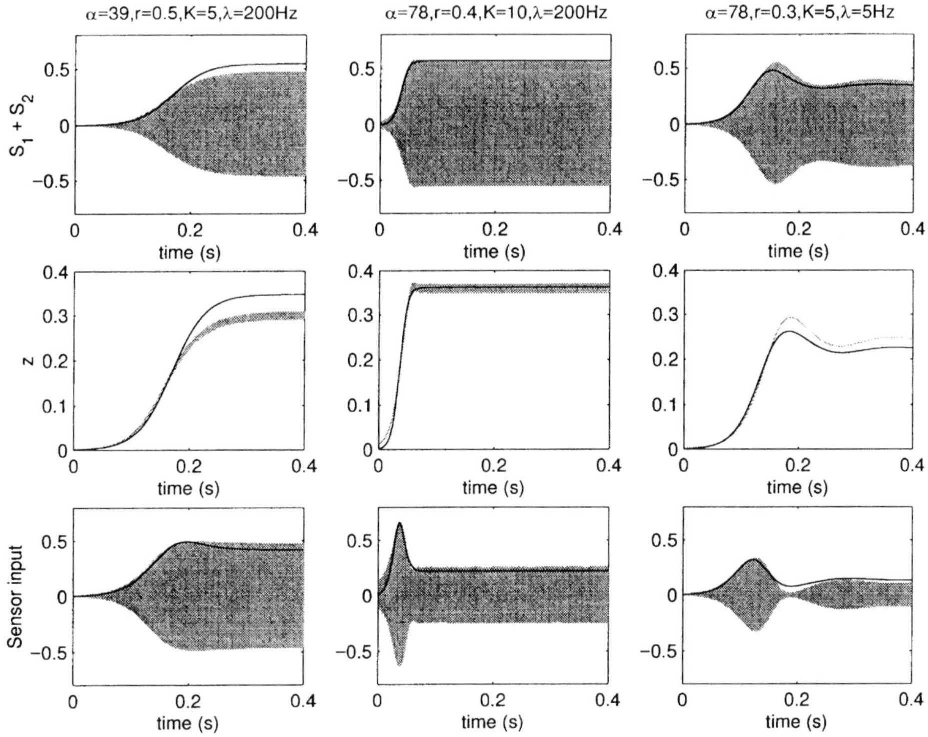

Figure 8. Closed-loop response (gray) and averaged equation simulation (black).

steady-state amplitude. Each column of sub-plots represents one experiment with a fixed set of AGC control parameters (α, K, r, λ). These parameters are given in the title of the top sub-plots. Each sub-plot row represents a particular variable in the closed-loop system. The first row represents the measurement $S_1 + S_2$ which is the signal that the AGC receives ('y' in Figure 4). The units are in volts normalized with respect to the A/D range of ± 10 volts, i.e., 1 represents 10 volts, 0.5 represents 5 volts, etc. The second row represents the state of the filter, z, used in the AGC. There are no physical units attached to this quantity since it represents a variable within the computer. The third row represents the signal that drives the sensor ('u' in Figure 4). Symbolically this is written as $D_1 = D_2 = K(r - z)(S_1 + S_2)$. The units are volts normalized with respect to the D/A range of ± 10 volts.

The black traces in Figures 8 and 9 are the variables from time simulations of the averaged model in Equation (9) with the same control parameters as its corresponding experiment and with Ω, α and c as determined above. The black traces in the first row of sub-plots represent the amplitude, $\alpha\Omega a(t)$, of the measured velocity. The black traces in the second rows are $z(t)$. Finally, the black traces in the third row of sub-plots are amplitudes of the oscillator input: $|K(r - z(t))\alpha\Omega a(t)|$. Thus, the black traces are the correct analogs from Equation (9) of the measured experimental variables. Figure 10 is the power spectral density of the constant amplitude portion of the sensor response from the top plot in column two of Figure 8. This figure confirms that the AGC is regulating the drive mode at 2431 Hz as it was designed to do (the PSDs of the other experiments look identical to this figure).

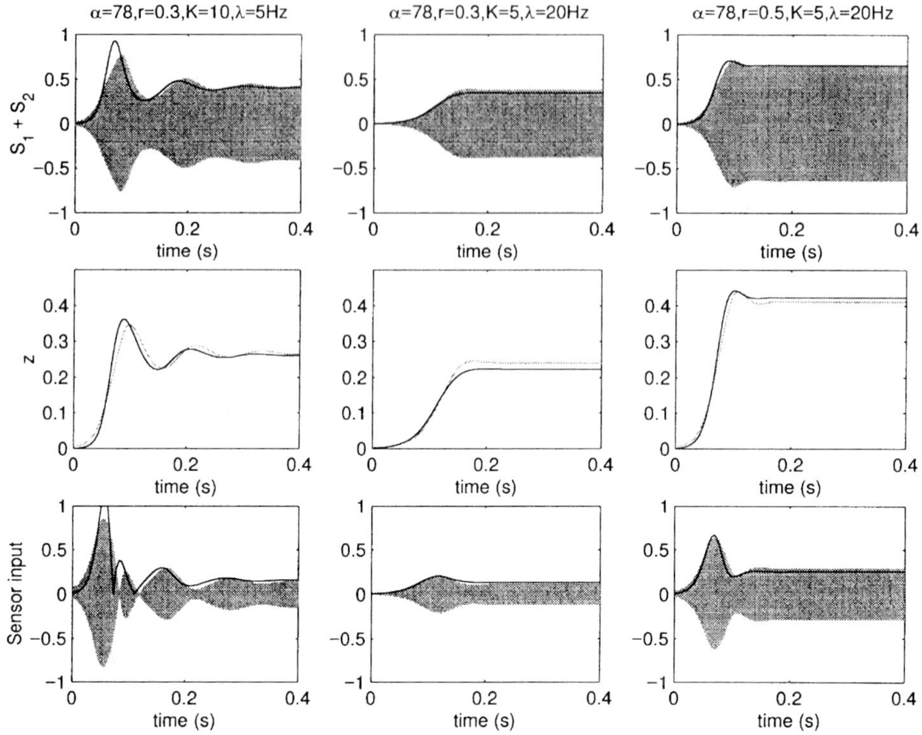

Figure 9. Closed-loop response (gray) and averaged equation simulation (black).

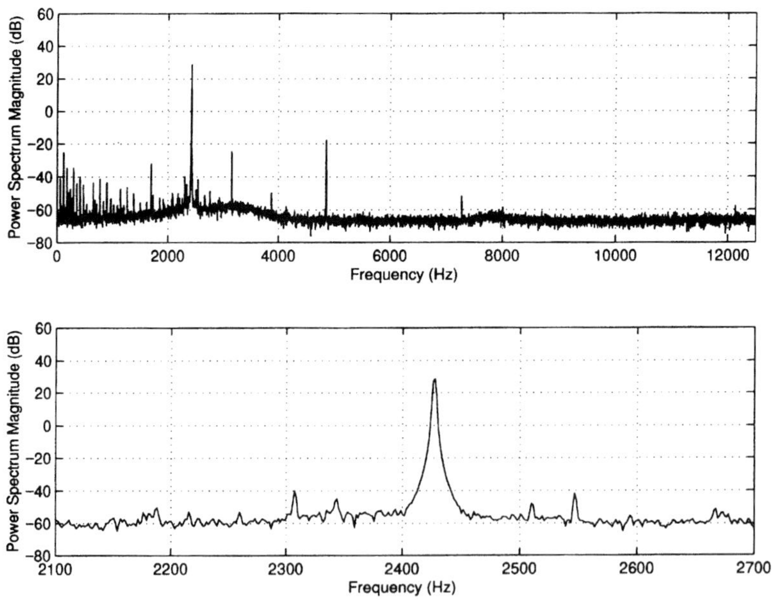

Figure 10. Power spectral density of the sensor output at equilibrium amplitude demonstrating excitation of the drive mode (top plot shows spectrum up to Nyquist frequency and the bottom plot is the detail in a neighborhood of the drive mode).

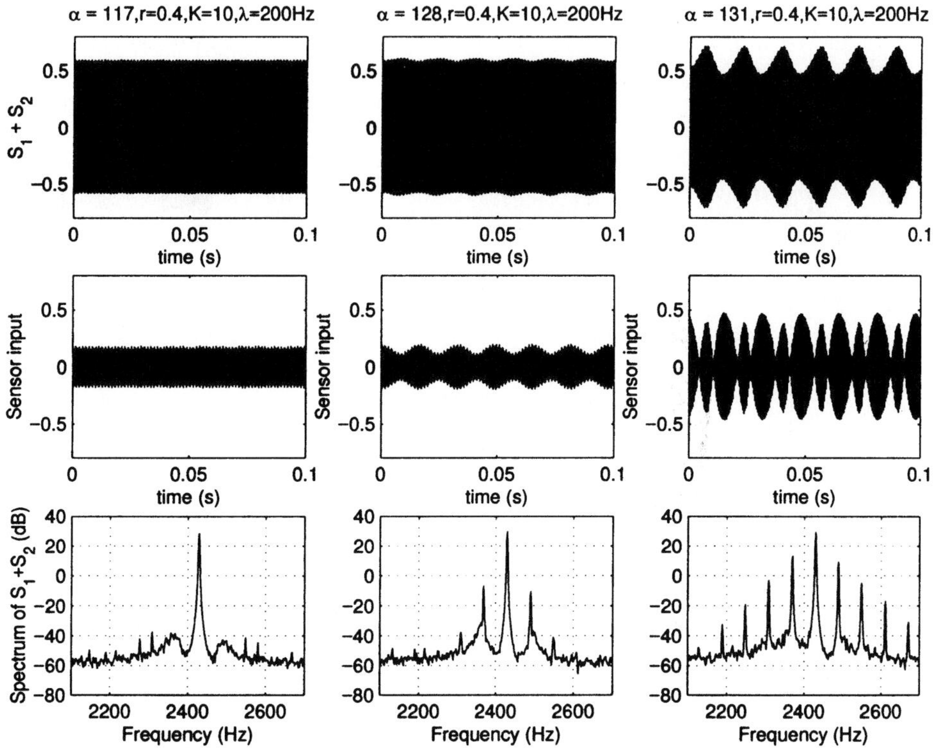

Figure 11. Quasi-periodic response of sensor for different values of sensor gain.

3.3. LIMITATIONS OF THE AGC/SDOF OSCILLATOR MODEL

The robustness of the AGC scheme is experimentally tested by perturbing control system parameters and this section demonstrates that a modest increase in the sensor transfer function gain from u to y in Figure 4 can produce complex behavior in the closed-loop system. In particular we perturb the gain parameter $\alpha = 78$ in Equation (11) to $\alpha = \{117, 128, 131\}$. These perturbations represent an increase in the transfer function gain by the approximate factors $\{1.5, 1.64, 1.68\}$. These gain changes are implemented by modifying a parameter in the signal processing software so even though the gain of the *physical* device is only known approximately, we can modify the parameters in the controller with high precision. The reference signal, AGC gain, and low pass filter corner frequency are fixed at $r = 0.4$, $K = 10$, and $\lambda = 200$ Hz. Figure 11 shows a transition from the stable limit cycle oscillation of the drive mode to a stable quasi-periodic response as α is increased. The sensor response power spectrum in the left-most column in Figure 11 corresponds to $\alpha = 117$ and shows that the sense mode is exited to a very small amplitude (the bump at $\Omega_{sen} \approx 2492$ Hz). The bump near 2370 Hz is a harmonic whose frequency is related to Ω_{drv} and Ω_{sen} according to $2\Omega_{drv} - \Omega_{sen}$. As α is further increased we observe a strong modulation in the sensor input and output signals with period equal to the difference between Ω_{drv} and Ω_{sen} (approximately 61 Hz). Power spectra show the presence of harmonics at $(m + 1)\Omega_{drv} - m\Omega_{sen}$ and $(m + 1)\Omega_{sen} - m\Omega_{drv}$, $m = 1, \ldots, 4$, in the sensor output.

These experimental results indicate that nonlinear active localization (that is, the desired mode of sensor operation) is possible only for certain ranges of parameters. This nonlinear

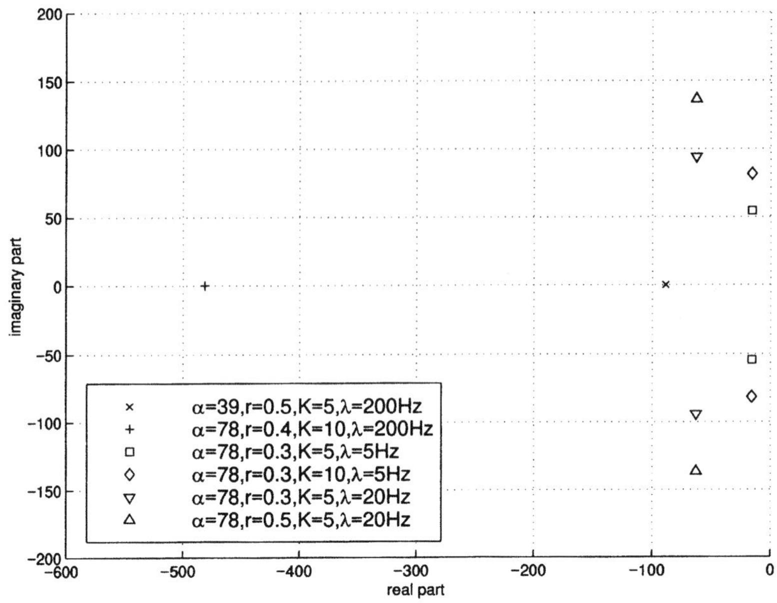

Figure 12. Locus of eigenvalues of closed-loop system linearized about equilibria.

localization is destroyed through what appears to be a Hopf bifurcation of the closed-loop dynamics that leads to quasi-periodic, nonlocalized response. Similar elimination of passive nonlinear localization through bifurcation has been reported in earlier work [3, 4].

4. Discussion

Comparison of the experimental data and the simulations in Figures 8 and 9 reveal that the averaged two state model in Equation (9) accurately predicts the response of the physical device. We conclude that the SDOF model in Equation (1), though simple, captures the salient features of the automatic gain control loop for the range of parameter values considered. The close fit of the simulations to the experimental sensor data is further evidence that neglecting the dynamics of the axial and sense modes is justified so long as the design of the drive and sense signal filters yields a Nyquist plot similar to Figure 6. In other words, the sensor input and output filters are designed so that the mode to be regulated by the AGC encircles -1 on the Nyquist plot while the other modes avoid encirclement for all likely values of feedback gain.

The AGC control parameters are chosen to provide different responses depending on the eigenvalues of the linearization in Equations (10). The eigenvalues of $D_x f(\bar{a}_1, \bar{z}_1)$ are are plotted in Figure 12 for the six different sets of gains considered in the experiments. Even though the simulations in Figures 8 and 9 are of the nonlinear model in Equation (9), the eigenvalues of the linearizations are a good indicator of oscillations in the response envelope. For example, the sensor response in the first two columns of Figure 8 exhibit over-damped envelopes; the eigenvalues of the linearizations for these cases are the '$\times$' and '$+$' in Figure 12 where the $\times$ corresponds to the first column with the longer rise time (the other eigenvalues corresponding to these cases are negative, real, and lie much further in the left half-plane and hence do not dominate the response). The response with the most oscillation is the experiment in the first column of Figure 9. The eigenvalues of the linearization for this case are the $\diamond$'s in Figure 12 – note that the ratio of imaginary-to-real parts is the largest for this parameter combination. The reader may confirm that the remaining experiments correlate well the linearizations given in Figure 12.

The robustness of this control method is tested through modification of the sensor transfer function gain. It is always desirable to design the controller such that the closed-loop response is insensitive to reasonable perturbations in the dynamics of the controlled device. Perturbations in the gain characteristics of the sensor are not uncommon since electronic components age, and changes in the operating environment, such as ambient pressure and temperature, can modify the sensor dynamics. Experiments reveal that an increase in the sensor gain by less than a factor of two over the nominal cases produces a quasi-periodic response that involves both the sense and drive modes. The quasi-periodic response is very undesirable and renders the sensor useless for its intended task. This set of experiments reveals the limitations of the AGC/SDOF oscillator model since the nonlinear control loop clearly excites the sense mode in addition to the drive mode. Thus, nonlinear localization through active control in the sensor is effective only for a carefully designed range of the control parameters and modest deviations in the sensor dynamics from the design condition can eliminate active localization and produce quasi-periodic nonlocalized response.

A topic of current research considers a two-degrees-of-freedom model with AGC where the objective is to explore how the sensor and controller parameters influence the stability of the drive mode limit cycle. A bifurcation study of the multi-mode system can be used to design a controller that avoids or delays the onset of the quasi-periodic response thereby producing a more robust system.

Acknowledgments

The authors thank Tony Tang, Vatche Vorperian, and Chris Stell at the Jet Propulsion Laboratory, Dorian Challoner at Hughes Space and Communications, and Jaime Sipila at UCLA. This work is supported by Hughes Electronics and UC MICRO Grant No. 96113, and the California Institute of Technology Grant No. PF-433.

References

1. Tang, T. K., Gutierrez, R. C., Wilcox, J. Z., Stell, C., Vorperian, V., Calvet, R., Li, W., Charkaborty, I., and Bartman, R., 'Silicon bulk micromachined vibratory gyroscope', in *Solid-State Sensor and Actuator Workshop*, Hilton Head, SC, June 3–6, 1996, pp. 288–293.
2. Tang, T. K., Gutierrez, R. C., Stell, C. B., Vorperian, V., Arakaki, G. A., Rice, J. T., Li, W. J., Charkaborty, I., Shcheglov, K., and Wilcox, J. Z., 'A packaged silicon MEMS vibratory gyroscope for microspacecraft', in *IEEE Proceedings of The Tenth Annual International Workshop on Micro Electro Mechanical Systems*, Nagoya, Japan, January 26–30, 1997, pp. 500–505.
3. Vakakis, A. F., Manevitch, L. I., Mikhlin, Y. V., Pilipchuk, V. N., and Zevin, A. A., *Normal Modes and Localization in Nonlinear Systems*, Wiley Interscience, New York, 1996.
4. Vakakis, A. F. and Salenger, G., 'Generation of localization in a discrete chain with periodic boundary conditions: Numerical and analytical results', *SIAM Journal on Applied Mathematics* **58**(6), 1998, 1730–1747.
5. Emad, J., Vakakis, A. F., and Miller, N., 'Experimental nonlinear localization in a periodically forced repetitive system of coupled magnetoelastic beams', *Physica D* **137**, 2000, 192–201.
6. King, M. E. and Vakakis, A. F., 'A very complicated structure of resonances in a nonlinear system with cyclic symmetry: Nonlinear forced localization', *Nonlinear Dynamics* **7**, 1995, 85–104.

7. King, M. E, Aubrecht, J., and Vakakis, A. F., 'Experimental nonlinear steady-state localization in coupled beams with active nonlinearities', *Journal of Nonlinear Science* **5**(6), 1995, 485–502.
8. Aubrecht, J., Vakakis, A. F., Tsao, T.-C., and Bentsman, J., 'Experimental study of nonlinear transient motion confinement in a system of coupled beams', *Journal of Sound and Vibration* **195**(4), 1996, 629–648.
9. Hale, J., *Ordinary Differential Equations*, Krieger, Melbourne, FL, 1980.
10. Jordan, D. W. and Smith, P., *Nonlinear Ordinary Differential Equations*, Oxford Applied Mathematics and Computing Science Series, Oxford University Press, Oxford, 1987.
11. Andronov, A. A., Vitt, A. A., and Khaikin, S. E., *Theory of Oscillators*, Dover, New York, 1966.
12. Lawrence, A., *Modern Inertial Technology: Navigation, Guidance, and Control*, second edition, Springer-Verlag, New York, 1992.
13. M'Closkey, R. T., Gibson, J. S., and Hui, J., 'Input/output dynamics of the JPL microgyroscope,' in *1998 IEEE Conference on Decision and Control*, Tampa, FL, December 7–10, 1998, pp. 4328–4332.
14. M'Closkey, R. T., Gibson, J. S., and Hui, J., 'Model identification of the JPL microgyroscope,' *Proceedings of the ASME, Dynamic Systems and Control* **64**, 1998, 801–805.

Nonlinear Dynamics **25:** 237–253, 2001.

Transition of Energy to a Nonlinear Localized Mode in a Highly Asymmetric System of Two Oscillators

OLEG V. GENDELMAN
Institute of Chemical Physics, RAS, Kosygin str., 4, 117977 Moscow, Russia

(Received: 11 February 1999; accepted: 31 August 1999)

Abstract. Redistribution of energy in a highly asymmetric system consisting of coupled linear and highly nonlinear damped oscillators is investigated. Special attention is paid to the excitation of a nonlinear normal mode while the energy is initially stored in other modes of the system. The transition proceeds via the mechanism of subharmonic resonance which is possible because of the strong nonlinearity of the system. The conditions of the energy transition to NNM being effective are revealed and guidelines to design such a systems are formulated in detail.

Keywords: Nonlinear normal mode, forced nonlinear oscillator, subharmonic resonance, transition of energy.

1. Introduction

The concept of nonlinear normal modes as additional types of energy distribution in dynamical systems with respect to the spectrum of ordinary linear modes is well established [1]. Usually these normal modes reveal themselves either if they are excited directly or as a result of the breakdown of some cooperative regimes of motion [2]. The question discussed in this paper is whether the nonlinear normal modes in dynamical systems may be excited while energy is initially stored in another stable vibration mode.

This question is of practical importance since the excitation of nonlinear normal modes is usually undesirable in real dynamical systems for safety reasons. On the other hand, the excitation of a nonlinear normal localized mode would prevent undesirable propagation of the energy through the system and absorb energy from 'dangerous' modes of motion.

For the case of linear modes such a process of energy transition between modes is obviously impossible since the modal distribution of energy is constant. The same situation holds for nonlinear integrable systems [3]. Then, it is doubtful to expect the effects of energy 'pumping' between different nonlinear modes if the system is highly symmetric or close to an integrable one.

Therefore, it seems reasonable to consider the highly asymmetric system of two coupled oscillators, where one oscillator is highly nonlinear with a small linear frequency, while the other one is linear with a high linear frequency:

$$\begin{aligned} &\ddot{y}_1 + \lambda_1 \dot{y}_1 + \omega_1^2 y_1 + C y_1^n \varepsilon (y_1 - y_2) = 0, \\ &\ddot{y}_2 + \lambda_2 \dot{y}_2 + \omega_2^2 y_2 + \varepsilon (y_2 - y_1) = 0. \end{aligned} \tag{1}$$

In this system of equations, y_1 and y_2 are, respectively, the displacements of highly nonlinear and linear oscillators, λ_1 and λ_2 are their damping coefficients, ω_1 and ω_2 the linear frequencies, C the intensity of the nonlinear term, ε the linear coupling parameter, n is an odd number greater than unity.

The frequency of the second oscillator and the intensity of the nonlinear term are considered to be in the order of unity. The linear frequency of the first oscillator, the damping terms in both equations and the coupling are supposed to be small in order to ensure high asymmetry and to allow the nonlinear normal mode localized at the first oscillator to appear. We suppose all the small coefficients (including squared frequencies) to be of the order ε.

Further analysis requires us to reveal the modal spectrum of the system. The modal analysis is provided for a conservative system, obtained from Equation (1), by eliminating the damping terms:

$$\begin{aligned} &\ddot{y}_1 + \omega_1^2 y_1 + C y_1^n + \varepsilon(y_1 - y_2) = 0, \\ &\ddot{y}_2 + \omega_2^2 y_2 + \varepsilon(y_2 - y_1) = 0. \end{aligned} \tag{2}$$

Linearized system (2) (with $C = 0$) obviously has two linear normal modes. Their frequencies and eigenvectors may be calculated using the standard procedure [4]. By using the substitution $y_i = A_i \exp(i\Omega t)$, where A_i denotes the appropriate component of the eigenvector and Ω the eigenfrequency, we easily get the expressions for two eigenfrequencies:

$$\begin{aligned} \Omega_1 &= \sqrt{\omega_1^2 + \varepsilon - \frac{\varepsilon^2}{\omega_2^2 - \omega_1^2}} + O(\varepsilon^3), \\ \Omega_2 &= \sqrt{\omega_2^2 + \varepsilon + \frac{\varepsilon^2}{\omega_2^2 - \omega_1^2}} + O(\varepsilon^3), \end{aligned} \tag{3}$$

while the respective eigenvectors may be introduced by the amplitude ratio $\kappa = A_1/A_2$:

$$\begin{aligned} \kappa_1 &= \frac{\omega_2^2 - \omega_1^2}{\varepsilon} + O(1), \\ \kappa_2 &= \frac{\varepsilon}{\omega_2^2 - \omega_1^2} + O(\varepsilon^2). \end{aligned} \tag{4}$$

Both modes are highly asymmetric and localized on the first and second oscillators, respectively. This localization occurs due to the linear structure of the system rather then because of nonlinear effects.

The linear modes determined by Equations (3–4) are characterized by certain orders of all parameters with respect to small parameter ε, so an analysis of the nonlinear system (2) may be also provided by establishing the correct orders of parameters. If proposing the evaluations of the orders of magnitude with respect to the small parameter for the time derivative and dependent variables in the form $\partial/\partial t \approx \varepsilon^a$, $y_i \approx \varepsilon^{b_i}$ (see, e.g., [5]), it is easy to obtain sets of values of powers a and b_i which lead to different asymptotic regimes of motion of the system of Equations (2). These sets are presented below:

$$\{1\} \quad a = \frac{1}{2}, \quad b_1 > \frac{1}{n-1}, \quad b_2 = 1 + b_1;$$

$$\{2\} \quad a = \frac{1}{2}, \quad b_1 = \frac{1}{n-1}, \quad b_2 = \frac{n}{n-1};$$

$$\{3\} \quad a = \frac{(n-1)b_1}{2}, \quad 0 < b_1 < \frac{1}{n-1}, \quad b_2 = 1 + b_1;$$

$$\{4\} \quad a = 0, \quad b_1 = 0, \quad b_2 = 1;$$

$$\{5\} \quad a = \frac{(n-1)b_1}{2}, \quad b_1 < 0, \quad b_2 = 1 - b_1(n-2);$$

$$\{6\} \quad a = 0, \quad b_1 > 1 + b_2, \quad b_2 > -1;$$

$$\{7\} \quad a = 0, \quad b_1 = 0, \quad b_2 = -1;$$

$$\{8\} \quad a = 0, \quad nb_1 > 1 + b_2, \quad b_2 < -1. \tag{5}$$

It is easy to see that the sets {1} and {6} describe weakly nonlinear regimes of motion similar to the linear normal modes revealed in Equations (3–4) – this fact may be checked by a simple comparison of the orders of parameters. Set {8} defines the mode localized at the second (linear) oscillator. However, in this case the amplitude of vibrations of the first oscillator is large enough to ensure that the regime will be essentially anharmonic. This fact is rather obvious since the orders of magnitude of variables depend on the power of nonlinearity n. Regimes {3} and {5} correspond to different essentially nonlinear modes localized at the first oscillator. Sets {2}, {4} and {7} correspond to transition points between the respective modes.

Equation (5) allows us to easily obtain the asymptotic equations describing the corresponding regimes [5]. The first approximation always may be obtained from the remaining terms with the highest order in Equations (2) and eliminating the others. The following orders of approximation may be obtained by one of the usual methods, while the correct orders of all the values are known from the estimations (5).

It should be mentioned that the analysis of the sets (5) does not constitute a complete description of the nonlinear normal modes of the system (2), which requires more cumbersome methods (see, e.g., [1]). Still, the mode described by sets {3} and {5} resembles a nonlinear normal mode in the accepted sense, since the energy distribution between the oscillators grows with the growth of the energy of the system. However, the mode described by set {8} is not a 'classical' NNM since the oscillator with a higher amplitude is completely linear. This regime is similar to the resonance vibrations of a highly nonlinear oscillator. Later in the paper, this statement will become more obvious. The mode discussed will be referred to below as a 'localized resonant mode'.

Now it is possible to start discussing the question of whether the energy may be 'pumped' between the different modes of the system. If the energy is initially concentrated at the second oscillator, then the only localized mode which may appear is the one described by the sets {7} and {8}. While the frequency of this mode satisfies the condition $a = 0$, $b_1 \leq 0$, the nonlinear normal mode described by sets {4} and {5} may be 'related' to this one, since it is characterized by the same asymptotic values of these two parameters. However, the value of b_2 is very different and therefore the transition described may occur only if there exists a way of reducing the value of the amplitude of the second oscillator with the help of damping without 'ruining' the structure of the localized mode.

Now it is possible to formulate explicitly the requirements for the 'pumping' process in the system (2) to occur in the desired way:

1. The localized mode described by set {8} should be excited.

2. The vibrations of the second oscillator should be damped faster than the first one, although the damping terms may be of the same order due to the physical essence of the system.

The first condition is analyzed in the second section of the paper. The second condition is analyzed in the third section and is accompanied by some examples of 'high-performance' systems. General requirements are presented in the concluding part of the paper.

2. Analysis of the Localized Resonant Mode

The purpose of this section is to analyze the conditions of excitation of the localized resonant mode, which is characterized by the sets of values {7} and {8} from the system (5). First, the system (2) without damping will be investigated. According to the general statements of the previous section, the first approximation describing this mode may be obtained from the system (2) by eliminating the terms having higher orders of magnitude in accordance with the values of orders of variables. The appropriate equations will have the following form:

$$\begin{aligned} &\ddot{y}_1 + C y_1^n - \varepsilon y_2 = 0, \\ &\ddot{y}_2 + \omega_2^2 y_2 = 0. \end{aligned} \tag{6}$$

The initial conditions 'natural' for the process investigated may be introduced as

$$\begin{aligned} y_1(0) &= 0, \quad y_1(0) = 0, \\ y_2(0) &= \alpha, \quad y_2(0) = \beta, \end{aligned} \tag{7}$$

where α and β denote the initial amplitude and velocity of the second oscillator respectively. After solving the second equation of the system (6), and substituting the solution to the first equation, we finally obtain the following system:

$$\begin{aligned} &\ddot{y}_1 + C y_1^n = \varepsilon A \sin(\omega_2 t + \theta), \\ &y_1(0) = 0, \quad y_1(0) = 0, \quad A = \sqrt{\alpha^2 + \left(\frac{\beta}{\omega_2}\right)^2}, \quad \theta = \arctan \frac{\alpha\omega_2}{\beta}. \end{aligned} \tag{8}$$

The problem we are interested in is to establish the conditions for the parameters A and θ (or α and β) which ensure that the oscillator described by Equation (8) with zero initial conditions will move in the resonant regime with a high value of energy. The stationary regimes of such a highly nonlinear oscillator may be successfully described by the harmonic balance method [6]. This method may be also applied for investigating the dependence of the stationary regime on the initial conditions [6] but it may be justified only if the nonlinearity is relatively small. It is obviously not the case for Equation (8) and the harmonic balance method should be 'adjusted' for the case of large nonlinearity. One way of doing this is proposed below.

Let us first consider the nonresonant regime of the oscillator described by Equation (8). The displacement may be presented as a combination of 'fast' forced vibrations with the frequency of the external force and small 'slow' vibrations caused by the nonlinear term. The possibility of such a presentation is based on the fact that the nonresonant regime is described by set {6} in solution (5) and therefore the inertia term determining the forced motion is the main one and must be balanced by the external force. The displacement is presented as follows:

$$y_1(t) = \gamma \sin(\omega_2 t + \theta) + \phi(t). \tag{9}$$

In this equation, the function $\phi(t)$ describes the free vibration of a highly nonlinear oscillator. By no means may it be described by the harmonic balance method and the 'adjustment' mentioned above should allow us to obtain the correct nonlinear equation for this function.

After substituting expression (9) into Equation (8) the following relationship is obtained:

$$-\gamma\omega_2^2 \sin(\omega_2 t + \theta) + \ddot{\phi}(t) + C(\gamma \sin(\omega_2 t + \theta) + \phi(t))^n = \varepsilon A \sin(\omega_2 t + \theta). \tag{10}$$

After opening the binomial, we have

$$-\gamma\omega_2^2 \sin(\omega_2 t + \theta) + \ddot{\phi}(t) + C\sum_{k=0}^{n} C_n^k \gamma^k \sin^k(\omega_2 t + \theta)\phi^{n-k}(t) = \varepsilon A \sin(\omega_2 t + \theta), \tag{11}$$

where $C_i^j = i!/(j!(i-j)!)$ denotes the binomial coefficients. All the terms under summation in Equation (11), besides the last one, contain the powers of the small function $\phi(t)$ and therefore may be neglected when determining the amplitude γ of the forced vibrations. Similarly, according to the harmonic balance method, only the component of the last term under the sum having the lowest frequency ω_2 should be taken into account. After some simple calculations, we obtain the following equation for the amplitude of the forced vibrations in the nonresonant regime:

$$-\gamma\omega_2^2 + C\gamma^n \frac{n!}{2^{n-1}((n+1)/2)!((n-1)/2)!} = \varepsilon A. \tag{12}$$

After extracting from Equation (11) the balanced harmonic terms having the frequency ω_2 according to Equations (9), (12), we get the equation for the function $\phi(t)$

$$\ddot{\phi}(t) + C\Bigg(\sum_{k=0}^{n} C_n^k \gamma^k \sin^k(\omega_2 t + \theta)\phi^{n-k}(t) \\ - \gamma^n \frac{n!}{2^{n-1}((n+1)/2)!((n-1)/2)!} \sin(\omega_2 t + \theta)\Bigg) = 0 \tag{13}$$

with the initial conditions necessary to balance the harmonic solution in the initial point:

$$\phi(0) = -\gamma \sin\theta, \quad \dot{\phi}(0) = -\gamma\omega_2 \cos\theta. \tag{14}$$

The function $\phi(t)$ varies slowly compared to the external force and therefore Equation (13) may be averaged with respect to the fast function $\sin\omega_2(t+\theta)$. After such averaging, only the terms with odd powers of $\phi(t)$ persist and finally the equation takes on the following form:

$$\ddot{\phi}(t) + C \sum_{p=0}^{(n-1)/2} \phi^{n-2p}(t) \left(\frac{\gamma}{2}\right)^{2p} \frac{n!}{(p!)^2(n-2p)!} = 0. \tag{15}$$

Equation (15) with initial conditions (14) constitutes the equation of motion of a free highly nonlinear oscillator and may be easily integrated. The frequency of the free vibrations of the oscillator described by Equation (15) Ω, depends on the initial conditions (14) and therefore on the parameters of the external force in Equation (8).

Excitations of the resonant regimes of the forced oscillator described by Equation (8) *become possible if the frequency Ω of free vibrations approaches the values ω_2/m, where m is an odd, $m \leq n$. In this case, the terms with the power $\phi^m(t)$ in Equation* (13) *obviously provide additional terms with the frequency of external force (via mechanism of frequency multiplying) which allow the resonant motion.*

The condition of the resonance introduced above allows us to establish whether a given external force with given initial conditions will lead to a resonant or nonresonant regime of vibrations. Clearly, this method is not restricted by zero initial conditions of the initial equation (8). Various resonant regimes of motion may be studied by the harmonic balance method [6] and are not discussed here in details since we are interested mainly in the conditions of transition from non-resonant to resonant regime.

As was mentioned above, the equation of free vibration (15) may be integrated with any power n, thus providing the integral for determining the frequency Ω. For the case $n = 3$, however, this integral may be calculated in terms of well-known special functions. For the sake of simplicity, the calculations will be performed for the case $\theta = 0$, $C = 2$. The expression for the period of vibrations of the oscillator is written as

$$\frac{T}{4} = \int_0^{p_1} \frac{d\phi}{\sqrt{(p_1^2 - \phi^2)(p_2^2 + \phi^2)}} = \frac{1}{\sqrt{p_2^2 + p_2^2}} K\left(\frac{p_1}{\sqrt{p_1^2 + p_2^2}}\right),$$

$$p_{1,2}^2 = -\frac{\pm\sqrt{9\gamma^4 + 4\gamma^2\omega_2^2} - 3\gamma^2}{2}, \tag{16}$$

where $K(z)$ denotes the complete elliptic integral of the first type. The resonance condition is then expressed as

$$T = \frac{2m\pi}{\omega_2} = \frac{4}{\sqrt[4]{9\gamma^4 + 4\gamma^2\omega_2^2}} K\left(\sqrt{\frac{\sqrt{9\gamma^4 + 4\gamma^2\omega_2^2} - 3\gamma^2}{2\sqrt{9\gamma^4 + 4\gamma^2\omega_2^2}}}\right). \tag{17}$$

This condition constitutes the transcendent equation for calculating the parameter γ which may be solved numerically. Denoting $\mu = \gamma/\omega_2$, Equation (17) is simplified:

$$\frac{m\pi}{2} = \frac{1}{\sqrt[4]{9\mu^4 + 4\mu^2}} K\left(\sqrt{\frac{\sqrt{9\mu^4 + 4\mu^2} - 3\mu^2}{2\sqrt{9\mu^4 + 4\mu^2}}}\right). \tag{18}$$

The critical amplitude of the resonant vibrations may be calculated with the help of the relationship (12) as

$$\varepsilon A_* = \omega_2^3\left(\frac{3}{2}\mu_*^3 - \mu_*\right), \tag{19}$$

where μ_* denotes the solution of Equation (18). The fact that the critical amplitude is proportional to the cube of the frequency of external force is clear from the symmetry properties of Equation (8) with $n = 3$.

Table 1. Critical amplitudes at various possible orders of resonance, $n = 3$.

m	μ_*	A_*
3	0.073	0.73
1	0.149	1.43

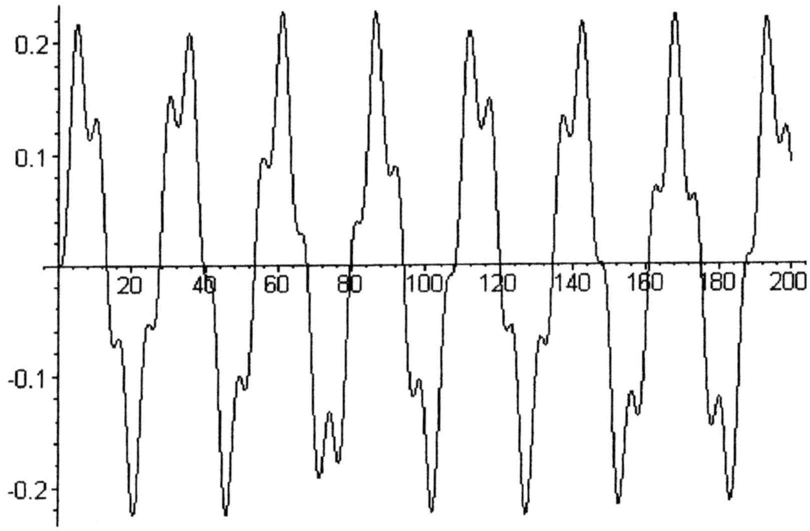

Figure 1a. Excitation of the resonant mode in a highly nonlinear oscillator, Equation (8), $n = 3$, $C = 2$, $\varepsilon = 0.1$, $\theta = 0$, $A = 0.4$, $\omega_2 = 1$.

A numerical solution of Equation (18) has been found for $m = 3$ and $m = 1$ and is summarized in Table 1 ($\omega_2 = 1$, $\varepsilon = 0.1$).

It is well known that the highly nonlinear oscillator of the type described by Equation (8) may have more than one resonant regime of motion [6]. Namely, various subharmonic and harmonic regimes of resonant vibrations may persist. For our purposes we are interested in the excitation of harmonic rather then subharmonic regimes, since the former has sufficiently higher energy of vibrations.

The analysis of the conditions of resonance performed above does not allow us to distinguish what resonant regime exactly will be excited. A numerical solution of Equation (8) allows us to clarify this question and is presented in Figures 1a–1c for various values of A. All numerical simulations in the present paper are accomplished with the help of MAPLE software.

It may be easily seen that for $A < 0.7$ resonant vibrations in the harmonic regime do not occur at all. The plot (Figure 1a) demonstrates low-frequency and low-amplitude anharmonic vibrations which are slightly perturbed by the harmonic function (it is possible to state that the approximation used in Equation (9) is thus justified for this regime). For the limit value $A = 0.73$, the harmonic regime may be immediately detected but needs a lot of time to be excited and is extremely unstable. Unstable harmonic vibrations are detected for intermediate

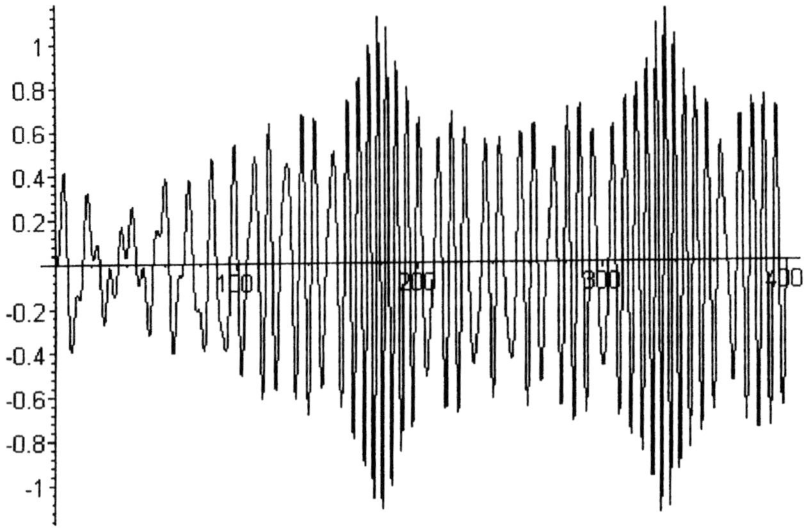

Figure 1b. Excitation of the resonant mode in a highly nonlinear oscillator, Equation (8), $n = 3$, $C = 2$, $\varepsilon = 0.1$, $\theta = 0$, $A = 1.0$, $\omega_2 = 1$.

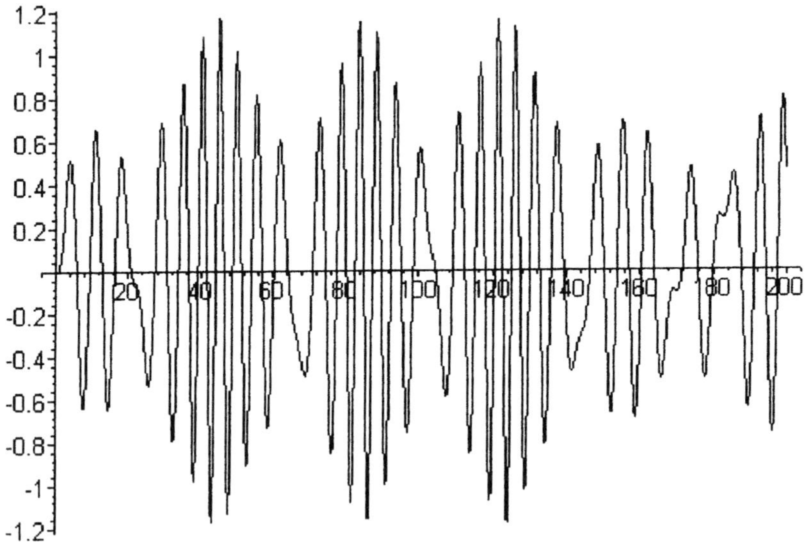

Figure 1c. Excitation of the resonant mode in a highly nonlinear oscillator, Equation (8), $n = 3$, $C = 2$, $\varepsilon = 0.1$, $\theta = 0$, $A = 1.43$, $\omega_2 = 1$.

values $0.72 < A < 1.42$ (Figure 1b), while the time of excitations decreases and the stability increases with the growth of A. Finally, for $A = 1.43$ (Figure 1c) harmonic resonant regime starts immediately and is fairly stable.

An immediate start of the harmonic resonant regime is a necessary condition for effective absorption of energy. Therefore, the limit $A = 1.43$ may be considered as the lowest amplitude of external force in Equation (8) which is suitable for the purpose of 'energy pumping'. It must be recalled that this limit value corresponds to the resonance of order $m = 1$.

A similar investigation may be performed numerically for the case $n = 5$ (the details will be presented elsewhere because of their awkwardness). The main conclusion is that in this system the case of $m = 3$ corresponds to the limit of excitation of the stable harmonic resonant regime. The corresponding value of the amplitude of the external force is $A = 2.05$.

The problem of computing the limit amplitudes for higher values of n is rather cumbersome from the viewpoint of numerical simulations and will be discussed in detail elsewhere. The working hypothesis derived from the investigations presented above will be that the required harmonic resonant regime starts for any n at the value $m = n - 2$.

It is clear that the conditions of excitation of a harmonic regime as well as the critical value of the external force depend strongly on the initial conditions for the variable y_1 and the phase θ of the external force. For instance, for the resonance $m = 1$ with $\theta = \pi/2$ and zero initial conditions, the critical amplitude was computed to be $A = 2.86$. This strong dependence on the initial conditions allows us to control the excitation of the nonlinear vibrations by slightly varying of the state of the nonlinear oscillator.

It should be mentioned that the stationary harmonic regime of Equation (8) constitutes vibrations with the frequency of external force, but their amplitude appears to be modulated. It is a common situation for stationary vibration regimes in conservative highly nonlinear systems and may be described by a modification of the harmonic balance method [6].

Let us recall that the initial system (1) has also damping terms. The system of equations for coupled damped oscillators moving in the regime described by the set of estimations {7} from system (5) in the first approximation takes the following form:

$$\begin{aligned} &\ddot{y}_1 + \lambda_1 \dot{y}_1 + C y_1^n - \varepsilon y_2 = 0, \\ &\ddot{y}_2 + \lambda_2 \dot{y}_2 + \omega_2^2 y_2 = 0. \end{aligned} \tag{20}$$

Quite naturally, the equation governing the excitation of a localized resonant mode (similar to Equation (8) for the system without damping) is written as follows:

$$\begin{aligned} &\ddot{y}_1 + \lambda_1 \dot{y}_1 + C y_1^n = \varepsilon \tilde{A} \exp\left(-\frac{\lambda_2 t}{2}\right) \sin(\tilde{\omega}_2 t + \tilde{\theta}), \\ &y_1(0) = 0, \quad \dot{y}_1(0) = 0, \quad \tilde{\omega}_2 = \sqrt{\omega_2^2 - \left(\frac{\lambda_2}{2}\right)^2}, \\ &\tilde{A} = \sqrt{\alpha^2 + \left(\frac{\beta + \frac{\lambda_2}{2}\alpha}{\tilde{\omega}_2}\right)^2}, \quad \tilde{\theta} = \arctan \frac{\alpha \tilde{\omega}_2}{\beta + \frac{\lambda_2}{2}\alpha}. \end{aligned} \tag{21}$$

To the best of our knowledge, the equations of a highly nonlinear oscillator with decreasing external force like Equation (21) have not been studied previously. The most essential complication is that this equation obviously cannot have any stationary regime. The process of interest here is the excitation of nonlinear vibrations having a frequency close to the frequency of the external force. This excitation will be followed by the damping of highly nonlinear vibrations due to the damping term of the nonlinear oscillator, whereas the effect of external force vanishes. Still, the excitation of this regime of nonlinear vibrations may be detected under the same resonance condition $m = 1$. The corresponding critical amplitude may be computed with the help of a procedure similar to one described in Equations (9–19), although the calculations are more bulky and are not presented here in detail. The result for particular

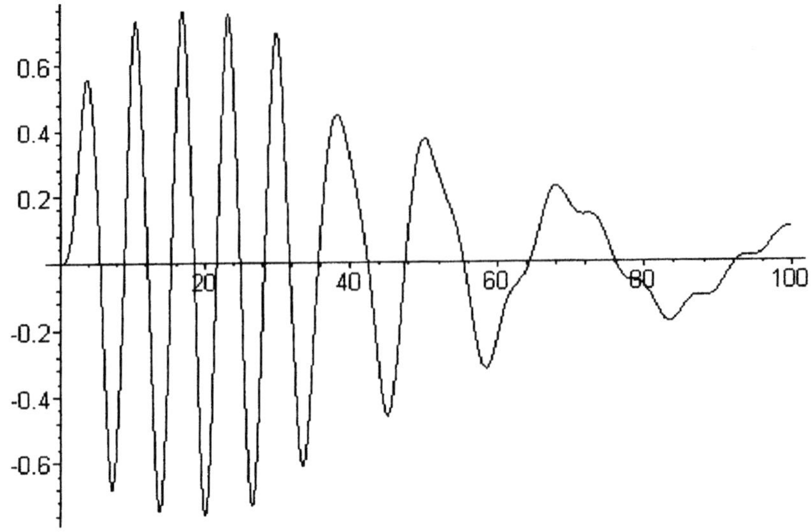

Figure 2. Excitation of the resonant mode in a highly nonlinear oscillator with damping, Equation (8), $n = 3$, $C = 2$, $\varepsilon = 0.1$, $\lambda_1 = \lambda_2 = 0.05$, $\tilde{\theta} = 0$, $\tilde{A} = 1.9$, $\omega_2 = 1$.

case of initial conditions similar to the case without damping described above for the oscillator with $n = 3$ and $C = 2$ is presented in Figure 2. For $\lambda_1 = \lambda_2 = 0.05$ and $\tilde{\theta} = 0$ the critical amplitude $\tilde{A} = 1.9$. Clearly, it is more than in the case without damping.

Characteristic feature of the regime plotted in Figure 2 is that the effect of modulation in this case is rather weak (unlike the regime of Figure 1c). This difference has an obvious physical reason – the characteristic 'decay time' of the external force is of the same order as the period of modulation. In other terms, the external source is 'detached' from the oscillator before it returns the energy due to modulation. The importance of this effect for our purposes is discussed in the next section.

3. Analysis of the Damping Process in the System of Two Oscillators and Conditions for 'Efficient' Energy Transfer

The second requirement for effective 'energy pumping' listed at the end of the first section is related to faster damping at the second (linear) oscillator than at the first (nonlinear) one. The second equation of system (21) is an ordinary equation of a damped linear oscillator. Its energy is dissipated at the following rate:

$$\frac{\partial E_2}{\partial t} \sim \exp(-\lambda_2 t). \tag{22}$$

The rate of the energy dissipation for the first oscillator may be estimated if we recall that this oscillator undergoes the transition to free damped nonlinear vibrations after the 'detaching' of the external force. Recently, it was proved [7] that for sufficiently high energy of vibrations, the solution of a free highly nonlinear oscillator may be estimated as $(C = (n+1)/2)$:

$$y_1 \approx Y_0 \exp\left(-\frac{2\lambda_1 t}{n+3}\right) \mathrm{Cam}_{(n)} \left(\frac{(n+3)Y_0^{(n-1)/2}}{\lambda_1(n-1)} \exp\left(-\frac{(n-1)\lambda_1 t}{n+3}\right) + \Psi\right), \tag{23}$$

where Y_0 and Ψ depend on the initial conditions and $\mathrm{Cam}_{(n)}$ is a special periodic function defined in the following way [8]:

$$du = \int \frac{\mathrm{d}\,\mathrm{Cam}_{(n)}(u)}{\sqrt{1 - \mathrm{Cam}_{(n)}^{n+1}(u)}}.$$

The dissipation rate of energy for the solution (23) obeys the estimation

$$\frac{\partial E_1}{\partial t} \sim \dot{y}_1^2 \sim \exp\left(-\frac{4\lambda_1 t}{n+3}\right). \tag{24}$$

Energy dissipation rates (22) and (24) allow us to formulate the necessary condition for the transfer from the mode described by sets {7} and {8} of Equation (5) to sets {4} and {5}, as required at the end of the first section. This condition is written as

$$\lambda_2 > \frac{4\lambda_1}{n+3}. \tag{25}$$

Summarizing the results of the second and third sections, it is possible to formulate the conditions of an efficient energy transfer to the nonlinear normal mode if the energy is initially localized at the linear oscillator:

1. The initial impact and amplitude must be sufficient for the localized resonant mode to be excited.
2. The damping coefficients must obey the relationship (25).

It should be specially emphasized that both conditions are necessary but not sufficient. They provide certain guidelines for designing the system of coupled oscillators with effective pumping but further analysis is needed to establish the exact conditions of this process for each concrete system.

Some additional features of the process of the energy pumping may be revealed by means of numerical simulation.

Let us consider the system of two oscillators (1) with two sets of parameters, the only difference being the value of the damping coefficients (see Figures 3a, 3b and Figures 4a, 4b and the figure captions). The plots in Figures 3b and 4b demonstrate the energy distribution between the oscillators. The energy distribution is characterized below by the ratio:

$$Q(y_1, y_2, t) = \ln \frac{\dot{y}_1^2 + \omega_1^2 y_1^2 + \frac{2C}{n+1} y_1^{n+1}}{\dot{y}_2^2 + \omega_2^2 y_2^2}. \tag{26}$$

Characterization of the energy distribution by relation (26) is not exact since it neglects the damping terms and the energy of coupling between the oscillators. The approximate nature of this criterion will lead to strange 'back and force energy pumping' (Figures 3b–7b) at the end of the process when the regime of vibrations must be nearly linear and therefore the modal distribution of energy will be constant, as was mentioned above. Still, it may be used as qualitative characteristics for the case of highly nonlinear vibrations which is of main interest.

The situations in Figures 3 and 4 are similar but different. Both cases correspond to the minimum of the excitation energy (the case in Figure 3 corresponds to numerical example of the excitation of the damped oscillator presented at the end of the previous section with a

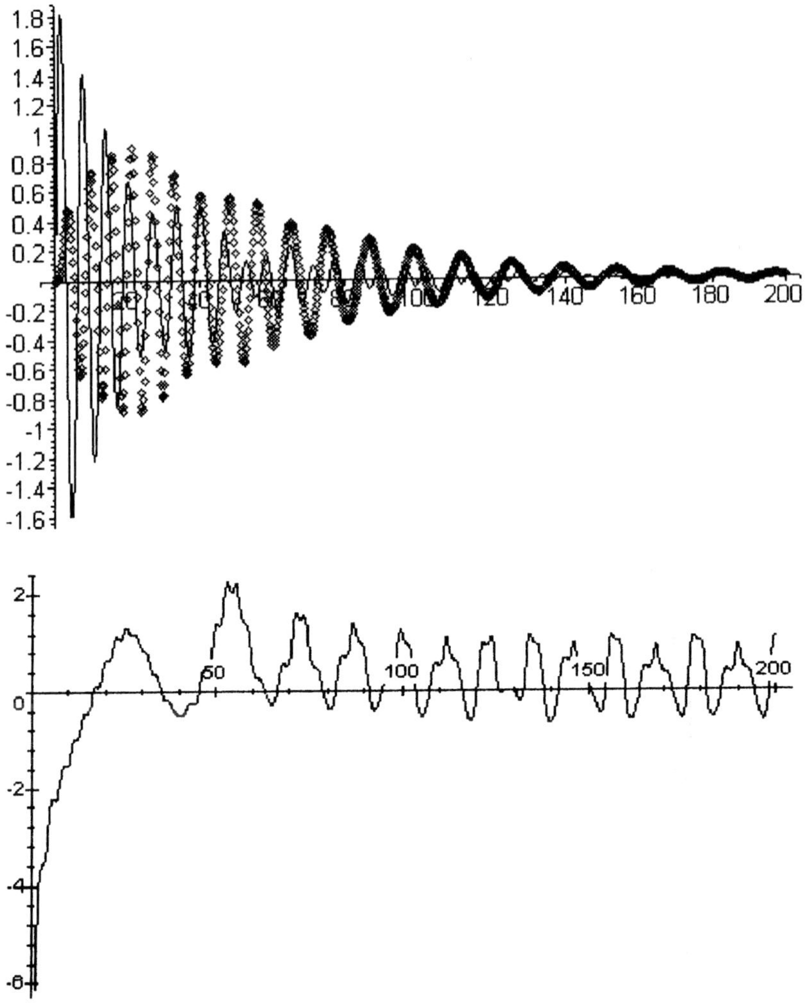

(a)

(b)

Figure 3. (a) Transition of energy in a coupled system of two oscillators via excitation of the resonant mode (displacements *versus* time). The displacement of the first oscillator is depicted by dots and the second by a solid line; $n = 3$, $C = 2$, $\lambda_1 = \lambda_2 = 0.05$, $\omega_1^2 = 0.1$, $\omega_2^2 = 0.9$, $\alpha = 0$, $\beta = 1.885$, $\varepsilon = 0.1$. (b) Energy distribution between oscillators, $Q(y)$ *versus* time.

small correction due to the nonzero linear frequency of the first oscillator) and the damping coefficients obey condition (25). In both cases, the energy is pumped to the first oscillator (the initial conditions in both cases correspond to impact of the second oscillator). However, in the system presented in Figure 3 no less than 50% of the energy of the system remains stored at the first oscillator after the initial transition whereas in the system presented in Figure 4, the energy is almost completely returned to the second oscillator.

The physical reason of this difference is rather obvious. Thorough analysis of the plots 3a and 4a show that in the former case, the smooth damping of the amplitude of the first oscillator begins 'immediately' after the excitation of the harmonic regime. In the latter case, the damping starts after the cycle of modulated vibrations of the first oscillator. So, as was mentioned above, the return of energy to the second oscillator occurs because of amplitude

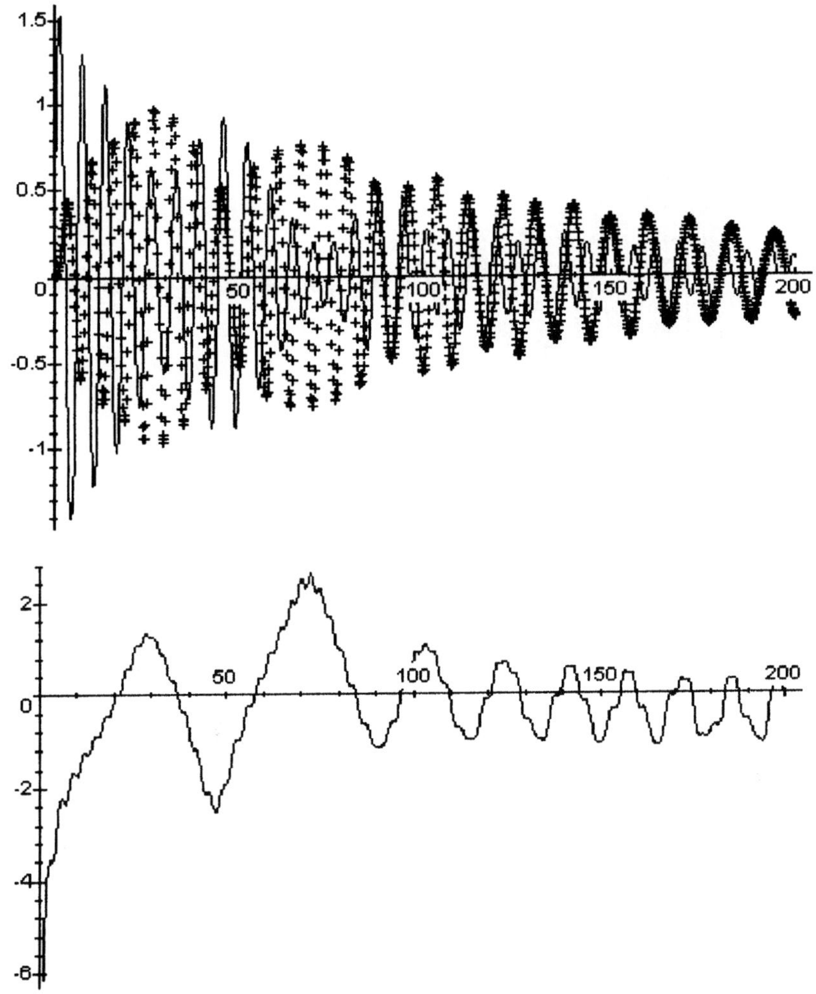

(a)

(b)

Figure 4. (a) Transition of energy in a coupled system of two oscillators via excitation of the resonant mode (displacements *versus* time). The displacement of the first oscillator is depicted by dots and the second by a solid line; $n = 3$, $C = 2$, $\lambda_1 = \lambda_2 = 0.02$, $\omega_1^2 = 0.1$, $\omega_2^2 = 0.9$, $\alpha = 0$, $\beta = 1.55$, $\varepsilon = 0.1$. (b) Energy distribution between oscillators, $Q(y)$ *versus* time.

modulation of the harmonic vibrations. In the former case, the characteristic time of damping is less than the half-period of modulation and in the latter case, the situation is opposite.

Both situations have their advantages and shortcomings from the viewpoint of applications. In the former case, the localization of energy in the nonlinear oscillator is very good but the damping is comparatively large and the characteristic time of 'pumping' (which is the time point when the condition $Q(y) = 0$ holds at the first time) is more than the characteristic time of damping (inverse damping coefficient). In other terms, the energy is dissipated quicker than when pumped to the first oscillator.

In the latter case the situation is opposite. The best performance should be expected from the system where both effect (damping and modulation) are balanced. Thus, the additional condition may be formulated as:

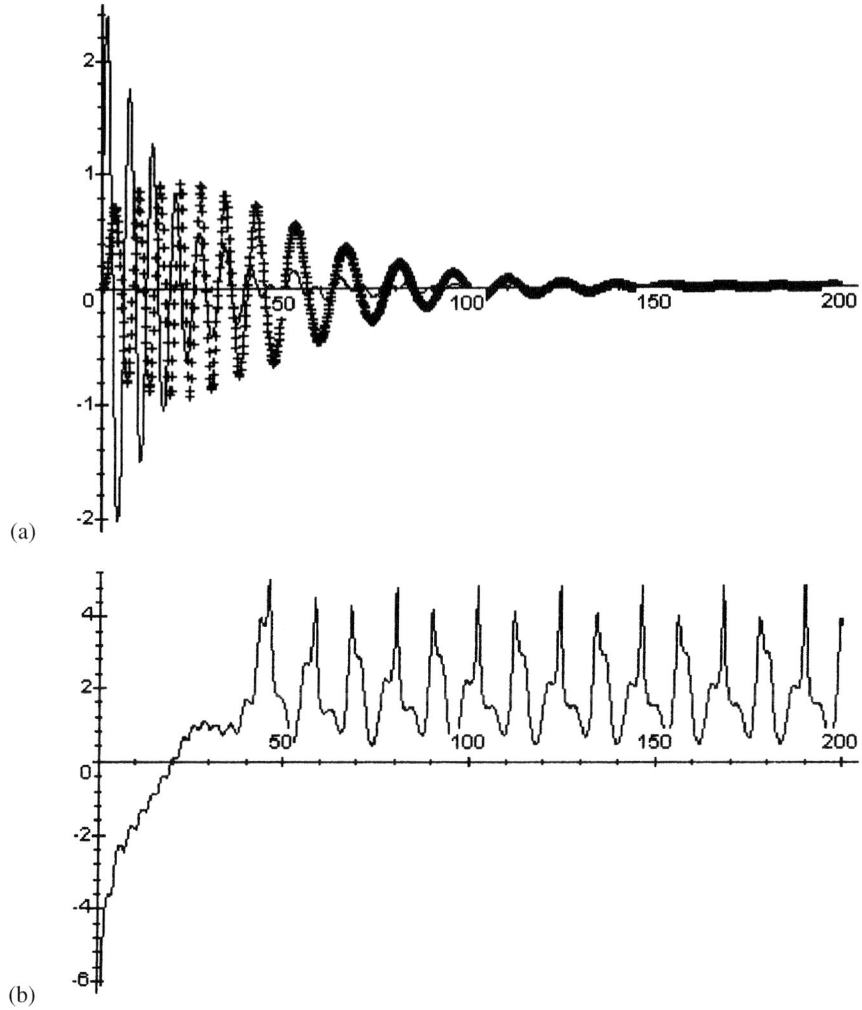

Figure 5. (a) Transition of energy in a coupled system of two oscillators via excitation of the resonant mode (displacements *versus* time). The displacement of the first oscillator is depicted by dots and the second by a solid line; $n = 7$, $C = 4$, $\lambda_1 = \lambda_2 = 0.07$, $\omega_1^2 = 0.1$, $\omega_2^2 = 0.9$, $\alpha = 0$, $\beta = 2.52$, $\varepsilon = 0.1$. (b) Energy distribution between oscillators, $Q(y)$ *versus* time.

3. The characteristic times of damping and modulation are to be close.

This condition is, of course, neither necessary nor sufficient, unlike conditions (1) and (2) formulated above. However, this condition should be also taken into account when designing the concrete system.

One more possibility for a good design of the system is to use the higher powers of the nonlinear term in order to improve the inequality in condition (25). Two additional examples plotted in Figures 5a, 5b and 6a, 6b are designed according to conditions and guidelines (1–3). They are characterized by some rather fast pumping of energy and rather good storage at the first oscillator and may be thus considered as 'high-performance' systems.

(a)

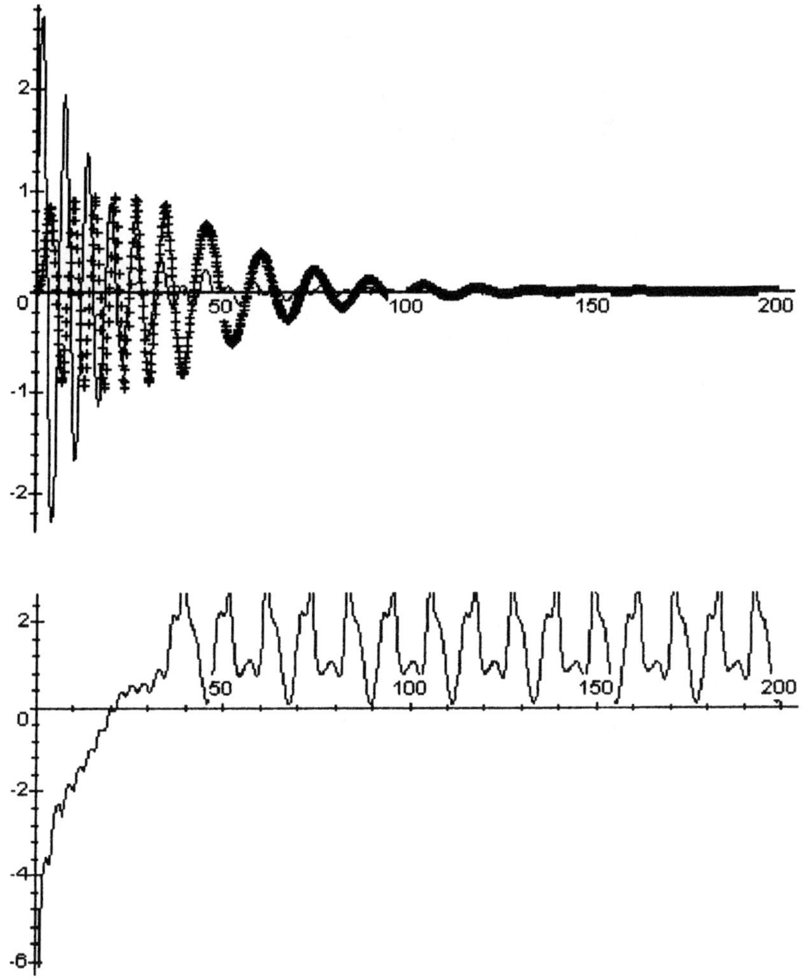

(b)

Figure 6. (a) Transition of energy in a coupled system of two oscillators via excitation of the resonant mode (displacements *versus* time). The displacement of the first oscillator is depicted by dots and the second by a solid line; $n = 15$, $C = 8$, $\lambda_1 = \lambda_2 = 0.08$, $\omega_1^2 = 0.1$, $\omega_2^2 = 0.9$, $\alpha = 0$, $\beta = 2.90$, $\varepsilon = 0.1$. (b) Energy distribution between oscillators, $Q(y)$ *versus* time.

4. Discussion and Concluding Remarks

The results of the previous sections allow us to conclude that the process of energy 'pumping' between two localized modes in a highly asymmetric system of coupled linear and nonlinear oscillators has been investigated. The conditions of an effective energy transfer to a localized normal mode were revealed and guidelines for a good design of such a system were established. The results provided by analytical investigations are in reasonable agreement with the numerical simulation data.

It should be mentioned that the use of a highly nonlinear oscillator as a vibration absorber (which seems to be most straightforward application of the process investigated above) has an interesting peculiarity, which is not typical for linear or weakly nonlinear absorbers. It is frequency-independent in the sense that it may be effective at any external frequency if the

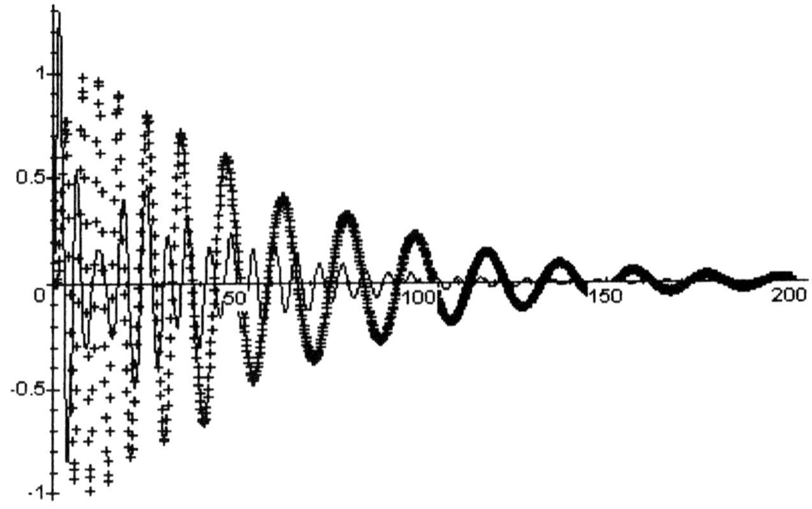

(a)

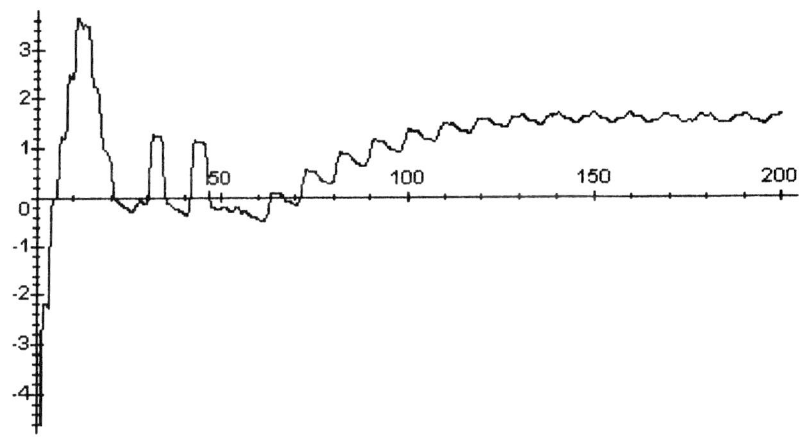

(b)

Figure 7. (a) Transition of energy in a coupled system of two oscillators via excitation of the resonant mode – the case of parametric coupling, Equation (28) (displacements *versus* time). The displacement of the first oscillator is depicted by dots and the second by a solid line; $n = 9$, $C = 5$, $\lambda_1 = \lambda_2 = 0.05$, $\omega_1^2 = 0.1$, $\omega_2^2 = 1$, $\alpha = 0$, $\beta = 1.47$, $\varepsilon = 0.27$. (b) Energy distribution between oscillators, $Q(y)$ *versus* time.

amplitude is sufficient enough to excite the highly nonlinear vibrations. In fact, all numerical examples investigated above used pure impact (including all frequencies) as the initial condition.

One important shortcoming of the passive system of coupled oscillators used as a vibration absorber lays in the fact that it is not active below the critical amplitude and its effectiveness is reduced while the amplitude grows above the critical resonant regime (simply because the nonlinear oscillator cannot absorb more than a certain amount of energy at a certain frequency). The way to overcome this shortcoming may be active control. Such control may effect the parameters of the system in order to get closer to the critical regime. The parameters of the system which have to be controlled may include the order of nonlinearity, the coupling and the initial conditions for the first oscillator – as it was demonstrated, the goal may be achieved by reasonably changing these values. The realization of such control schemes may be an interesting challenge.

Another challenge is the investigation of another mechanism of coupling between the oscillators, namely the parametric mechanism of energy transfer which usually occurs to be more effective than ordinary linear coupling. The system of equations which 'realizes' this mechanism in a most simple manner may be written as

$$\begin{aligned} &\ddot{y}_1 + \lambda_1 y_1 + \omega_1^2 y_1 + C y_1^n + \varepsilon(y_1 - y_2)^3 = 0, \\ &\ddot{y}_2 + \lambda_2 y_2 + \omega_2^2 y_2 + \varepsilon(y_2 - y_1)^3 = 0. \end{aligned} \tag{27}$$

The coupling terms are cubic and therefore the main mechanism of the energy transfer should be a nonlinear parametric resonance. System (28) may be much more effective than the system with linear coupling investigated above, as it is demonstrated in Figures 7a, 7b. Still, the conditions of energy 'pumping' in such a system are not yet sufficiently understood.

Finally, the problems of real applications of a highly nonlinear vibration absorber require the investigation of more complicated systems, especially a nonlinear oscillator coupled to the system with more than one degree of freedom.

Acknowledgments

The author is very grateful to Professor Alexander Vakakis from the University of Illinois, Urbana-Champaign, and to Professor Leonid Manevitch from the Institute of Chemical Physics, Moscow, for useful discussions and suggestions. The author is also grateful to Russian Foundation of Basic Research for partial financial support of this work (grant 98-03-33366a).

References

1. Vakakis, A. F., Manevitch, L. I., Mikhlin, Y. V., Pilipchuk, V. N., and Zevin, A. A., *Normal Modes and Localization in Nonlinear Systems*, Wiley Interscience, New York, 1996.
2. Kosevitch, A. M. and Kovalev, A. S., *Introduction to Nonlinear Physical Mechanics*, Naukova Dumka, Kiev, 1989 [in Russian].
3. Ablowitz, M. J. and Segur, H., *Solitons and the Inverse Scattering Transform*, SIAM, Philadelphia, PA, 1981.
4. Landau, L. D. and Lifshits, E. M., *Theoretical Physics, Vol. 1: Mechanics*, Nauka, Moscow, 1988 [in Russian].
5. Manevitch, L. I. and Cherevatsky, B. P., 'On the approximate description of normal vibrations in the nonlinear 2DOF system', in *Problems of Toughness, Reliability and Fracture of Mechanical Systems*, Dnepropetrovsk, 1969, pp. 26–34 [in Russian].
6. Hayashi, C., *Nonlinear Oscillations in Physical Systems*, McGraw-Hill, New York, 1964.
7. Salenger, G., Vakakis, A. F., Gendelman, O., Manevitch, L., and Andrianov, I., 'Transitions from strongly- to weakly-nonlinear motions of damped nonlinear oscillators', *Nonlinear Dynamics* **20**(2), 1999, 99–114.
8. Rosenberg, R., 'The Ateb(h)-functions and their properties', *Quarterly of Applied Mathematics* **21**(1), 1963, 37–47.

Nonlinear Dynamics **25:** 255–274, 2001.

Application of Nonlinear Normal Mode Analysis to the Nonlinear and Coupled Dynamics of a Floating Offshore Platform with Damping

JEFFREY M. FALZARANO, ROBERT E. CLAGUE*, and RAVIKIRAN S. KOTA**
School of Naval Architecture, University of New Orleans, New Orleans, LA 70148, U.S.A.

(Received: 10 March 1999; accepted: 17 November 1999)

Abstract. The nonlinear dynamics of ships and floating offshore platforms has attracted much attention over the last several years. However the topic of multiple-degrees-of-freedom systems has almost been completely ignored with very few exceptions. This is probably due to the complexity of analyzing strongly nonlinear and coupled systems. It turns out that coupling may be particularly important for certain critical dynamics such as the dynamics of a floating offshore platform about its diagonal axis. In a previous work, Kota et al. [1] applied the recently developed nonlinear normal mode technique to analyze the coupled nonlinear dynamics of a floating offshore platform. Although this previous work was restricted to unforced and undamped systems, in this work a comparison of the two alternative nonlinear normal mode analysis techniques was completed. Considering the relative practical importance of damping *versus* external forcing for this system, in the present work, we utilize just one of the two major techniques available [2] to analyze damped multiple-degrees-of-freedom nonlinear dynamics. Specifically, we investigate the effect of nonlinearity, and non-proportionate damping. Our results show that this technique allows one to simply consider the effect of nonlinearity and general damping on the resulting normal modes. This technique is particularly powerful because it allows one to visualize the modes in a geometric fashion using the invariant manifold concept from dynamical systems.

Keywords: Nonlinear dynamical systems, nonlinear normal modes, MODU, dynamical stability, floating platform motion.

1. Introduction

The methods of nonlinear normal modes (NNMs) have found widespread use in engineering mechanics predominantely for the study of nonlinear structural vibrations. The use herein is therefore a somewhat unique application because it involves the study of coupled multiple-degrees-of-freedom (MDOF) rigid body dynamics of a floating body. The application is that of a floating semi-submersible column stabilized offshore oil drilling rig rotating about its critical (approximately diagonal) axis.

The global nonlinear dynamics have been studied extensively in a series of previous papers [3]. However, this paper focuses on continuing the analysis begun in a recent previous paper [1], where NNMs were used to study the coupled nonlinear dynamics of an actual drilling rig. However in that previous work forcing and damping were neglected. In that paper, the two major NNM analysis approaches, i.e., [2, 4], where simultaneously applied to study the coupled global nonlinear dynamics in an effort to gain experience with and to compare the two methods. Our previous experience [5] suggests that although the forcing might be neglected

* Currently at Friede and Goldman Ltd., Houston, TX 77079, U.S.A.
** Currently at Brown and Root Deepwater Div., Houston, TX 77072, U.S.A.

in practical application, the damping probably should not be ignored. Physically this is due to the fact that for a vessel of this type most of the hull volume is submerged far below the free surface and relatively unaffected by the external wave excitation. The damping is not only important with regards to decaying free oscillations and limiting resonant amplitudes but also in modifying the size of the safe basin [5]. The experience gained during the previous investigation where the two alternative approaches were simultaneously applied to analyze the exact same physical problem has led us to believe that the so called Invariant manifold approach is most easily extendable to consider systems with damping. As such, this paper will exclusively utilize this method. Essentially this paper will provide results for how the NNMs are affected by various forms and amounts of damping in order to provide insight into how the coupled nonlinear dynamics are effected by the damping.

This paper being part of a special issue dealing exclusively with NNMs and localization this paper will limit its general discussion of the NNM technique and instead refer the reader to the other papers in this issue, the textbook written by Vakakis et al. [4] and our previous work in this area [1, 6]. However in order for this paper to be relatively self-contained we will describe the physical modeling as completely as possible.

2. Background

This study is the result of a series of investigations which have analyzed the MDOF nonlinear dynamics of marine vehicles using modern geometric methods over the last several years. These investigations have only been possible due to the developments outside the marine dynamics field in applied mathematics and mechanics [7, 8]. These previous investigations have considered steady state MDOF ship motions at various headings [9, 10] and transient global nonlinear dynamics of single degree of freedom (SDOF) [11] and transient global nonlinear dynamics of ship roll motion coupled with sway and yaw [3, 12]. This work was subsequently extended to consider the nonlinear dynamics of floating offshore platforms [5]. Most recently our work focused on the coupled nonlinear dynamics of floating offshore platforms about their critical axis [13]. In that work it was realized that the coupling of the out-of-plane righting arm although small at small angles continuously increased and became quite large as one approached the in-plane angle of vanishing stability. It was realized at that time that the powerful methods of NNMs being developed outside the marine field by Vakakis et al. [4], and Shaw and Pierre [2] might provide valuable insight into this important problem. This paper summarizes the most significant results of this investigation; more complete details are described by Kota and co-workers [1, 6].

3. Problem Formulation

3.1. Description of the Explicit Physical Model

The partially linearized 6-DOF equations of motion of a vessel, in the vessel coordinate system with nonlinear restoring force/moment [14], can be written in a vector form as

$$(\mathbf{M} + \mathbf{A}(\omega))\underline{\ddot{x}} + \mathbf{B}(\omega)\underline{\dot{x}} + \mathbf{C}\underline{x} + \underline{g}(\underline{x}) = \underline{F}(t). \tag{1}$$

In the above equation, the vector $\mathbf{x}$ represents the six degrees of motion, $\mathbf{M}$ is the 6×6 physical mass/inertia and dynamic coupling (in air) matrix, $\mathbf{A}$ and $\mathbf{B}$ are the hydrodynamic added mass and damping matrices, **C** is the linear part of the hydrostatic restoring force/moment, and the vector **g** is the nonlinear component of the restoring force/moment. Note that **A** and **B** are frequency dependent. The system is now modeled as a forced, nonlinear spring mass damper system which can be analyzed for response to any external input (initial conditions or forcing) using various techniques.

The focus of attention in this study is the dynamics of tilt about an arbitrary horizontal axis at an angle of orientation α with respect to the vessel's principal longitudinal axis. So, the roll and pitch equations are isolated from the general 6-DOF equation above, assuming that neither is coupled, to linear approximation, with any other degree of freedom statically or inertially. Thus, we have the following equations:

$$\text{Roll}: \qquad [I_{44} + A_{44}(\omega)\ddot{\varphi} + B_{44}(\omega)\dot{\varphi} + \Delta GZ(\varphi) = F_4 \cos(\omega t + \gamma_4), \tag{2a}$$

$$\text{Pitch}: \qquad [I_{55} + A_{55}(\omega)]\ddot{\theta} + B_{55}(\omega)\dot{\theta} + \Delta GZ(\theta) = F_5 \cos(\omega t + \gamma_5). \tag{2b}$$

In these equations the variables, φ and θ are the roll ($i = 4$) and pitch ($i = 5$) coordinates, while I_{ii} and Δ represent physical mass moment of inertia and displacement, respectively. A_{ii} and B_{ii} are the frequency dependent hydrodynamic added moment of inertia and radiated wave damping respectively, $GZ(\cdot)$ is the nonlinear polynomial approximations to the static restoring arm. F_i is the amplitude and γ_i the phase of wave forcing at a frequency of ω.

Along the same lines, an SDOF equation of motion for general tilt angle dynamics about an orientation angle a can be expressed as

$$\text{Tilt}: \quad [I_\alpha + A_\alpha(\omega)]\ddot{\xi} + B_\alpha(\omega)\dot{\xi} + \Delta GZ(\xi) = F_\alpha \cos(\omega t + \gamma_\alpha). \tag{3}$$

In this equation, ξ is the tilt coordinate. I_α, A_α and B_α are the physical mass moment of inertia, hydrodynamic added inertia and radiated wave damping respectively, evaluated about the tilt orientation axis, α. F_α and γ_α are the amplitude and phase of the wave force (moment) again evaluated about the orientation axis. This equation is obtained from (2) using a simple coordinate transformation

When the coupling of the out-of-plane rotation to the in-plane rotation is considered, a 2-DOF analysis is essential. So, the two principal coordinates are – the in-plane rotation, ξ; and the out-of-plane rotation, ζ. This 2-DOF system is nonlinear and coupled dynamically and statically since the axis of orientation and axis of tilt need not necessarily be in the planes of symmetry of the vessel. Thus, the 2-DOF equations of motion in vector form are

$$(\mathbf{I}_\alpha + \mathbf{A}_\alpha(\omega)\ddot{\underline{\mathbf{z}}} + \mathbf{B}_\alpha(\omega)\dot{\underline{\mathbf{z}}} + \Delta \underline{GZ}(\underline{z}) = \underline{F_\alpha} \cos(\omega t + \underline{\gamma_\alpha}). \tag{4}$$

In the equation above, the matrices, shown in bold, are full with non-zero cross-diagonal terms while the vector $\underline{GZ}(z)$ is nonlinear and coupled. The vector $\underline{z}$ is given by

$$\underline{z} = \begin{pmatrix} \xi \\ \zeta \end{pmatrix}. \tag{5}$$

This completes the problem formulation for the tilt motions of a semi-submersible about an arbitrary tilt orientation axis.

The damping for a floating body such as is being considered herein arises from two sources, i.e., the energy dissipated due to radiated waves and the energy dissipated due to viscous and rotational fluid effects. The linear radiated wave damping is accurately calculated using state of the art potential flow based ship/platform computer programs. However, it is well known by ship motion practioneers, that under certain circumstances when the radiated wave damping is small that an accurate prediction of the viscous and rotational flow damping is required. Unfortunately, the prediction of the non-ideal flow damping is not as well established as is the prediction of potential flow radiated wave damping. Various methods are used in practice which include performance of specialized small scale physical model tests, the use of empirical data and the direct numerical computation using computational fluid mechanics (CFD).

The model testing option is by far the most common in practice but unfortunately is fundamentally unsound for this specific purpose. The problem with model testing arises due to the conflicting demands of two different scaling laws, i.e., Froude number $V/\sqrt{gL}$ and Reynolds number VL/ν. Where V is the characteristic speed, L is a characteristic length, g is the gravitational acceleration and ν is the fluid kinematic viscosity. The satisfaction of Froude scaling requires that the full scale prototype and the small scale model test have this same Froude number which suggests that the ratio of the inertial and gravity forces are the same for the model and full scale. The satisfaction of Reynolds scaling requires that the model and full scale Reynolds numbers be the same and this assures that the ratio of the inertial and viscous forces will be the same. Ship and offshore platform model testing typically satisfies the Froude scaling due to the importance of gravity water waves. Unfortunately, it is then generally impossible to simultaneously satisfy the Reynolds scaling. The error introduced by incorrectly satisfying the Reynolds number and therefore incorrectly predicting the viscous forces is generally referred to as scale effects.

Another option is to include the viscous effects using empirical component-wise data; this method suffers from the fact that much of the empirical data have been obtained from improperly scaled model tests or exists only over a limited range. A more accurate alternative is to directly predict the damping using CFD. Few successful attempts have been made in this area to date although this promises to be the most attractive alternative in the future because of its ability to eliminate scale effects errors [15].

However,it should be noted here that the only nonlinearities considered as those due to the static restoring moments. This is essentially due to the fact that only linear hydrodynamics were available and the insignificance of the nonlinear inertial terms. Moreover, this implies that this analysis is essentially a low frequency analysis which is what ship/platform dynamics typically are. The insignificance of the nonlinear inertial terms in the body fixed equations of motion for ship/platform motions is in direct contrast to certain flight dynamics maneuvers where the nonlinear inertial terms are in fact dominate [16].

4. Theoretical Background

The methods of normal mode analysis are well established for linear MDOF discrete and continuous vibratory systems. However, the logical question to ask is what happens when the systems are nonlinear? Are the same techniques applicable and are the same or similar results realized? A powerful technique used to analyze normal modes in nonlinear systems is *nonlinear normal modes* (NNMs). In the area of NNMs, two alternative analysis techniques have evolved, the first is more compact and mathematically elegant and easily extendable to continuous systems [4] while the second is more heuristic yet possibly more generally applicable to a wider class of problems including damped and possibly forced systems [2].

In this section, the analytical techniques implemented to analyze the coupled dynamics of a MODU about its critical axis are described. These techniques are the theory of NNMs which is the main emphasis of this paper. After an introductory overview, the second method of analysis for MDOF systems using NNMs is described. The first methodology is described and utilized in a previous paper [1].

4.1. Nonlinear Normal Mode Analysis

An efficient approach to understand the dynamics of MDOF nonlinear systems is by the use of NNMs. The idea of using NNMs for such systems was first proposed by Rosenberg [18]. Many researchers have extended his theory to seek normal modes of mainly discrete, homogenous, conservative nonlinear oscillators. Certain unique properties, like mode bifurcations and localizations, of NNMs were demonstrated by Vakakis et al. [4]. Shaw and Pierre [2] developed a fundamentally different definition and approach to obtain NNMs which is also applicable to non-conservative systems. The emphasis of this paper is on the application of the methodology suggested by Shaw and Pierre [2] to the problem of platform motions about their critical axis. Rosenberg's definition of NNMs and his classification of two qualitatively distinct types of normal modes is presented followed by a brief description of the aforementioned methodology. A more complete description is included in [6]. Results are discussed in the next section.

The existence of normal modes and their useful application in obtaining 'natural' free vibrations and steady state forced vibrations is well established for linear conservative systems with many degrees of freedom. This approach is also applicable to non-conservative systems for the special case of proportional damping [17]. Having known the normal modes of motion for an N-DOF linear system, the transient response of the system can easily be expressed as a linear combination of the N normal modes using the modal analysis technique.

The central importance of normal modes in simplifying the analysis of coupled N-DOF linear systems has motivated the search for a similar class of solutions for nonlinear systems. It has been shown by Rosenberg [18] and Vakakis et al. [4] that although the principle of superposition does not hold for nonlinear systems, certain physical phenomena like resonance occur in the neighborhood of normal mode vibrations regardless of the system being linear or nonlinear. However, for nonlinear systems it is not possible to obtain normal modes in the same manner as for linear systems. To obtain the normal modes of a linear system, expressed in vector form as

$$(\ddot{x}) + [A](x) = (0), \tag{6}$$

the eigenvalues and the corresponding eigenvectors of the fundamental system matrix (A) are determined. These eigenvalues physically represent the squares of the natural frequencies of the system. Then, the eigenvector associated with each eigenvalue represents the normal mode of vibration of the system at that frequency. However, this procedure of obtaining eigenvalues and hence obtaining eigenvectors (normal modes) is only valid for linear systems. Demonstration of existence of eigenvalues has its roots in the theory of quadratic forms, and its application to vibration problems requires that the potential energy be a quadratic form [18]. For a general nonlinear system, the potential energy need not be a quadratic form. Hence, a new definition of normal modes is necessary which would not only be applicable to a general nonlinear system but is equally valid for linear systems.

According to Rosenberg [18], normal mode vibrations of an autonomous MDOF system are the 'vibrations-in-unison' of that system. This means that during normal mode vibration all masses of the system undergo synchronous periodic motions attaining their maxima, minima and equilibria at the same instant of time [18]. When the system is linear, the motion of every mass during a normal mode vibration is a simple harmonic function of time having the same frequency and the same zeroes as every other, i.e.

$$x_i(t)/x_j(t) = A_i \cos(\omega t)/A_j \cos(\omega t) \equiv c_{ij}, \text{ const.} \tag{7}$$

It was shown that symmetric 2-DOF systems [4] and homogenous n-DOF systems [18] are such that their normal mode vibrations satisfy (7) no matter how strongly nonlinear the systems or how large the displacements [18]. When the displacements of any two masses m_i and m_j during normal mode vibrations are such that

$$x_i(t)/x_j(t) \not\equiv \text{const}, \tag{8}$$

then the normal mode vibration is defined to be *non-similar*.

4.2. Invariant Manifold Approach

A useful and general methodology to obtain the NNMs was proposed by Shaw and Pierre [2]. In this approach, a normal mode motion is defined as one where all displacements and velocities are related to a single reference pair of displacement and velocity. This is in contrast to the other methodolgies wherein only the displacement of the reference point is the dependent variable. As a consequence of considering two dependent variables, the modal trajectory is no longer a curve but a manifold. A key feature of this method is its applicability to a broad class of problems including nonlinear and general non-conservative systems. The applicability of this method to general damped and nonlinear systems is particularly attractive since classical normal modes techniques are not applicable to such systems. Prior to the development of this technique, various specialized or approximate methods where proposed to analyze non-classically damped linear systems.

Recall for linear systems that the normal modes are mutually orthogonal. This leads to the *invariance property* of normal modes, i.e., vibrations started on any one of these modes will remain on that mode for all time with no participation what-so-ever from any of the other modes. This property is also found to be valid for nonlinear systems. The *invariance property* of modal subspaces and the theory of invariant manifolds of dynamic systems form the basis for the definition of NNMs as invariant manifolds in the phase space of the equations of motion governing the dynamical system. An invariant manifold, by definition, is a surface contained in the phase space of a dynamical system which has the property that orbits starting on the surface remain on the surface throughout the course of their dynamical evolution [8].

Consider the N-DOF nonlinear autonomous system expressed as

$$\ddot{x}_i + f_i(\underline{x}, \dot{\underline{x}}) = 0, \quad i = 1, 2, \ldots, n, \tag{9}$$

where $\underline{x}$ is a $(n \times 1)$ vector of displacement and f a vector nonlinear function of displacement and velocity. The governing equations can be expressed in Cauchy standard first-order form as

$$\dot{x}_i = y_i, \quad \dot{y}_i = -f_i(x_i, y_i), \quad i = 1, 2, \ldots, n. \tag{10}$$

Now, we introduce, the as yet unknown functions x and y which are functions relating the displacement and velocity at any point to a single reference pair, say, $(x_1, y_1) = (u, v)$, respectively, i.e.

$$x_i = \hat{x}_i(u, v), \quad y_i = \hat{y}_i(u, v), \quad i = 2, 3, \ldots, n. \tag{11}$$

Using these functional relationships in the equations of motion, we obtain

$$\dot{x}_i = \frac{\partial \hat{x}_i}{\partial u}\dot{u} + \frac{\partial \hat{x}_i}{\partial v}\dot{v}, \quad \dot{y}_i = \frac{\partial \hat{y}_i}{\partial u}\dot{u} + \frac{\partial \hat{y}_i}{\partial v}\dot{v}, \quad i = 2, 3, \ldots, n. \tag{12}$$

Also, for the dynamics of the reference pair (u, v), we obtain

$$\dot{u} = v, \quad \dot{v} = -f_1(u, v; \hat{x}_2, \hat{x}_3, \ldots, \hat{x}_n; \hat{y}_2, \hat{y}_3, \ldots, \hat{y}_n). \tag{13}$$

Substituting (11–13) into (10) for $i = 2, 3, \ldots n$ and thus eliminating time from the equations, we obtain a pair of equations, in the unknown functions $\hat{x}(u, v)$ and $\hat{y}(u, v)$, for each mass point i.

We assume a homogenous power series in u and v for $\hat{x}(u, v)$ and $\hat{y}(u, v)$ with unknown coefficients. By matching each power of u and v term by term starting from the lowest order, a system of simultaneous algebraic equations is obtained for each power. This can be solved for the unknown coefficients and the solution used to obtain the next higher-order coefficients and so on. The functions $\hat{x}(u, v)$ and $\hat{y}(u, v)$ then represent the invariant modal manifolds for the given nonlinear system. Having obtained the solutions for the functions $\hat{x}(u, v)$ and $\hat{y}(u, v)$, (13) can be used to generate the modal dynamics of each mode. If there are no internal resonances or mode bifurcations, the number of real solutions coincides with the number of degrees of freedom of the system. What we have done in the present investigation is to consider the effect of various types and amounts of damping on the resulting NNMs. It is well known from linear MDOF systems that only when the damping matrix is proportionate do classical normal modes exist. Proportionate or classically damped refers to damping matrix being a scalar multiple of either the mass/inertia matrix or the restoring force matrix. Typically damping for MDOF systems is not exactly proportionate in which case classical linear normal modes do not exist. However, since this is a relatively common problem in engineering practice, various methods have evolved to consider the problem and to estimate the error involved in approximating the general system by the uncoupled system [19]. However, it turns out that the invariant manifold approach used here is not restricted to either non-proportionately damped or nonlinear systems.

5. Results

The geometric methods described in the earlier section are used to study the capsizing mechanism of a column-stabilized semi-submersible Mobile Offshore Drilling Unit (MODU) – *Ocean Ranger* [20]. Using an SDOF model, the existence of the critical rotation axis was established [13]. In the present study, a 2-DOF analysis is performed, specifically about the critical axis, in order to investigate the importance of the coupling of the out-of-plane rotations to the in-plane rotations. To study the dynamics of this 2-DOF model, it is essential to decouple the two modes. This is achieved by using the NNM technique described in an earlier section using the invariant manifold approach. Then, the decoupled modes are individually analyzed as SDOF systems to study the dynamics.

The vessel is geometrically and inertially asymmetric about the critical axis. This suggests that an SDOF analysis might not be adequate in determining the overall vessel dynamics and stability. For every orientation, at every tilt angle, there are two righting moments, i.e., an in-plane moment and an out-of-plane moment. At small tilt angles, the magnitude of out-of-plane moment is comparatively small. However, from intermediate to large angles of tilt, its

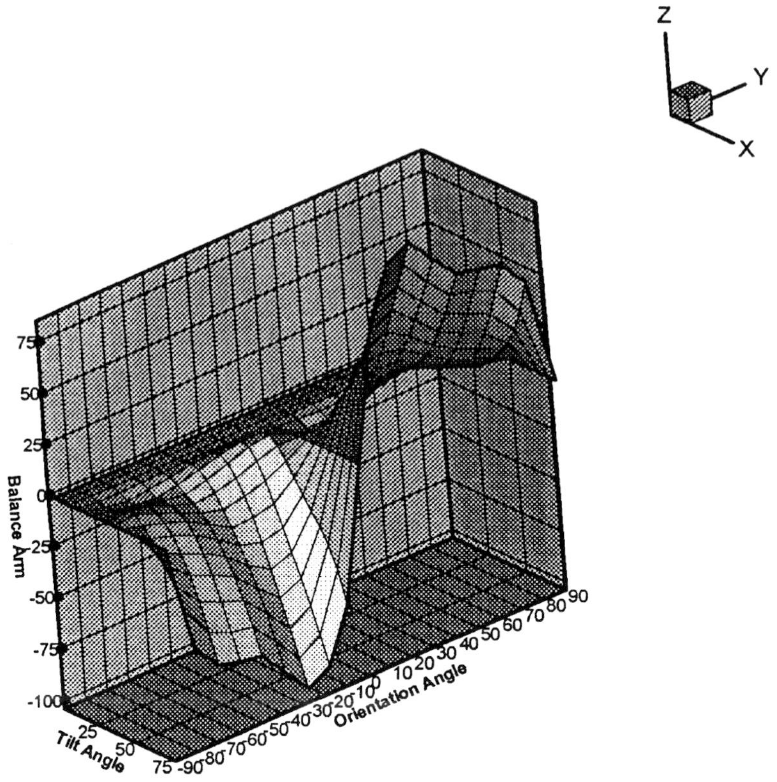

Figure 1. Variation of out-of-plane moment arm with tilt axis orientation.

magnitude rapidly increases while the in-plane decreases. However, both degrees of freedom are strongly coupled statically and dynamically (Figures 2a and 2b). The variation of the out-of-plane moment arm with tilt axis orientation is shown in Figure 1. Moreover, in the present study we will consider damping which is not even approximately proportional.

In the 2-DOF analysis presented here, the effects of damping has now been considered although the effects of forcing have not yet been considered. The NNM technique is used to analyze nonlinear coupled oscillators, as discussed in the last section. Recall that the analysis methodology – the Invariant Manifold approach [2] used to obtain the normal modes of a coupled nonlinear oscillator, were already discussed in the previous section.

First, the dependence of the in-plane righting moment arm and the out-of-plane moment arm with the in-plane tilt angle and out-of-plane tilt angle were obtained. This was achieved by computing the moments against in-plane tilt angle at different fixed out-of-plane tilt angles. These out-of-plane tilt angles are vector sums of appropriate heel and trim. Euler angles were used for the transformations. Then, a homogenous cubic polynomial in two variables, i.e., the in-plane tilt and out-of-plane tilt angles, is used to obtain a surface fit approximation to model the stiffness in the equations of motion. A similar surface is obtained for the out-of-plane moment arm (Figures 2a and 2b).

The invariant manifold approach is implemented to analyze the system. In this approach, the reference pair of coordinates are the in-plane displacement and velocity. As a result, the modal subspace is a surface known as an invariant modal manifold by virtue of its invariance property described in the previous section.

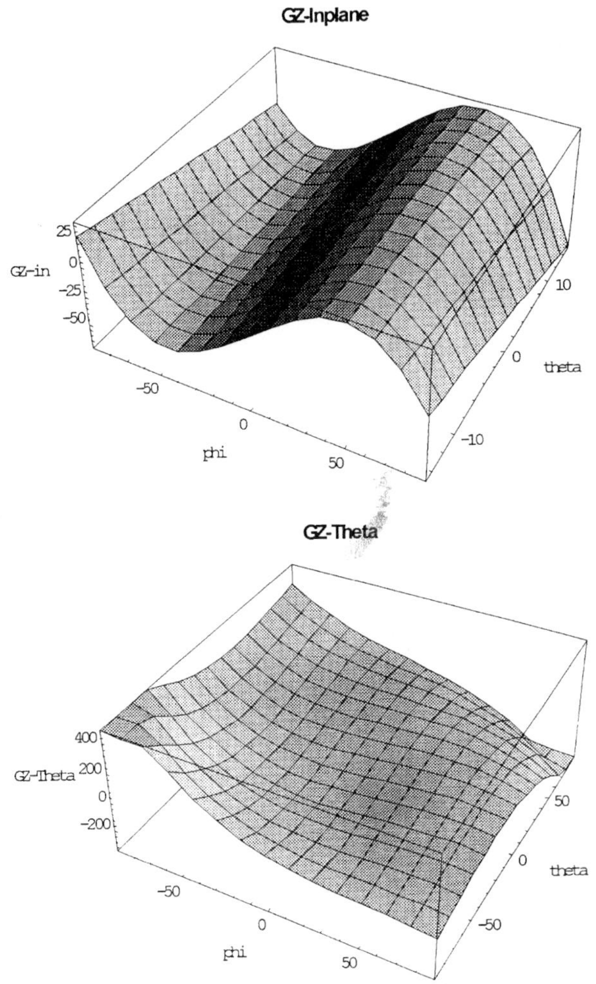

Figure 2. Surfaces of (a) in-plane and (b) out-of-plane, restoring arms.

As described in the last section, the second-order system is expressed in Cauchy standard form, thereby introducing a velocity-dependence. Equations (12) and (13) form the basis of the formulation. In the present problem, the system possesses 2-DOF so that $n = 2$ in Equations (12) and (13). The unknown functions x_1 and y_1 are chosen as bi-cubic polynomials in u and v consistent with the order of the polynomial approximations used for the restoring terms. For the general undamped case, solving the system of algebraic equations yielded two real and four complex solution sets for the coefficients implying two real NNMs. Figure 3 shows the two modes obtained.

Note that for undamped systems such as was considered in the previous investigation [1] and illustrated in Figure 3, that despite the system being undamped there was a velocity dependence in the functions representing the invariant modal manifolds. This dependence, however, only appeared as even powers which could be eliminated using conservation of energy. However, in the present analysis where damping has been included, this is no longer the case and the velocity dependent terms no longer need to appear as even powers but appear

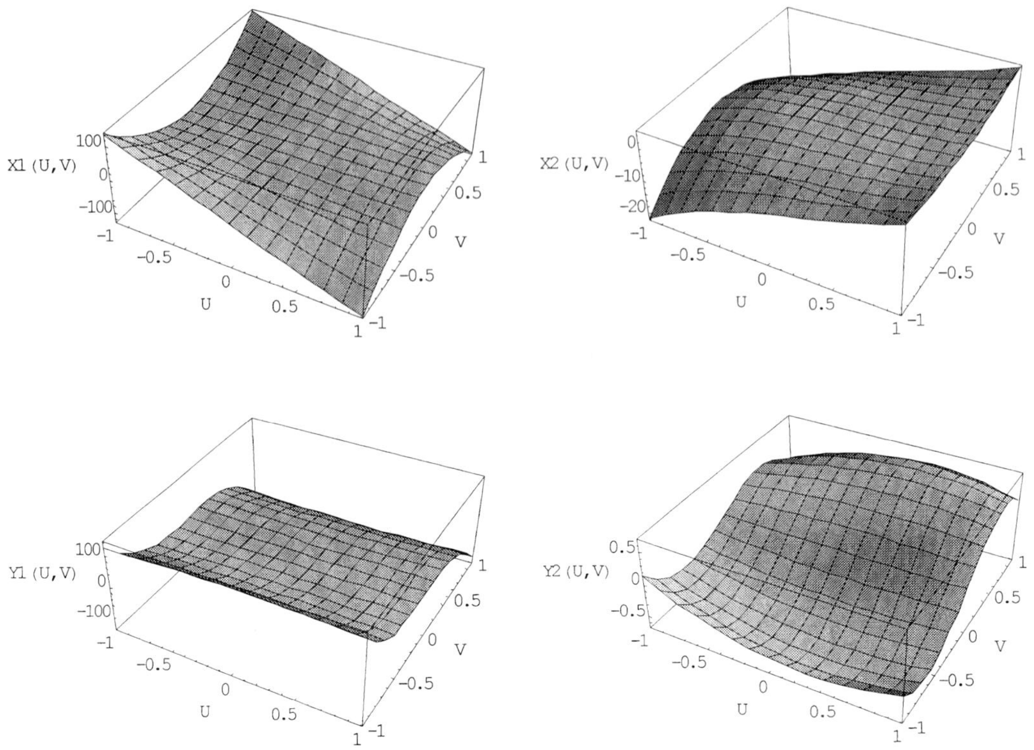

(mode 1) (mode 2)

Figure 3. Invariant modal manifolds for the undamped system.

in a more general dissipative form. Also included are the results (Figure 4) for the linearzized system. One can note that the NNMs are tangent to the linearized system.

The focus of the present work is on a systematic study of the simultaneous effect of non-linearity and general damping on the resulting dynamics. In order to accomplish this goal we ultimately consider the actual system parameters as predicted using a state of the art linear ship hydrodynamics computer program and a general hydrostatics computer program. In order to observe a qualitative change in behavior, we consider various fractions of the total actual damping both less than and equal to the actual radiated wave damping. In order to determine the effects of neglecting the non-proportionate damping, we determine the linear proportionate damping and include various fractions of the non-proportionate damping. We believe that this work is an important step in completely understanding the dynamics of this system.

For comparison, we repeat the results without damping of the previous investigation [1] (see Figure 3). For additional comparison, we include the linearized results in Figure 4. We found for the uncoupled roll and pitch equations that the roll damping was 75% of critical damping while the pitch damping was 200% of critical damping. Moreover, we found that although the in-plane righting arm was restoring, the out-of-plane was not. This resulted in unstable motion in the out-of-plane mode around the zero equilibrium. In order to understand the behavior of the system as effected by damping, we include three sets of systematic results. These results are for proportionate damping only (Figure 5), proportionate plus 50% of non-proportionate damping (Figure 6) and the actual damping (Figure 7) which includes both the

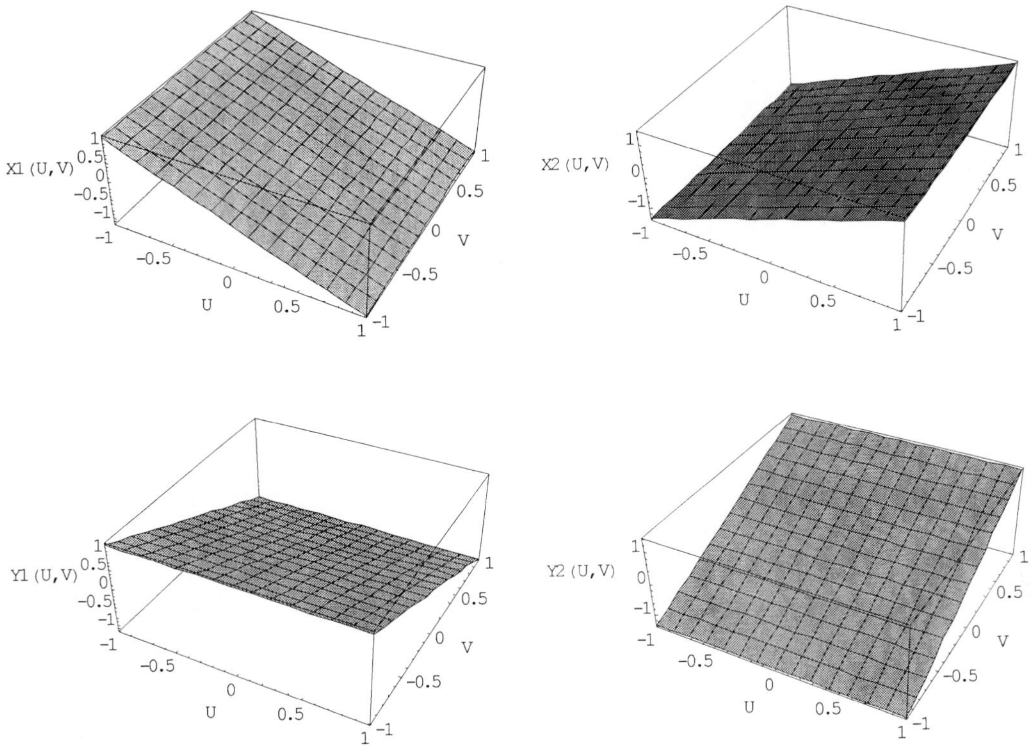

(mode 1) (mode 2)

Figure 4. Invariant modal manifolds for the linearized undamped system.

proportionate and 100% of the non-proportionate damping. For all three sets of results, only a small change in behavior of the modes occurred for damping in range of 25 to 50% of the total. However, in the range of 50 to 75% of the damping, significant qualitative and quantitative changes in behavior of the modes occurred. The actual damped case (Figure 7) experiences dramatic changes in behavior at a relatively small percentage of damping. The 50% of non-proportionate damping case (Figure 6) requires a bit higher percentage of damping for this change. Finally, the just proportionately damped case (Figure 5) requires the highest relative percentage of damping. This makes sense if one considers that the total damping in the system effects this behavior and the proportionate case has the least amount of damping available while the actual damped case has the greatest amount of damping available. The first set of systematic results (Figure 5) are for 0% of non-proportionate damping and various amounts of proportionate damping. These only classically damped results include 25, 50, 75 and 100% of the proportionate damping. The results are not physical but are used to demonstrate the relative importance of neglecting the non-proportionate damping which is typically done in engineering analysis. It should be clear from these results that the error that would exist if the non-proportionate damping was neglected is substantial. From the results given herein and additional results not included, we observed that a significant change in the mode behavior occurs at approximately 89% of the proportionate damping. This change results in a significant qualitative and quantitative modification of first mode's characteristics and a significant quantitative change in the second mode's character. In addition, similar to the behavior found by

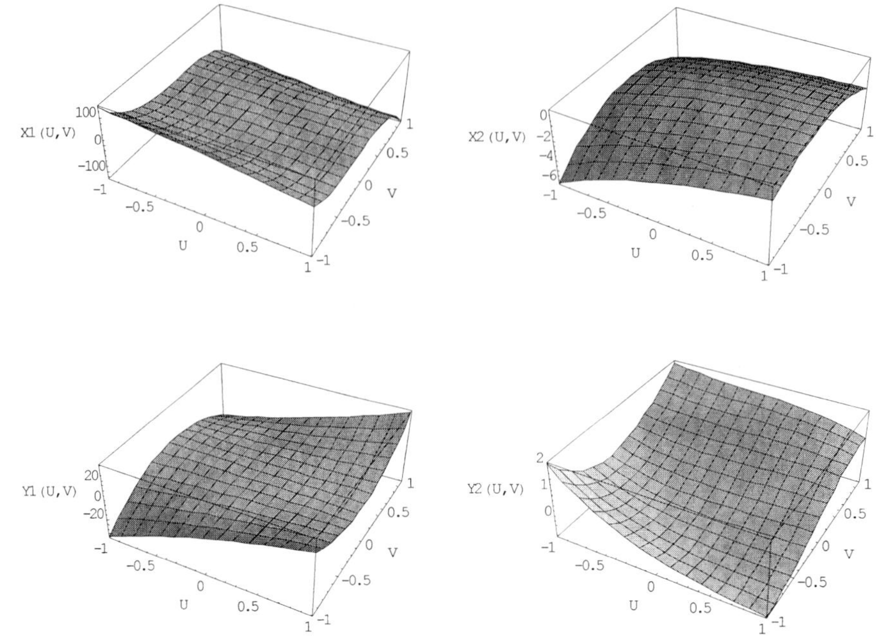

(mode 1) (mode 2)

Figure 5a. Invariant modal manifolds for the system with 25% proportionate damping and 0% non-proportionate damping.

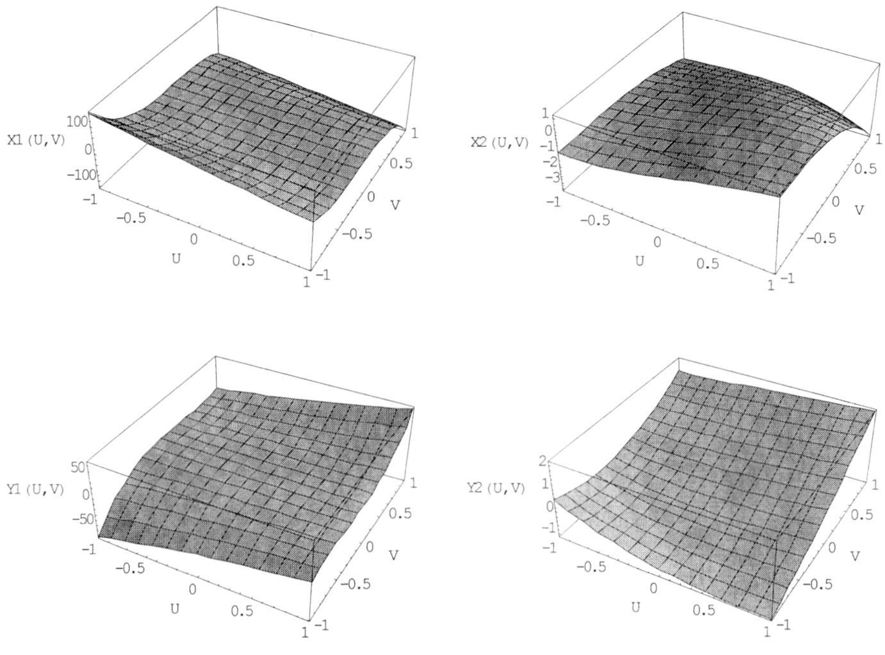

(mode 1) (mode 2)

Figure 5b. Invariant modal manifolds for the system with 50% proportionate damping and 0% non-proportionate damping.

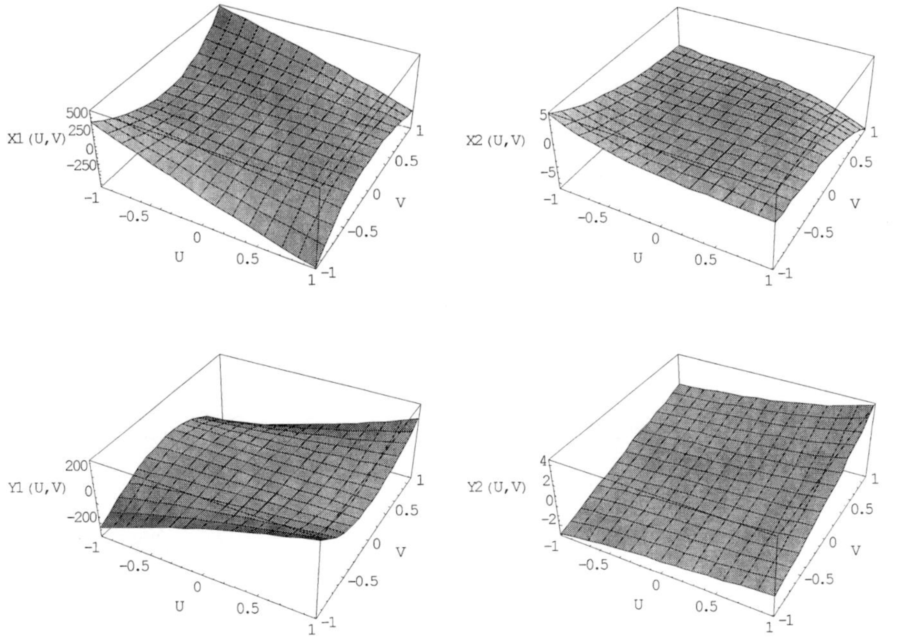

(mode 1) (mode 2)

Figure 5c. Invariant modal manifolds for the system with 75% proportionate damping and 0% non-proportionate damping.

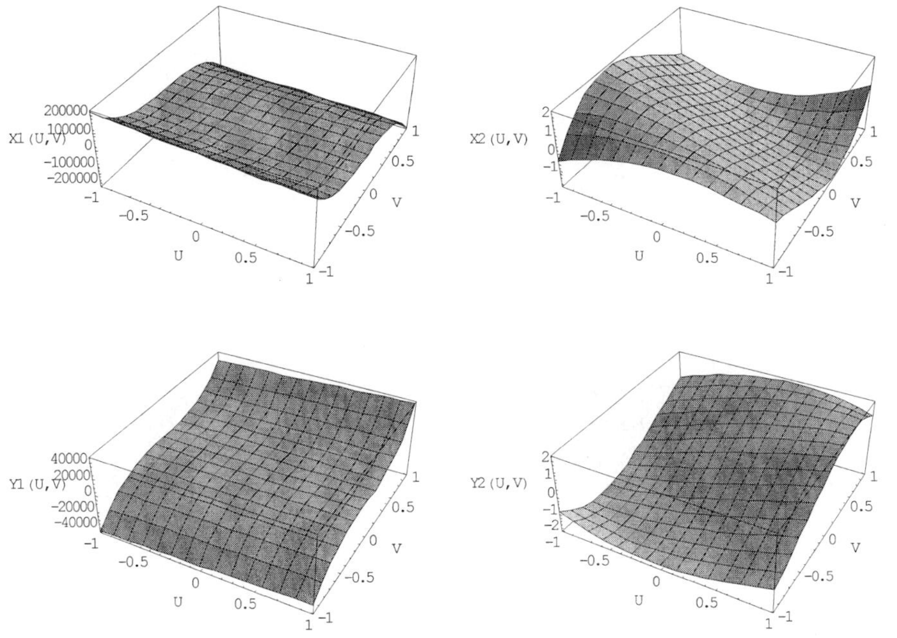

(mode 1) (mode 2)

Figure 5d. Invariant modal manifolds for the system with 100% proportionate damping and 0% non-proportionate damping.

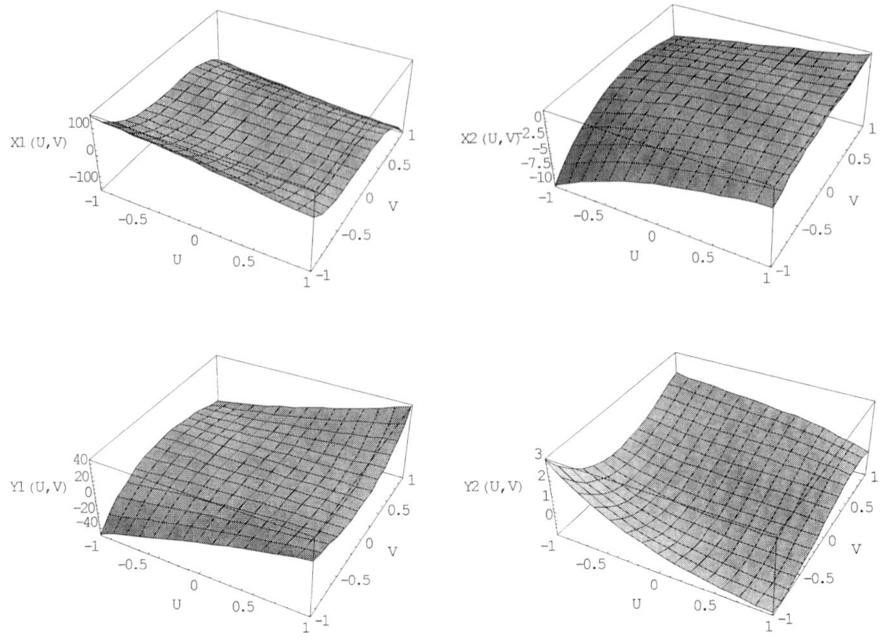

(mode 1) (mode 2)

Figure 6a. Invariant modal manifolds for the system with 25% proportionate damping and 50% non-proportionate damping.

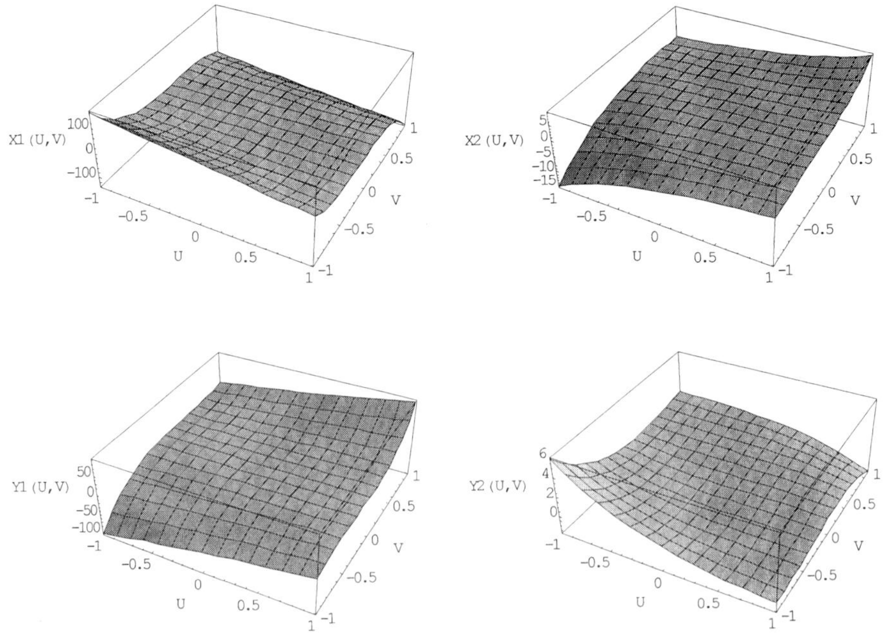

(mode 1) (mode 2)

Figure 6b. Invariant modal manifolds for the system with 50% proportionate damping and 50% non-proportionate damping.

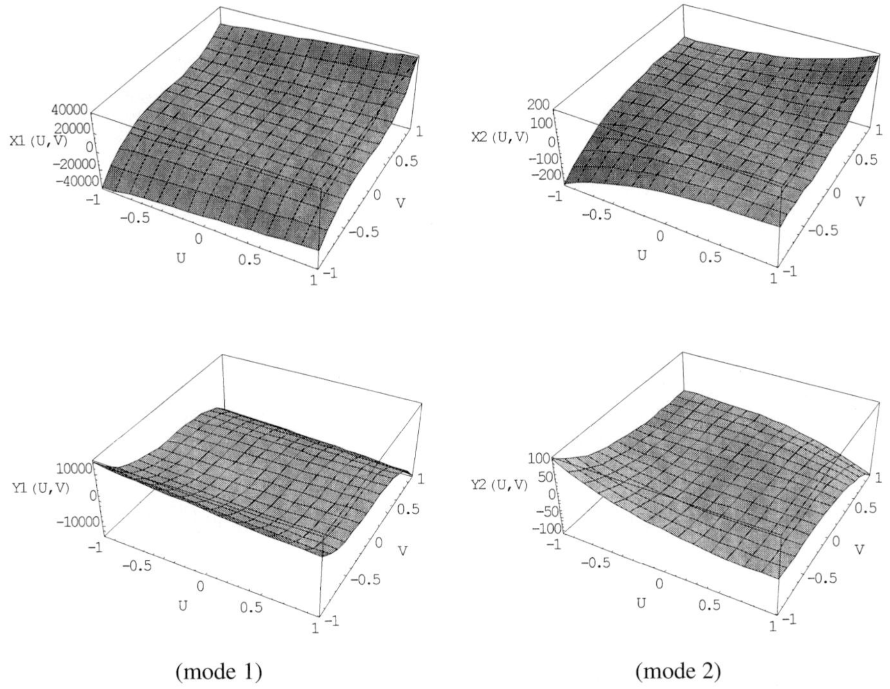

Figure 6c. Invariant modal manifolds for the system with 75% proportionate damping and 50% non-proportionate damping.

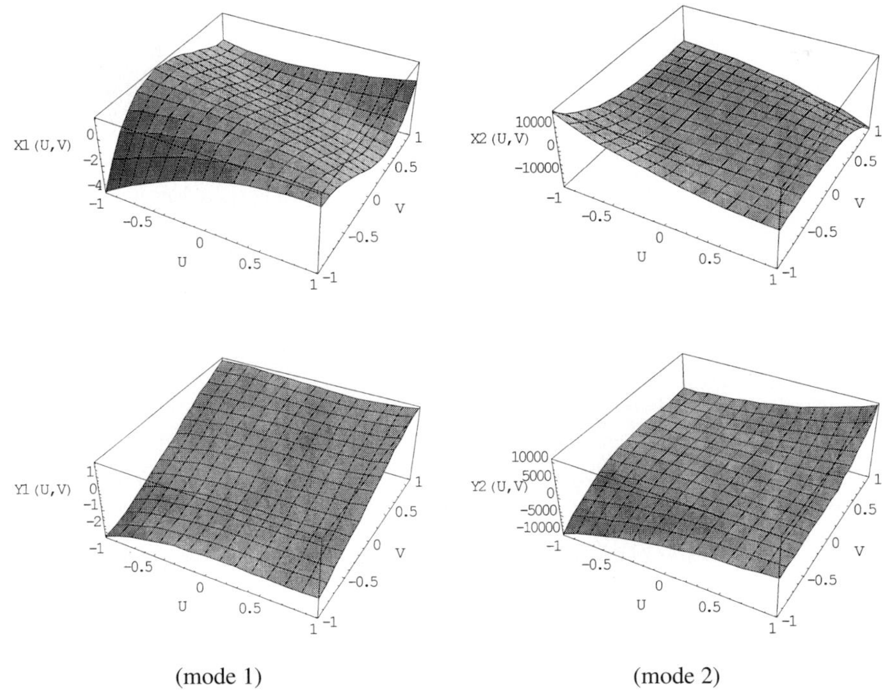

Figure 6d. Invariant modal manifolds for the system with 100% proportionate damping and 50% non-proportionate damping.

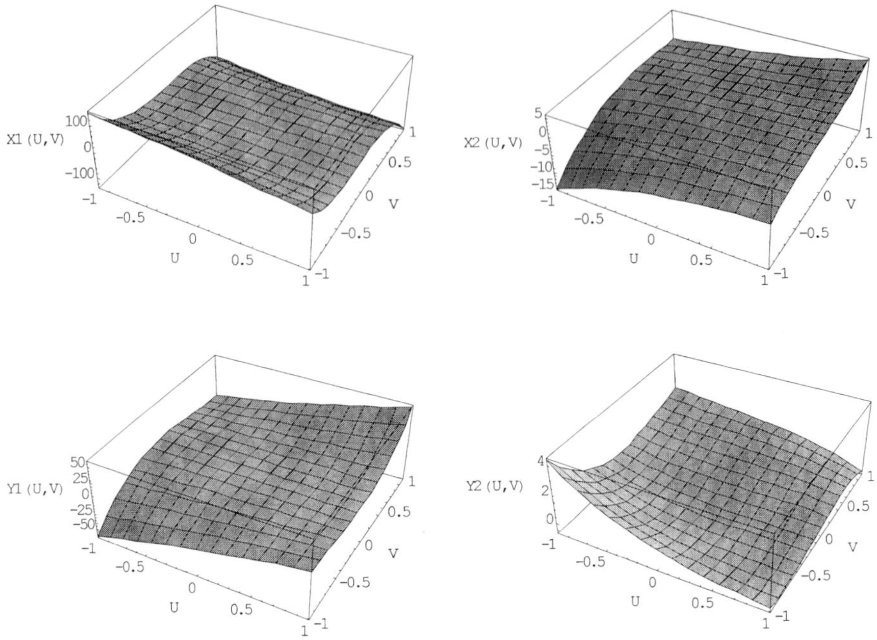

(mode 1) (mode 2)

Figure 7a. Invariant modal manifolds for the system with 25% of actual damping.

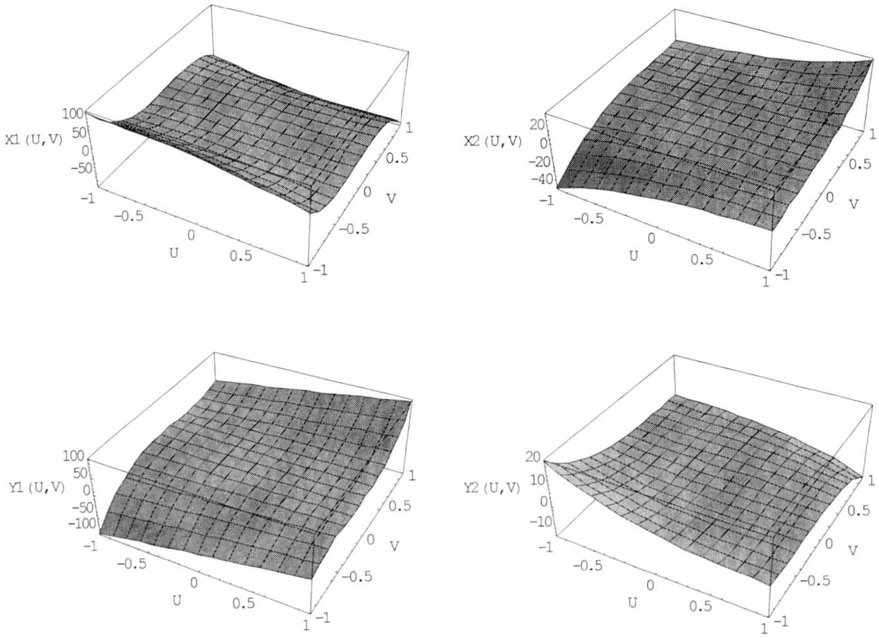

(mode 1) (mode 2)

Figure 7b. Invariant modal manifolds for the system with 50% of actual damping.

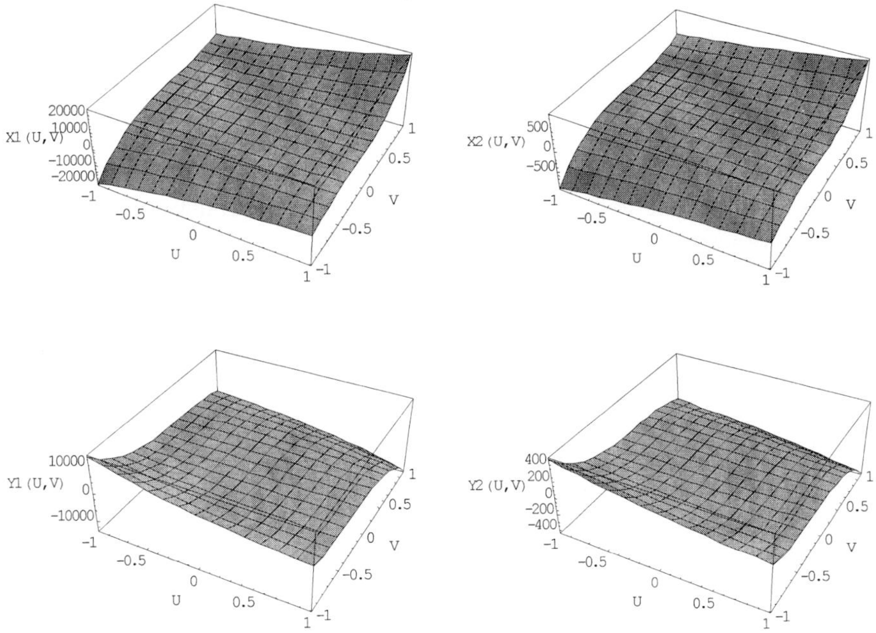

(mode 1) (mode 2)

Figure 7c. Invariant modal manifolds for the system with 75% of actual damping.

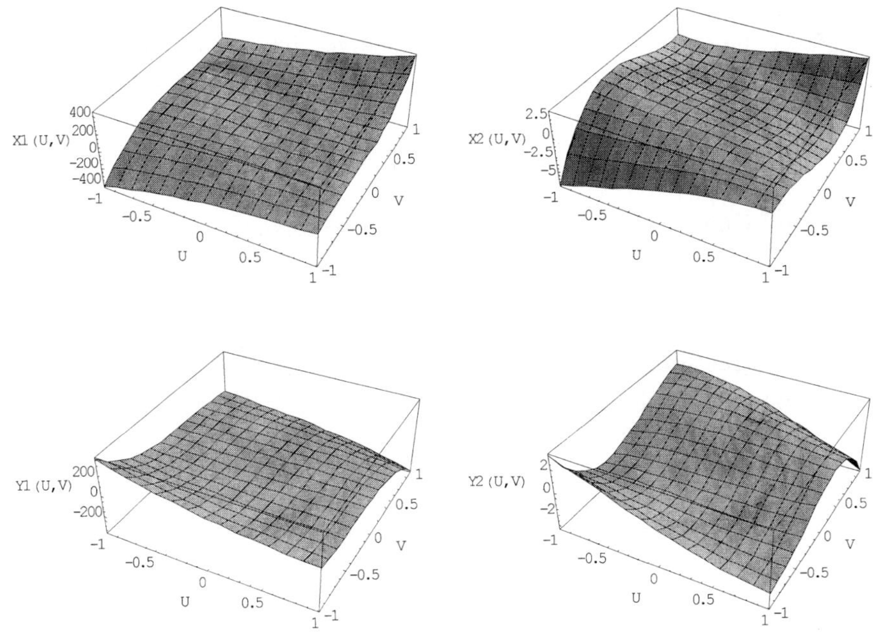

(mode 1) (mode 2)

Figure 7d. Invariant modal manifolds for the system with 100% of actual damping.

Shaw and Pierre [2], the number of real modes changes from two to six at 94% proportionate damping for this system.

The next set of results (Figure 6) are for the proportionate and the 50% of non-proportionate damping. Various amounts of proportionate and non-proportionate damping were considered in the analysis. In addition to viewing the results as arranged herein, one could also view the results across figures and for example viewing Figures 5a, 6a and 7a and observe the effect of increasing the amount of non-proportionate damping from zero to 100% of actual on the resulting modes. From the results given herein and additional results not included, we observed that a significant change in the mode behavior occurs at approximately 66% of the damping. This change results in a significant qualitative and quantitative modification of both the first and second modes' characteristics. Similar to what was found in the proportionate case, the number of real modes changes from two to six at 97% damping of this case.

The last set of systematic results (Figure 7) are for the actual predicted radiated wave damping matrix with various fractions thereof the total. We then include the manifolds for 25, 50, 75, 100% of the radiated wave damping. From the results given herein and additional results not included, we observed that a significant change in the mode behavior occurs at approximately 55% of the actual damping. This change results in a significant qualitative and quantitative modification of the first mode's characteristics and a significant quantitative change in the second mode's character. Similar to what was found in the two previous cases, the number of real modes changes from two to six at 116% of the actual damping for this system.

It should be clear that damping has a dramatic effect on the character of the NNMs. The lower percentages of damping are used to understand the behavior of the system. Higher percentages of damping would have been used to model the inclusion of additional viscous damping however the system experienced a qualitative change in behavior at full radiated wave damping. Therefore additional radiated wave damping did not need to be considered. Similar to what was observed by Shaw and Pierre [2], at this critical damping level, all six solutions became real and an ambiguity exists with regards to the NNMs.

6. Summary and Conclusions

The results shown in the previous section represent the two normal modes (the in-plane and the out-of-plane) of oscillation of the vessel at the critical orientation angle and around the zero in-plane tilt angle. It can be inferred from the results that around the zero equilibrium, the vessel executes small amplitude in-plane oscillations. For the damped and forced system, recalling the saddle type behavior of the out-of-plane oscillation, it is predicted that these out-of-plane oscillations will either aggravate or limit the eventual stability of the vessel at this critical in-plane angle of vanishing stability. Analysis of NNMs allowed us to capture the nonlinear interaction between the two modes which could not be captured using a linearized approach. Specifically, the nonlinear interaction of the saddle type orbit can not be captured using linear analysis. For this vessel, near the stable orbit at the zero in-plane tilt angle the out-of-plane dynamics was of the saddle type and the out-of-plane restoring was relatively small. However, for the saddle orbit near the in-plane angle of vanishing stability, the out-of-plane restoring becomes dominate in a relative sense. Moreover, the NNMs allow us to infer what will happen under external wave excitation. In-plane and out-of-plane resonances are expected which can not be accurately predicted using a linearized and uncoupled analysis.

For all three sets of systematic results, between 25 and 50% damping relatively small changes in mode behavior occurs. However, between 50 and 75% damping, significant changes in mode behavior occurs. In this damping range, the proportionately damped case does not change as dramatically as the other two cases. It is not until the damping range of 75 to 100% that the proportionately damped case experiences dramatic changes as occurred with the non-proportionately damped cases. As general rule, for the dynamical systems considered herein, we found that the rate of mode behavior change was proportional to the relative amount of damping present in the system. We believe that this result may be true for all types of dynamical systems and not just the specific one studied herein.

The results of this study emphasize the importance of using a dynamical systems approach to study the critical tilt dynamics of a marine vehicle. It was verified in previous studies that for hull forms similar to a semi-submersible, the effect of wave forcing is negligible (e.g., [5]). However, damping appears to be as important as for conventional hull forms, particularly about this critical axis. Although, in this paper, viscous damping has not been explicitly calculated, the significant increase in vessel stability due to damping is clear from what was demonstrated in previous studies [13]. This may justify the need to know viscous damping explicitly and that was the subject of a related study [15].

The existence of a critical axis in the proximity of the principal diagonal of the waterplane was predicted and verified in previous work [13]. Physically, this means that a tilting moment about the critical axis is likely to cause capsize more readily than moments about vessel longitudinal (i.e., roll) or transverse (i.e., pitch) axes. This is so because the out-of-plane motion is unstable. This seems to have been the mechanism of capsize of the MODU *Ocean Ranger*. High waves and strong steady wind in an approximately quartering direction caused a dangerous list of the vessel resulting in the flooding of one of its chain lockers [20].

For a more in-depth study of the motions about the critical axis, a 2-DOF analysis was performed. Preliminary stability analyses indicated strong static and dynamic coupling of the in-plane motions to the out-of-plane motions and vice versa. Though in this study, the effects of forcing have not been considered, the decoupling procedures seem to be readily applicable to forced systems [2]. However, a major drawback of the procedure lies in the degree of computational complexity involved in solving the system of algebraic equations even for a simple cubic polynomial approximation for stiffness. It is well known that at least a fifth-order polynomial is necessary to model the righting arm curve accurately [1]. The invariant manifolds approach was utilized and the results compared for various amounts and types of damped systems. Work is already under way extending these analyses to more general systems, possibly even systems which include forcing. The invariant manifold approach seems to be a general approach capable of analyzing a broad class of systems of practical engineering relevance.

Acknowledgements

The authors would like to acknowledge the past and current support of the National Science Foundation's Dynamic Systems and Control Program and its current program manager Dr. Alison Flatau. The first author would also like to acknowledge the partial support of the Office of Naval Research, MOB Science and Technology program and its program manager Gene Remmers, and the NFESC MOB project team. The authors would also like to acknowledge Professor Alex Vakakis for his participation on previous part of this work.

References

1. Kota, R. S., Falzarano, J. M., and Vakakis, A., 'Survival analysis of a deep-water floating offshore platform about its critical axis including coupling', *Journal of International Society of Offshore and Polar Engineers* **8**(2), 1998, 492–499.
2. Shaw, S. W. and Pierre, C., 'Normal modes for nonlinear vibratory systems', *Journal of Sound and Vibration* **164**, 1993, 85–124.
3. Falzarano, J. M. and Zhang, F., 'Multiple degree of freedom global analysis of transient ship rolling motion', in *Nonlinear Dynamics of Marine Vehicles: Modeling and Applications*, ASME Winter Annual Meeting, New Orleans, LA, J. Falzarano and F. Papoulias (eds.), ASME, New York, 1993, pp. 57–72.
4. Vakakis, A. F., Manevitch, L. I., Mikhlin, Y. V., Pilipchuk, V. N., and Zevin, A. A., *Normal Modes and Localization in Nonlinear Systems*, Wiley Series in Nonlinear Science, Wiley, New York, 1996.
5. Falzarano, J. M., Kota, R. S., and Vishnubhotla, S., 'Survival analysis of deep-water floating offshore platforms', in *Proceedings Brasil Offshore '95*, Rio de Janeiro, Brasil, F. Carneiro et al. (eds.), Wiley, New York, 1995, pp. 139–152.
6. Kota, R. S., 'Survival analysis of deep-water floating offshore platforms about their critical axis', Master's Thesis, UNO, 1997.
7. Guckenheimer, J. and Holmes, P., *Nonlinear Oscillations, Dynamical Systems and Bifurcations of Vector Fields*, Applied Mathematical Sciences, Vol. 42, Springer-Verlag, New York, 1983.
8. Wiggins, S., *Introduction to Applied Nonlinear Dynamical Systems and Chaos*, Texts in Applied Mathematics, Vol. 2, Springer-Verlag, New York, 1990.
9. Falzarano, J. M., Steindl, A., Troesch, A. W., and Troger, H., 'Rolling motion of a ship treated as a bifurcation problem', in *Bifurcation, and Chaos: Analysis, Algorithms, Applications*, R. Seydel (ed.), Birkhäuser, Basel, 1991, pp. 117–121.
10. Taz ul Mulk, M. and Falzarano, J. M., 'Complete six degrees of freedom nonlinear ship rolling motion', *Journal of Offshore Mechanics and Arctic Engineering* **116**, 1994, 191–201.
11. Falzarano, J. M., Shaw, S. W., and Troesch, A. W., 'Application of global methods for analyzing dynamical systems to ship rolling and capsizing', *International Journal of Bifurcation and Chaos in Applied Science and Engineering* **2**(1), 1992, 101–115.
12. Zhang, F. and Falzarano, J. M., 'Multiple degree of freedom global transient ship rolling motion: Large amplitude forcing', in *Symposium on Stochastic Dynamics and Reliability of Nonlinear Ocean Systems*, Chicago, IL, R. A. Ibrahim and Y. K. Lin (eds.), ASME, New York, 1994, pp. 99–108.
13. Falzarano, J. M. and Kota, R. S., 'Survival analysis of deep-water floating offshore platforms: Critical axis', in *Proceedings, ASME Offshore Mechanics and Arctic Engineering Conference*, Florence, Italy, S. Chakrabarti et al. (eds.), ASME, New York, 1996, Vol. I-B, pp. 49–54.
14. Chakrabarti, S. K., *Hydrodynamics of Offshore Structures*, Springer-Verlag, New York, 1987.
15. Korpus, R. A. and Falzarano, J. M., 'Prediction of viscous ship roll damping using unsteady Navier–Stokes', in *Proceedings, ASME Offshore Mechanics and Arctic Engineering Conference*, Florence, Italy, S. Chakrabarti et al. (eds.), ASME, New York, 1996, Vol. I-A, pp. 241–248. Also in *ASME Journal of Offshore Mechanics and Arctic Engineering* **116**, 1997, 108–113.
16. Abzug, M. and Larrabee, E., *Airplane Stability and Control*, Cambridge University Press, Cambridge, UK, 1997.
17. Caughey, T. K., 'Classical normal modes in damped linear dynamic systems', *ASME Journal of Applied Mechanics* **32**, 1965, 503–509.
18. Rosenberg, R. M., 'On nonlinear vibrations of systems with many degrees of freedom', *Advances in Applied Mechanics* **9**, 1966, 155–242.
19. Hwang, J. H. and Ma, F., 'On the approximate solution of nonclassically damped linear system', *ASME Journal of Applied Mechanics* **60**, 1993, 695–701.
20. United States Coast Guard, 'MODU ocean ranger capsizing and sinking in the Atlantic Ocean on 15 February 1982 with multiple loss of life', Marine Casualty Report, 1983.

Nonlinear Dynamics **25:** 275–292, 2001.

Performance of Nonlinear Vibration Absorbers for Multi-Degrees-of-Freedom Systems Using Nonlinear Normal Modes

GREGORY S. AGNES
Department of Aeronautics and Astronautics, Air Force Institute of Technology, AFIT/ENY, 2950 P. St. Building 640, WPAFB, OH 45433-7765, U.S.A.

DANIEL J. INMAN
Center for Intelligent Materials and Structures, Virginia Tech, Virginia, MC 0261, Blacksburg, VA 24061, U.S.A.

(Received: 17 November 1998; accepted: 20 March 2000)

Abstract. Linear vibration absorbers are a valuable tool used to suppress vibrations due to harmonic excitation in structural systems. While limited evaluation of the performance of nonlinear vibration absorbers for nonlinear structures exists in the literature for single mode structures, none exists for multi-mode structures. Consequently, nonlinear multiple-degrees-of-freedom structures are evaluated. The theory of nonlinear normal modes is extended to include consideration of modal damping, excitation and small linear coupling, allowing estimation of vibration absorber performance. The dynamics of the $N + 1$-degrees-of-freedom system are shown to reduce to those of a two-degrees-of-freedom system on a four-dimensional nonlinear modal manifold, thereby simplifying the analysis. Quantitative agreement is shown to require a higher-order model which is recommended for future investigation.

Keywords: Vibration absorber, nonlinear normal modes.

1. Introduction

Linear vibration absorbers are a valuable tool used to suppress vibrations due to harmonic excitation in structural systems. A recent review of the theory and application of the linear vibration absorber [1] notes the mass and stroke length design limitations imposed by practical applications. Geometric considerations can limit the stroke length (the amount of travel) of the added mass. Since the mass added to a system must often be minimized, the tradeoff between mass and stroke is a crucial design constraint. Given a fixed mass and stroke length, an absorber is efficient for only a narrow range of driving force and frequency variations. Consequently, active and/or adaptive feedback control is topical [1, 2]. Using active control to minimize the response of the system, smaller absorber stroke lengths can be achieved, albeit with increased system complexity. A second solution reported in the literature is to use nonlinear absorbers.

Results on nonlinear (mechanical) vibration absorbers exist only for single-degree-of-freedom systems. Early work [3–5] considered a linear system with a nonlinear absorber and neglected damping entirely. Most commonly, a nonlinear spring is used in place of the linear spring when designing the absorber. In 1952, Roberson [3] reported that a broadening of the suppression bandwidth was possible and that the zero response frequency (design operational frequency in most cases) changed with force level and nonlinearity. (Note that suppression bandwidth refers to the frequency band in which the response of the system with an absorber is less than the non-resonant response of the system. Otherwise lower response could be achieved by simply shifting the modal frequency of the structure. This definition is standard in the vibration absorber literature.) Arnold [4] and Pipes [5] came to similar conclusions with slightly different systems. Miller and Gartner [6] presented a nonlinear design using pneumatic springs, while Hunt and Nissen [7] used Belleville washers and reported that a broadening of the suppression bandwidth is possible. Several authors [8–10] have applied optimization methods to absorber design, but considered system damping of ten percent or more. This is unrealistic for many applications. Each pointed to bandwidth improvements and several claimed higher harmonic behavior improved the design.

However, Shaw et al. [11] warned of the presence of quasiperiodic responses near the tuning frequency. Such motions were present in numerical simulations of lightly damped linear systems with nonlinear absorbers. These responses were also confirmed analytically in [12, 13]. Rice and McCraith [12] used the method of harmonic balance to numerically investigate the response of a linear system with a nonlinear absorber to narrowband excitation. Natsiavas [13] considered the case of the nonlinear system with a nonlinear absorber and demonstrated the adverse effects of system nonlinearity on absorber performance. Recently the authors [14–17] considered performance criteria for nonlinear vibration absorbers for single-degree-of-freedom nonlinear systems. The selection of absorber parameters will thus only briefly be discussed herein.

Having established performance criteria and a design approach for single-degree-of-freedom systems, multi-degrees-of-freedom systems were next considered and the results are reported herein. Motivated by the work of Slater [18, 19], an approach using nonlinear normal modes (NNM) was developed as follows:

1. Find the system equations of motion.
2. Use NNMs to transform the equations of motion to modal coordinates.
3. Design the vibration absorber in modal coordinates for the two-dimensional modal manifold being excited. (Note that it is assumed that the mode of interest is not in internal resonance.)
4. Evaluate the response by transforming the modal coordinates back to physical coordinates using nonlinear normal mode theory.

To illustrate this design process, an example of Shaw and Pierre [20] is used. The system (Figure 1), consists of two equal masses connected with one nonlinear and two linear springs. This example essentially outlines the procedure for absorber design for multi-degrees-of-freedom discrete (or discretized continuous) systems. The concepts of nonlinear normal modes will be reviewed, then applied to vibration absorber design and this example presented.

2. Nonlinear Normal Modes

Two definitions of NNMs are present in the literature: one due to Rosenberg [21–23] and another due to Shaw and Pierre [20, 24]. For this work the definition due to Shaw and Pierre and later extended by Slater [18, 19] is assumed. With this definition, a NNM is an invariant manifold tangent to the linear modal plane as the nonlinearity in the system decreases to zero. In the absence of internal resonance, each mode of the multi-degrees-of-freedom system

$$M_{ij}\ddot{x}_j = f_i(x_j, \dot{x}_j), \quad j = 1, \ldots, n \tag{1}$$

is a two-dimensional manifold parametrized as:

$$x_i = X_i(u, v), \quad \text{and} \quad y_i = \dot{x}_i = Y_i(u, v). \tag{2}$$

The choice of u and v is arbitrary. Shaw and Pierre use the first physical coordinate, (x_1, y_1) whereas Nayfeh and co-workers [25–27] first transforms the system using the linear eigenvectors and uses (q_k, p_k) where $p_k = \dot{q}_k$ to find the nonlinear normal mode tangent to the kth linear mode. Note that the discussion herein pertains to systems which are linearizable within the neighborhood of an equilibrium point and for which an eigensolution to the linear system of equations exists. It should also be noted that the form of the nonlinear stiffness on the manifold is dependent on the choice of reference coordinates. The dynamics described remain the same and transformation between definitions is possible. For the problem at hand, the modal coordinate approach proved simplest. Using standard tensor notation, the system equations

$$M_{ij}\ddot{x}_j + K_{ij}x_j + \varepsilon N_i(x_j) = 0 \tag{3}$$

undergo a linear modal transformation,

$$x_i = \Phi_{ij}q_j \tag{4}$$

where the eigenvectors, Φ_{ij}, are normalized such that

$$M_{ij}\Phi_{ir}\Phi sj = \delta_{sr} \quad \text{and} \quad K_{ij}\Phi_{ir}\Phi_{sj} = \delta_{sr}\omega^2_{(r)}. \tag{5}$$

The () on the index indicates summation is not performed on that term. The parameter ε has been introduced as a bookkeeping parameter to emphasize smallness scale the equations. The modal equations of motion are then

$$\ddot{q}_j + \omega^2_{(j)}q_j + \varepsilon G_j(q) = 0, \tag{6}$$

where

$$G_j(q) = g_{jrst}q_r q_s q_t \tag{7}$$

is assumed for the form of the nonlinearities considered for this study. Using the method of multiple scales, we can solve for the nonlinear normal modes as in [25–27]. Expanding q_j in power series of ε

$$q_j = q_{j0} + \varepsilon q_{j1} + \cdots \tag{8}$$

and defining time as

$$T_0 = t, \quad T_1 = \varepsilon t, \quad \ldots \tag{9}$$

and its derivative as

$$D_0 = \frac{\partial}{\partial T_0}, \quad D_1 = \frac{\partial}{\partial T_1}, \quad \ldots, \tag{10}$$

the equations may be ordered in ε to yield

$$D_0^2 q_{j0} + \omega^2_{(j)}q_{j0} = 0 \tag{11a}$$

$$D_0^2 q_{j1} + \omega^2_{(j)}q_{j1} = -2D_0D_1q_{j0} - g_{jrst}q_{r0}q_{s0}q_{t0}. \tag{11b}$$

To construct the nonlinear normal mode which tends to the kth mode as $\varepsilon \to 0$, we let

$$q_{j0} = \begin{cases} A_k \, \mathrm{e}^{i\omega_k T_0} + \mathrm{cc}, & j = k, \\ 0, & j \neq k. \end{cases} \tag{12}$$

in Equation (11b):

$$D_0^2 q_{j1} + \omega_{(j)}^2 q_{j1} = \begin{cases} -2D_0 D_1 q_{k0} - g_{kkkk} q_{k0}^3, & j = k, \\ -g_{jkkk} q_{k0}^3, & j \neq k. \end{cases} \tag{13}$$

Substituting Equation (12) into Equation (13) and eliminating secular terms, the response on the kth nonlinear normal mode is

$$q_j = \begin{cases} A_k \, \mathrm{e}^{i\omega_k T_0} + \varepsilon \frac{g_{kkkk}}{8\omega_k^2} A_k^3 \, \mathrm{e}^{3i\omega_k T_0} + \mathrm{cc} + O(\varepsilon^2), & j = k, \\ \varepsilon \left(\frac{3g_{jkkk}}{\omega_k^2 - \omega_j^2} A_k^2 \bar{A}_k \, \mathrm{e}^{i\omega_k T_0} + \frac{g_{kkkk}}{9\omega_k^2 - \omega_j^2} A_k^3 \, \mathrm{e}^{3i\omega_k T_0} \right) + \mathrm{cc} + O(\varepsilon^2), & j \neq k, \end{cases} \tag{14}$$

where

$$D_1 A_k = \frac{3i}{2\omega_k} g_{kkkk} A_k^2 \bar{A}_k \tag{15}$$

can be solved using standard techniques. To place this result in a more common form, we express (see [27])

$$q_j = \begin{cases} u_k, & j = k, \\ \Gamma_{1jk} q_k^3 + \Gamma_{2jk} q_k \dot{q_k}^2, & j \neq k, \end{cases} \tag{16}$$

with

$$\Gamma_{1jk} = \frac{7\omega_k^2 - \omega_j^2}{(9\omega_k^2 - \omega_j^2)(\omega_k^2 - \omega_j^2)} g_{jkkk} \tag{17}$$

$$\Gamma_{2jk} = \frac{6}{(9\omega_k^2 - \omega_j^2)(\omega_k^2 - \omega_j^2)} g_{jkkk}. \tag{18}$$

This solution procedure is valid unless the kth mode is in internal resonance. For the undamped case presented here, the manifold is parametrized by the single variable q_k. This will not be true for nonconservative systems. However, for lightly damped and forced systems, the response of the system to harmonic forcing near the modal frequency will tend to the undamped nonlinear mode defined above [28, 18]. Using this result, the forced response of the vibration absorber attached to a forced and lightly damped structure can be analyzed.

3. Multi-Degrees-of-Freedom Absorber Design

To model a vibration absorber coupled to a multi-degrees-of-freedom system, the same approach as in [15, 16] can be followed. The equations of motion in physical coordinates are

$$M_{ij} \ddot{x}_j + K_{ij} x_j + \varepsilon N_i(x_j) - \varepsilon A_i x_a = 0, \tag{19a}$$

$$m \ddot{x}_a + k x_a + \varepsilon N_a x_a^3 - \varepsilon B_i x_i = 0. \tag{19b}$$

Introducing the modal transformation similar to the previous section,

$$x_i = \Phi_{ij} q_j, \tag{20a}$$

$$x_a = \frac{1}{\sqrt{m}} q_a, \tag{20b}$$

the equations can be rewritten in modal coordinates as

$$\ddot{q}_j + \omega_{(j)}^2 q_j + \varepsilon g_{jrst} q_r q_s q_t - \varepsilon \alpha_j q_a = 0, \tag{21a}$$

$$\ddot{q}_a + \omega_a^2 q_a + \varepsilon g_a q_a^3 - \varepsilon \beta_j q_j = 0. \tag{21b}$$

The absorber is tuned to be near a resonance of the system. Thus, $\omega_a = (1+\varepsilon\delta)\omega_k$ and a one-to-one internal resonance is formed between the absorber and the kth mode. Note that $\beta_k = \alpha_k$ and δ is a small mistuning of the absorber frequency. Applying the method of multiple scales, the equations can be transformed into a series of partial differential equations. The $O(\varepsilon^0)$ equations are:

$$D_0^2 q_{j0} + \omega_{(j)}^2 q_{j0} = 0, \tag{22a}$$

$$D_0^2 q_{a0} + \omega_{(k)}^2 q_{a0} = 0. \tag{22b}$$

The solution for the kth mode is

$$q_{j0} = \begin{cases} A_k \, \mathrm{e}^{i\omega_k T_0} + \mathrm{cc}, & j = k, \\ 0, & j \neq k, \end{cases} \tag{23a}$$

$$q_{a0} = A_a \, \mathrm{e}^{i\omega_k T_0} + \mathrm{cc}. \tag{23b}$$

The $O(\varepsilon)$ equations are:

$$D_0^2 q_{j1} + \omega_{(j)}^2 q_{j1} = -2D_0 D_1 q_{j0} - g_{jrst} q_{r0} q_{s0} q_{t0} + \alpha_j q_{a0}, \tag{24a}$$

$$D_0^2 q_{a1} + \omega_{(k)}^2 q_{a1} = -2D_0 D_1 q_{a0} - g_a q_{a0}^3 + \alpha_j q_{j0} - 2\delta q_{a0}. \tag{24b}$$

Substituting Equations (23a)-(23b) into Equations (24a)-(24b),

$$D_0^2 q_{j1} + \omega_{(j)}^2 q_{j1} = \begin{cases} -2D_0 D_1 q_{k0} - g_{kkkk} q_{k0}^3 + \alpha_k q_{a0}, & j = k, \\ -g_{jkkk} q_{k0}^3 + \alpha_j q_{a0}, & j \neq k, \end{cases} \tag{25a}$$

$$D_0^2 q_{a1} + \omega_{(k)}^2 q_{a1} = -2D_0 D_1 q_{a0} - g_a q_{a0}^3 + \alpha_k q_{k0} - 2\delta q_{a0}. \tag{25b}$$

Therefore, for a vibration absorber coupled to a multi-degrees-of-freedom system, the nonlinear modal response for an absorber coupled to the kth linear mode is given by

$$q_j = \begin{cases} A_k \, \mathrm{e}^{i\omega_k T_0} + \varepsilon \frac{g_{kkkk}}{8\omega_k^2} A_k^3 \, \mathrm{e}^{3i\omega_k T_0} + \mathrm{cc}, & j = k, \\ \varepsilon \left(\frac{3g_{jkkk} A_k^2 \bar{A}_k - \alpha_j A_a}{\omega_k^2 - \omega_j^2} \, \mathrm{e}^{i\omega_k T_0} + \frac{g_{jkkk}}{9\omega_k^2 - \omega_j^2} A_k^3 \, \mathrm{e}^{3i\omega_k T_0} \right) + \mathrm{cc}, & j \neq k, \end{cases} \tag{26a}$$

$$q_a = A_a \, \mathrm{e}^{i\omega_k T_0} + \varepsilon \frac{g_a}{8\omega_k^2} A_a^3 \, \mathrm{e}^{3i\omega_k T_0} + \mathrm{cc}, \tag{26b}$$

where

$$D_1 A_k = \frac{3i g_{kkkk}}{2\omega_k} A_k^2 \bar{A}_k - \frac{i\alpha_k}{2\omega_k} A_a, \tag{27a}$$

$$D_1 A_a = \frac{3i g_a}{2\omega_k} A_a^2 \bar{A}_a - \frac{i\alpha_k}{2\omega_k} A_k + \frac{i\delta}{\omega_k} A_a. \tag{27b}$$

These equations are identical to the solution for a single-degree-of-freedom system with a vibration absorber presented in previous work [14–16] in the absence of external forcing and damping. The solution of the 'absorbed' mode can be used to find (A_k, A_a) which were denoted (A_1, A_2) in those references.

Since the forced and lightly damped response will tend towards the nonlinear normal mode, nonlinear normal mode theory can be modified to include forcing and modal damping. First Equations (21a–21b) are modified to include forcing and damping:

$$\ddot{q}_j + 2\varepsilon\omega_{(j)}\zeta_{(j)}\dot{q}_j + \omega_{(j)}^2 q_j + \varepsilon g_{jrst} q_r q_s q_t - \varepsilon\alpha_j q_a = \gamma_j q_f, \tag{28a}$$

$$\ddot{q}_a + 2\varepsilon\omega_{(a)}\zeta_a\dot{q}_a + \omega_a^2 q_a + \varepsilon g_a q_a^3 - \varepsilon\beta_j q_j = 0, \tag{28b}$$

$$\ddot{q}_f + \omega_f^2 q_f = 0. \tag{28c}$$

Note that in Equation (28c) the forcing has been written as a harmonic variable coupled (linearly) to each mode by γ_j (but not coupled to the force). This places the equations in the form of nonlinear normal mode theory discussed in the previous section. Forcing is simply treated as an additional coordinate of the system which is harmonically oscillating. Thus, the results of the previous section directly apply.

To proceed, we set $\omega_a = (1+\delta)\omega_k$ and $\omega_f = (1+\sigma)\omega_k$ to construct the response of the kth mode. Again $\beta_k = \alpha_k$ and δ is the frequency mistuning of the absorber. σ is the mistuning of the forcing from the modal frequency. Applying the method of multiple scales, the equations can be transformed into a series of partial differential equations. The $O(\varepsilon^0)$ equations are:

$$D_0^2 q_{j0} + \omega_{(j)}^2 q_{j0} = 0, \tag{29a}$$

$$D_0^2 q_{a0} + \omega_{(k)}^2 q_{a0} = 0, \tag{29b}$$

$$D_0^2 q_{f0} + \omega_{(k)}^2 q_{f0} = 0. \tag{29c}$$

The solution for the kth mode is

$$q_{j0} = \begin{cases} u_0 = A_k \,\mathrm{e}^{i\omega_k T_0} + \mathrm{cc}, & j = k, \\ 0, & j \neq k, \end{cases} \tag{30a}$$

$$q_{a0} = u_a = A_a \,\mathrm{e}^{i\omega_k T_0} + \mathrm{cc}, \tag{30b}$$

$$q_{f0} = f_0 = \frac{f}{2}\,\mathrm{e}^{i\omega_k T_0} + \mathrm{cc}. \tag{30c}$$

Note that the parameter u_0 has been introduced to emphasize it is the coordinate of the manifold. The $O(\varepsilon)$ equations are:

$$\begin{aligned} D_0^2 q_{j1} + \omega_{(j)}^2 q_{j1} = {} & -2D_0 D_1 q_{j0} - 2\zeta_{(j)}\omega_{(j)} D_0 q_{j0} \\ & - g_{jrst} q_{r0} q_{s0} q_{t0} + \alpha_j q_{a0} + \gamma_j q_{f0}, \end{aligned} \tag{31a}$$

$$D_0^2 q_{a1} + \omega_{(k)}^2 q_{a1} = -2D_0 D_1 q_{a0} - 2\zeta_a \omega_{(k)} D_0 q_{a0} - g_a q_{a0}^3 + \alpha_j q_{j0} - 2\delta\omega_{(k)} q_{a0}, \tag{31b}$$

$$D_0^2 q_{f1} + \omega_{(k)}^2 q_{f1} = -2D_0 D_1 q_{f0} - 2\sigma\omega_{(k)} q_{f0}, \tag{31c}$$

Substituting Equations (30a–30b) into Equations (31a–31c),

$$D_0^2 q_{j1} + \omega_{(j)}^2 q_{j1} = \begin{cases} -2D_0D_1u_0 - 2\zeta_k D_0u_0 - g_{kkkk}u_0^3 + \alpha_k u_a + \gamma_k f_0, & j = k, \\ -g_{jkkk}u_0^3 + \alpha_j u_a, & j \neq k, \end{cases} \quad (32a)$$

$$D_0^2 q_{a1} + \omega_{(k)}^2 q_{a1} = -2D_0D_1u_a - 2\zeta_a D_0u_a - g_a u_a^3 + \alpha_k u_0 - 2\delta\omega_{(k)}u_a, \quad (32b)$$

$$D_0^2 q_{f1} + \omega_{(k)}^2 q_{f1} = -2D_0D_1 f_0 - 2\sigma\omega_{(k)} f_0, \quad (32c)$$

the nonlinear normal mode which reduces to the kth linear mode can be found. Noting

$$2u_0^3 = \left[u_0^3 - \frac{u_0\dot{u}_0^2}{\omega_k^2}\right] + \left[u_0^3 + \frac{u_0\dot{u}_0^2}{\omega_k^2}\right], \quad (33)$$

where the first bracketed term is harmonic with frequency $3\omega_k$, while the second term is harmonic with frequency ω_k. The nonlinear normal mode can be expressed:

$$q_j = \begin{cases} u_0 + \frac{g_{kkkk}}{16\omega_k^2}\left(u_0^3 - \frac{u_0\dot{u}_0^2}{\omega_k^2}\right), & j = k, \\ \frac{\alpha_j}{(\omega_j^2-\omega_k^2)}u_a - \frac{g_{jkkk}}{2(\omega_j^2-9\omega_k^2)}\left(u_0^3 - \frac{u_0\dot{u}_0^2}{\omega_k^2}\right) & \\ \quad -\frac{3g_{jkkk}}{2(\omega_j^2-\omega_k^2)}\left(u_0^3 + \frac{u_0\dot{u}_0^2}{\omega_k^2}\right) + \frac{\gamma_j}{\omega_j^2-\omega_k^2}q_f, & j \neq k, \end{cases} \quad (34a)$$

$$q_a = u_a + \frac{g_a}{16\omega_k^2}\left(u_a^3 - \frac{u_a\dot{u}_a^2}{\omega_k^2}\right), \quad (34b)$$

$$q_f = \frac{f}{2}\mathrm{e}^{i\omega_k(T_0+\sigma T_1)+\phi_0} + \mathrm{cc}, \quad (34c)$$

where ϕ_0 is the phase of the excitation, determined from the initial conditions. The dynamics on the manifold are:

$$\ddot{u}_0 + 2\zeta_k\omega_k\dot{u}_0 + \omega_k^2 u_0 + g_{kkkk}u_0^3 - \alpha_k u_a = \frac{f}{2}\mathrm{e}^{i\omega_k(T_0+\sigma T_1)+\phi_0} + \mathrm{cc}, \quad (35a)$$

$$\ddot{u}_a + 2\zeta_a\omega_k\dot{u}_a + (1+2\delta)\omega_k^2 u_a + g_a u_a^3 - \alpha_k u_0 = 0. \quad (35b)$$

These are identical to the equations of a vibration absorber on a single-degree-of-freedom system derived in previous work [14–16].

Having solved for the kth mode solution, the lth mode solution is now derived to determine the effect of the vibration absorber on the other modes of the structure. The solution of Equations (28a–28c) for the lth mode is

$$q_{j0} = \begin{cases} u_l = A_l\,\mathrm{e}^{i\omega_l T_0} + \mathrm{cc}, & j = l \neq k, \\ 0, & j \neq l, \end{cases} \quad (36a)$$

$$q_{a0} = 0, \quad (36b)$$

$$q_{f0} = f_0 = \frac{f}{2}\mathrm{e}^{i\omega_k T_0} + \mathrm{cc}, \quad (36c)$$

The $O(\varepsilon)$ equations of motion thus are

$$D_0^2 q_{j1} + \omega_{(j)}^2 q_{j1} = \begin{cases} -2D_0D_1u_l - 2\zeta_l\omega_l D_0u_l - g_{llll}u_l^3 + \gamma_l f_0, & j = l \neq k, \\ -g_{jlll}u_l^3 + \gamma_j f_0, & j \neq l, \end{cases} \quad (37a)$$

$$D_0^2 q_{a1} + \omega_{(k)}^2 q_{a1} = \alpha_l u_l, \quad (37b)$$

$$D_0^2 q_{f1} + \omega_{(k)}^2 q_{f1} = 0. \quad (37c)$$

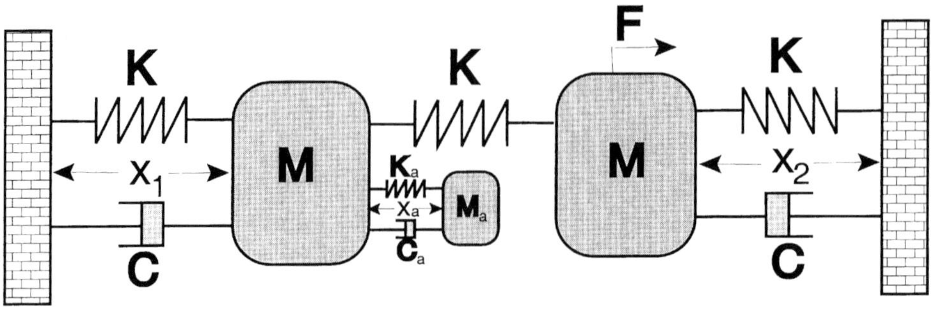

Figure 1. Two-degrees-of-freedom nonlinear system.

The dynamics of the lth nonlinear mode are thus unaffected by the presence of the vibration absorber tuned to the kth mode (in the absence of a k–l internal resonance). They represent the response of a nonlinear single-degree-of-freedom system

$$\ddot{u}_l + 2\zeta_l\omega_l\dot{u}_l + \omega_l^2 u_l + g_{llll}u_l^3 = \frac{f}{2}\,\mathrm{e}^{i\omega_k(T_0+\sigma T_1)+\phi_0} + \mathrm{cc}. \tag{38}$$

For steady-state solutions these solutions will decay to a non-resonant forced response for positive damping. *Therefore the design of a nonlinear vibration absorber for a nonlinear multi-degrees-of-freedom system reduces to that of the single-degree-of-freedom design presented in previous work* [14–16]. To verify this result, a nonlinear two-degrees-of-freedom system used in [20] was examined.

4. Example: Two-Degrees-of-Freedom Shaw and Pierre System

The two-mass, three-spring system shown in Figure 1 was used in [20] and later by subsequent authors [18, 19, 28, 29] to illustrate the concept of nonlinear normal modes.

Herein it is used to illustrate the design procedure developed in the previous section. The equations of motion of the two-degrees-of-freedom system with a vibration absorber attached to mass 1 and external forcing applied to mass 2 are:

$$M\ddot{X}_1 + C\dot{X}_1 - C_a\dot{X}_a + 2KX_1 - KX_2 + NX_1^3 - K_aX_a - N_aX_a^3 = 0, \tag{39a}$$

$$M\ddot{X}_2 + C\dot{X}_2 + 2KX_2 - KX_1 = F, \tag{39b}$$

$$M_a(\ddot{X}_a + \ddot{X}_1) + C_a\dot{X}_a + K_aX_a + N_aX_a^3 = 0. \tag{39c}$$

Following the methodology outlined in previous work [14–16], these equations can be nondimensionalized, scaled in ε and dropping terms of $O(\varepsilon^2)$ and higher, written

$$\ddot{x}_1 + 2\varepsilon c\dot{x}_1 + 2x_1 - x_2 + \varepsilon n x_1^3 - \varepsilon\alpha x_a - \varepsilon n_a x_a^3 = 0, \tag{40a}$$

$$\ddot{x}_2 + 2\varepsilon c\dot{x}_2 + 2x_2 - x_1 = f, \tag{40b}$$

$$\ddot{x}_a + 2\varepsilon c_a\dot{x}_a + (1 + 2\varepsilon\delta)x_a + \varepsilon n_a x_a^3 + 2\varepsilon\alpha x_1 - \varepsilon\alpha x_2 = 0, \tag{40c}$$

where the following parameters were defined:

$$t = \omega T, \quad c = \frac{C}{2\sqrt{KM}}, \quad \omega = \sqrt{\frac{K}{M}}, \quad n = \frac{N}{K},$$

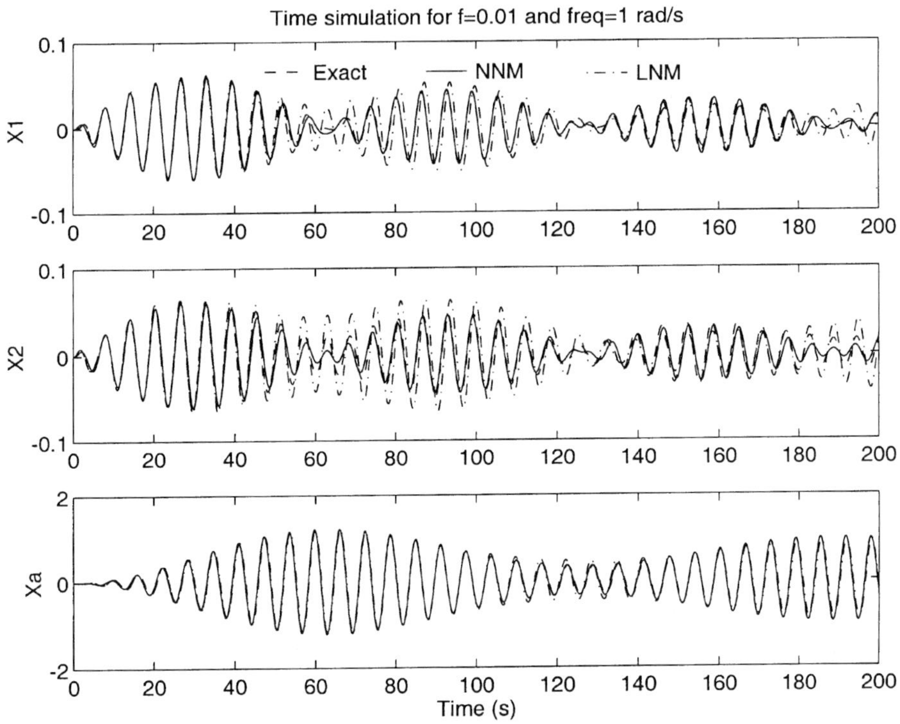

Figure 2. Transient response of the two-degrees-of-freedom system in physical coordinates for $f = 0.01$ and $\sigma = 0$.

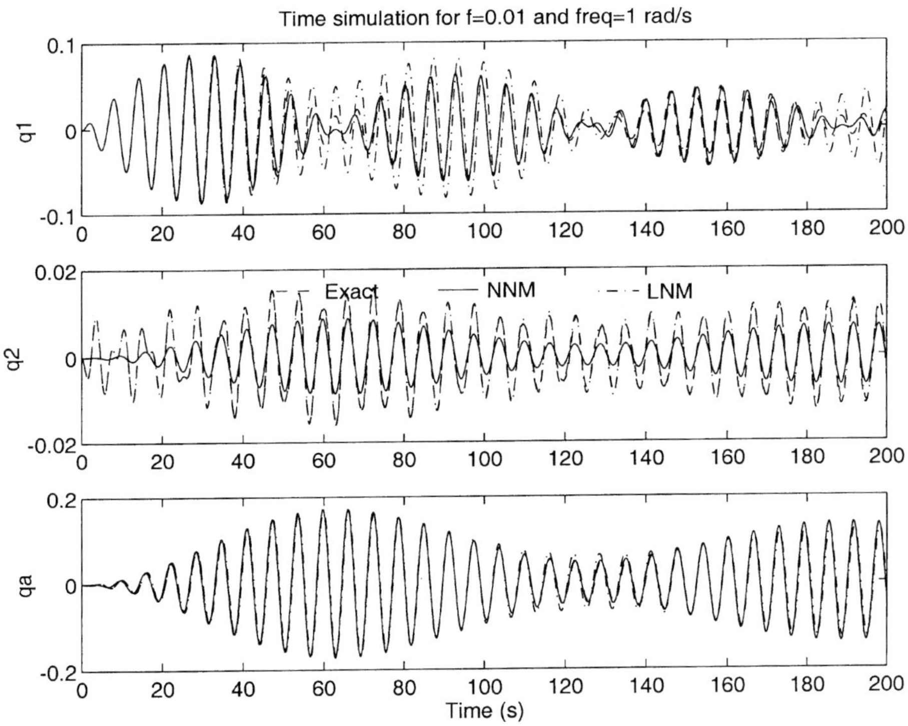

Figure 3. Transient response of the two-degrees-of-freedom system in modal coordinates for $f = 0.01$ and $\sigma = 0$.

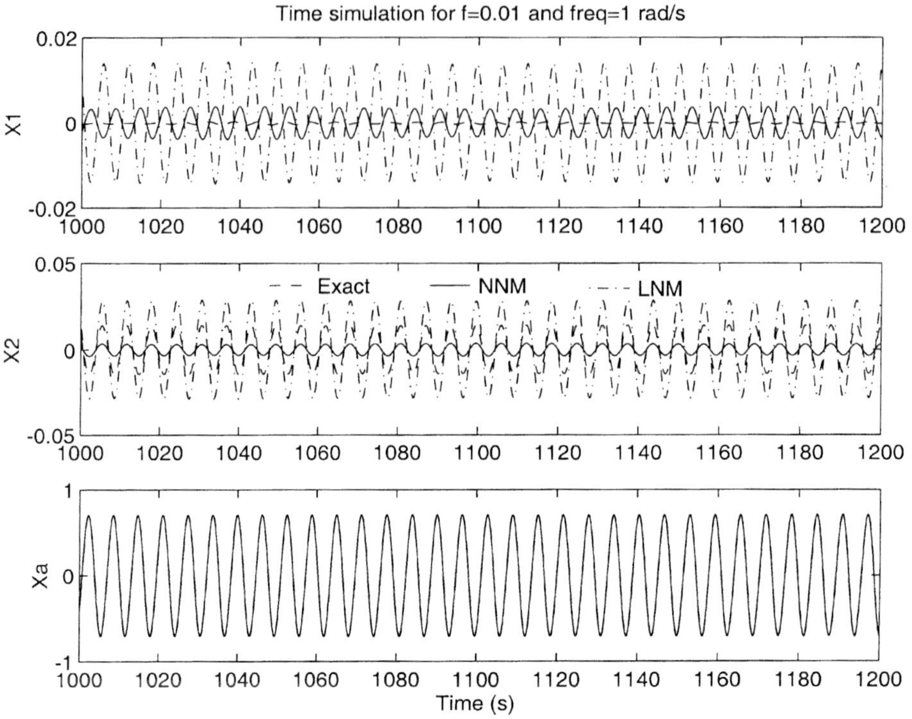

Figure 4. Steady-state response of the two-degrees-of-freedom system in physical coordinates for $f = 0.01$ and $\sigma = 0$.

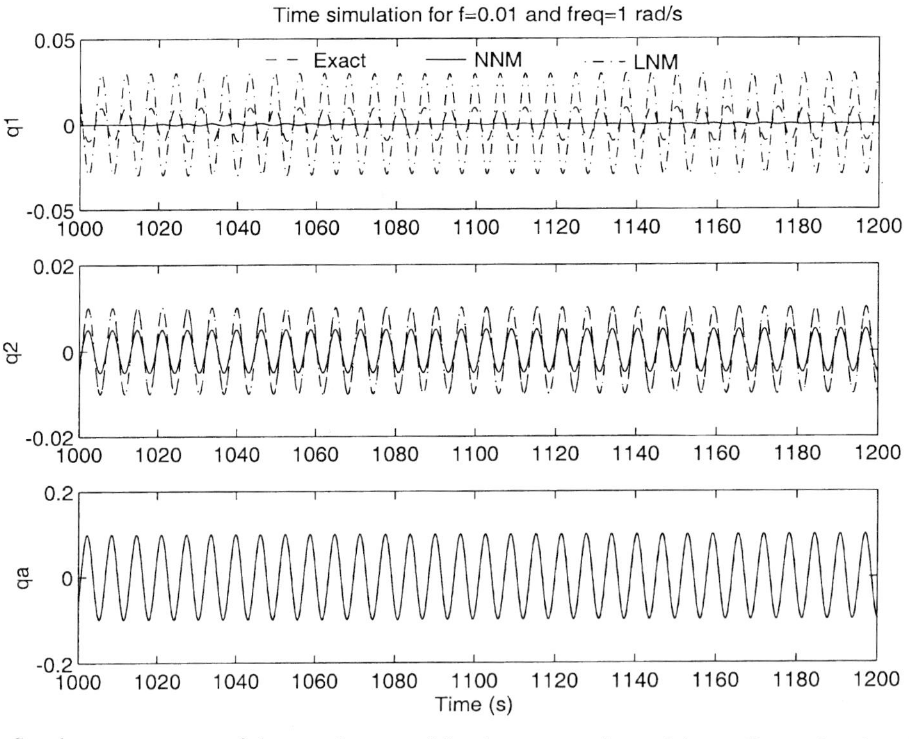

Figure 5. Steady-state response of the two-degrees-of-freedom system in modal coordinates for $f = 0.01$ and $\sigma = 0$.

$$\alpha = \sqrt{\frac{M_a}{M}}, \quad f = \frac{F}{K}, \quad c_a = \frac{C_a}{2\sqrt{K_a M_a}}, \quad \omega_a = \sqrt{\frac{K_a}{M_a}},$$

$$n_a = \frac{N_a}{\alpha^2 K}, \quad (1+\delta) = \frac{\omega_a}{\omega}, \quad x_i = X_i, \quad x_a = \alpha X_a, \tag{41}$$

and ε is a small bookkeeping parameter added to emphasize the scaling of the equations. Also differentiation is now with respect to non-dimensional time, T. These equations are of the form assumed in the previous section. Applying the modal transform

$$\begin{Bmatrix} x_1 \\ x_2 \\ x_a \end{Bmatrix} = \begin{bmatrix} \frac{1}{\sqrt{2}} & \frac{1}{\sqrt{2}} & 0 \\ \frac{1}{\sqrt{2}} & -\frac{1}{\sqrt{2}} & 0 \\ 0 & 0 & 1 \end{bmatrix} \begin{Bmatrix} q_1 \\ q_2 \\ q_a \end{Bmatrix}, \tag{42}$$

the equations are transformed as in the previous sections to (linear) modal coordinates:

$$\ddot{q}_1 + 2\zeta_1 \dot{q}_1 + q_1 - \varepsilon \frac{\alpha}{\sqrt{2}} q_a + \varepsilon \frac{n}{4}(q_1 + q_2)^3 = \frac{f}{\sqrt{2}}, \tag{43a}$$

$$\ddot{q}_2 + 2\sqrt{3}\zeta_2 \dot{q}_2 + 3q_2 - \varepsilon \frac{\alpha}{\sqrt{2}} q_a + \varepsilon \frac{n}{4}(q_1 + q_2)^3 = -\frac{f}{\sqrt{2}}, \tag{43b}$$

$$\ddot{q}_a + 2\zeta_a \dot{q}_a + (1 + 2\delta) q_a - \varepsilon \frac{\alpha}{\sqrt{2}} q_1 - \varepsilon \frac{3\alpha}{\sqrt{2}} q_2 + \varepsilon n_a q_a^3 = 0, \tag{43c}$$

where ζ_i are the modal damping ratios. Applying the results of the previous section, the nonlinear normal mode of interest is:

$$q_1 = u - \frac{n}{64}(u^3 - u\dot{u}^2), \tag{44a}$$

$$q_2 = \frac{\alpha}{2\sqrt{2}} u_a - \frac{n}{48}(u^3 - u\dot{u}^2) - \frac{3n}{16}(u^3 + u\dot{u}^2) - \frac{f}{2}, \tag{44b}$$

$$q_a = u_a + \frac{n_a}{16}(u_a^3 - u_a \dot{u}^2), \tag{44c}$$

with the dynamics on the manifold governed by

$$\ddot{u} + \zeta_1 \dot{u} + u + \varepsilon \frac{n}{4} u^3 - \varepsilon \frac{\alpha}{\sqrt{2}} u_a = \frac{f}{\sqrt{2}}, \tag{45a}$$

$$\ddot{u}_a + \zeta_a \dot{u}_a + (1+\delta) u_a + \varepsilon n_a u_a^3 - \varepsilon \frac{\alpha}{\sqrt{2}} u = 0. \tag{45b}$$

Here u and u_a represent the manifold coordinates. The solution to these equations via a perturbation method was examined in [15, 16]. Selection of the absorber parameters (δ and n_a) is not the focus of this discussion, but they are chosen to provide performance for the system at low excitation levels. Instead, we wish to compare the performance of the nonlinear normal mode approach for the response to the full equations of motion via time integration. Assuming $n = -0.04$, $n_a = 0.01$, $\zeta_1 = 0.01$, $\alpha = 0.1\sqrt{2}$, $\delta = \sigma = 0$, Equations (45a–45b) were integrated numerically using MATLAB®, and compared with direct integration of the equations of motion in physical coordinates and integration performed using the linear

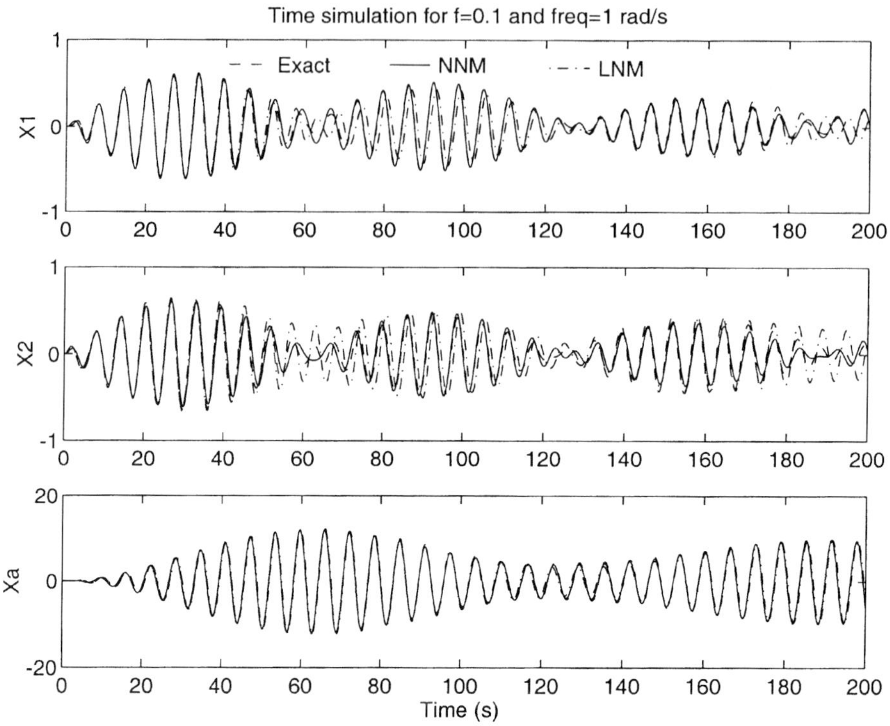

Figure 6. Transient response of the two-degrees-of-freedom system in physical coordinates for $f = 0.1$ and $\sigma = 0$.

modal transformation. The linear normal modal transformation consists of integrating Equations (43a) and (43c) only and setting $q_2 = 0$. These three cases are denoted NNM, exact, and LNM on the graphs. Harmonic forcing amplitudes of $f = 0.01, 0.1$, and 0.3 were considered for a frequency of 1.0 rad/s. Results are plotted in Figures 2–13 for both the calculated physical coordinates, X_i, and the linear modal coordinates, q_i. All integrations were performed from rest, and both the initial response and the steady state response are plotted as indicated in the figure captions.

Examining Figures 2–5, several observations can be made. First, the qualitative responses of the figures are similar. All responses appear harmonic. Second, the transient responses for all three cases agree well for both q_1 and q_a, but the present nonlinear normal mode response is off. This makes physical sense since the present approach considers the solution only within the manifold of the first mode. The initial conditions are not on that manifold since on the manifold the system is at rest only if $f = 0$. As time progresses, the three responses deviate. This results from the approximations made. Recall that the slower time scales in the application of the method of multiple scales were not present in this order of analysis. The transient results for the physical coordinates demonstrate similar results, except the motion of X_2 is in agreement. This validates the assumption that the second mode, q_2, does not participate in the response to a large extent.

Examining the steady state results between $t = 1000$ and 1200 s, the agreement of the three models is poor. Physically, the reason for this is that the width of the suppression bandwidth is an $O(\varepsilon^2)$ quantity (see [15, 16] for details). The nonlinear normal mode analysis predicts that the suppression bandwidth lies centered near 1 rad/s. Thus, examining Figure 5, the

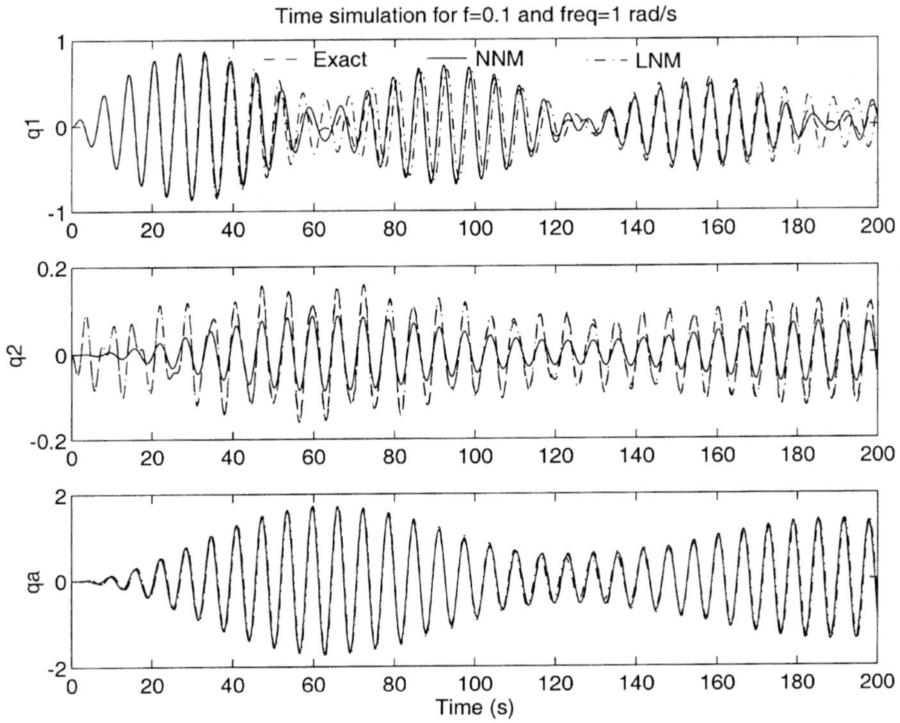

Figure 7. Transient response of the two-degrees-of-freedom system in modal coordinates for $f = 0.1$ and $\sigma = 0$.

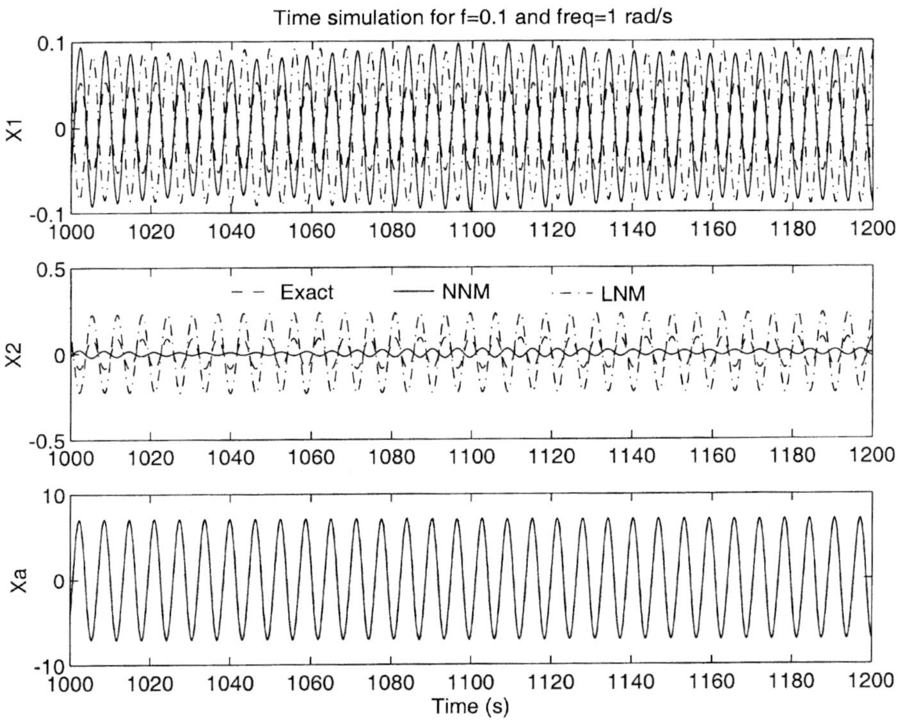

Figure 8. Steady-state response of the two-degrees-of-freedom system in physical coordinates for $f = 0.1$ and $\sigma = 0$.

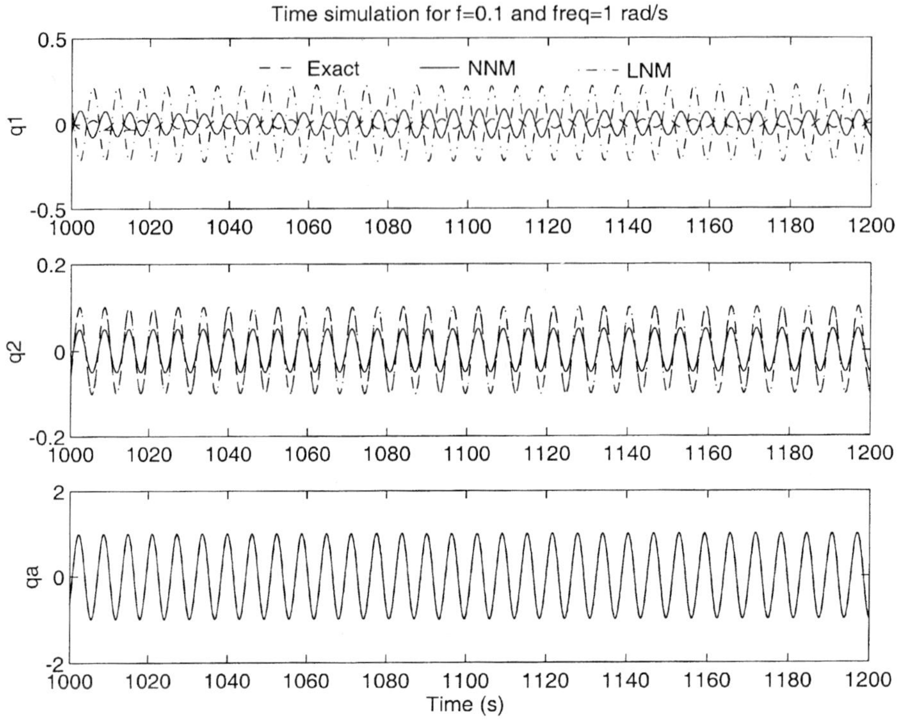

Figure 9. Steady-state response of the two-degrees-of-freedom system in modal coordinates for $f = 0.1$ and $\sigma = 0$.

NNM predicts suppressed response. However the presence of the second mode has shifted the suppression bandwidth slightly preventing 'perfect' absorption. This is not captured by the present analysis. The present analysis therefore only provides qualitative agreement with the physical model. To achieve quantitative agreement for multi-mode systems, an $O(\varepsilon^2)$ analysis would be required. Note that both the second mode and absorber displacements are in reasonable agreement. This is because these responses are not as sensitive to the slight shifting of the suppression bandwidth. Further discussion of these issues along with the perturbation analysis can be found in [15, 16]. The responses in physical coordinates shown in Figure 4 display similar information.

Increasing the forcing level to $f = 0.1$, the agreement between the present model and the physical model improves as shown in Figures 6–9. Again, the absorber response is in excellent agreement while the system response, although in better agreement, shows some improvement. It was demonstrated in [15, 16] as the excitation level increases, the suppression region shifts higher in frequency and away from the 1 rad/s (excitation frequency) responses plotted. The sensitivity of the system to higher-order effects decreases with increased distance from the suppression region, so improved results can be expected. Otherwise, similar observations can be made from these plots. Note that the steady state is periodic in all three models.

Increasing the forcing level further to $f = 0.3$ shows even better agreement as plotted in Figures 10–13. The agreement between the physical coordinates for both transient and modal coordinates is improved. Examining the steady state responses, the response in all three cases is quasiperiodic. A large phase difference between the exact and the NNM solution is

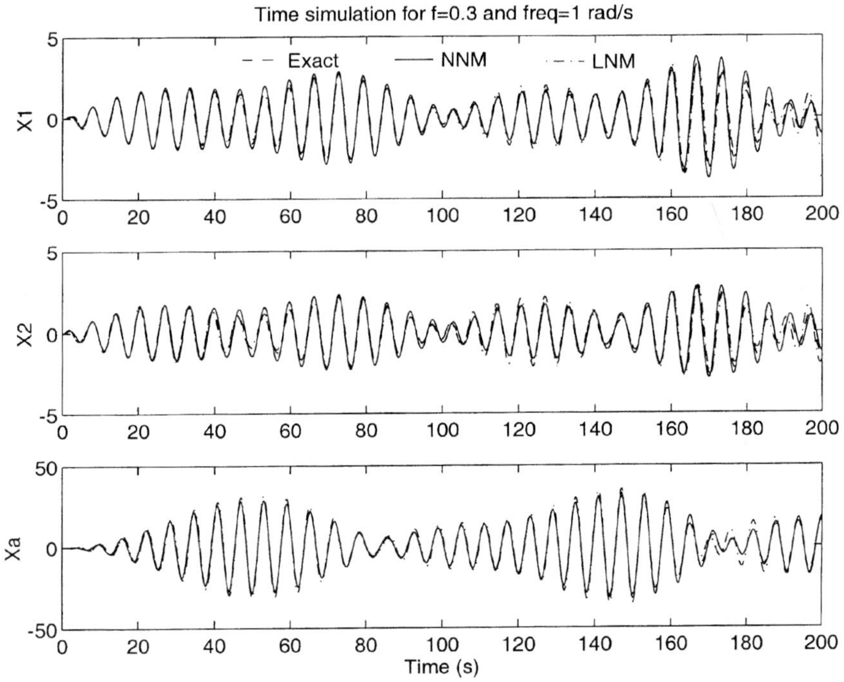

Figure 10. Transient response of the two-degrees-of-freedom system in physical coordinates for $f = 0.3$ and $\sigma = 0$.

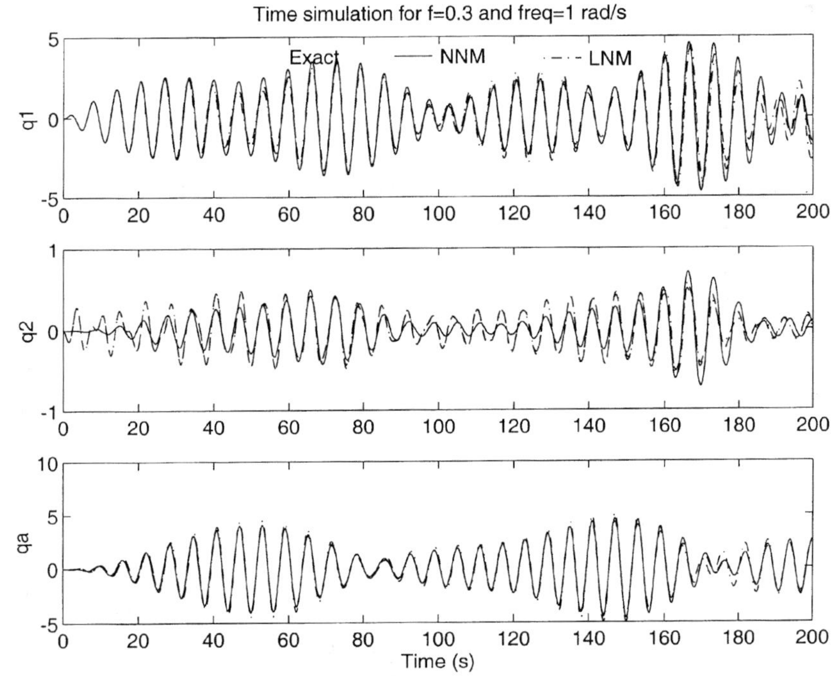

Figure 11. Transient response of the two-degrees-of-freedom system in modal coordinates for $f = 0.3$ and $\sigma = 0$.

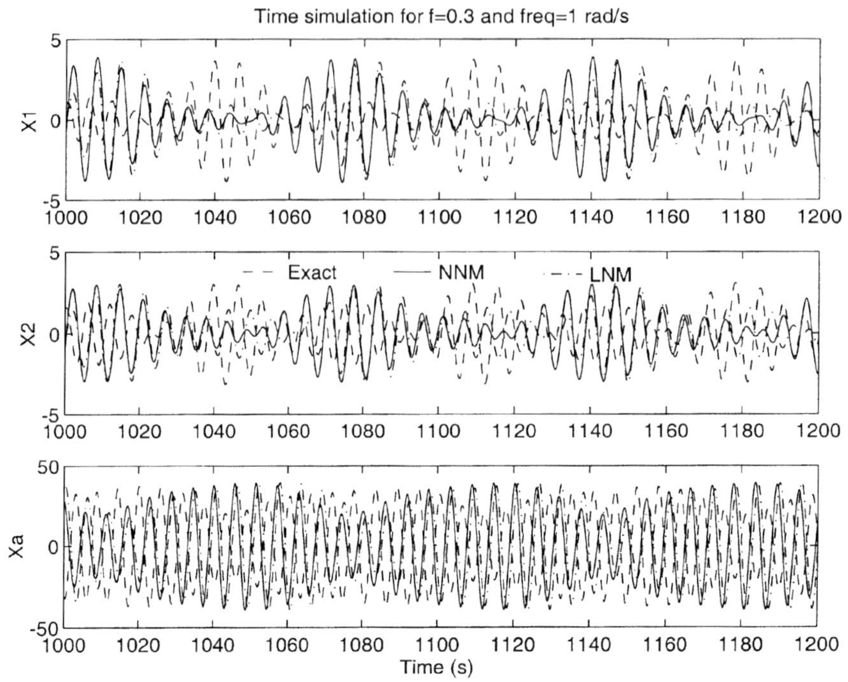

Figure 12. Steady-state response of the two-degrees-of-freedom system in physical coordinates for $f = 0.3$ and $\sigma = 0$.

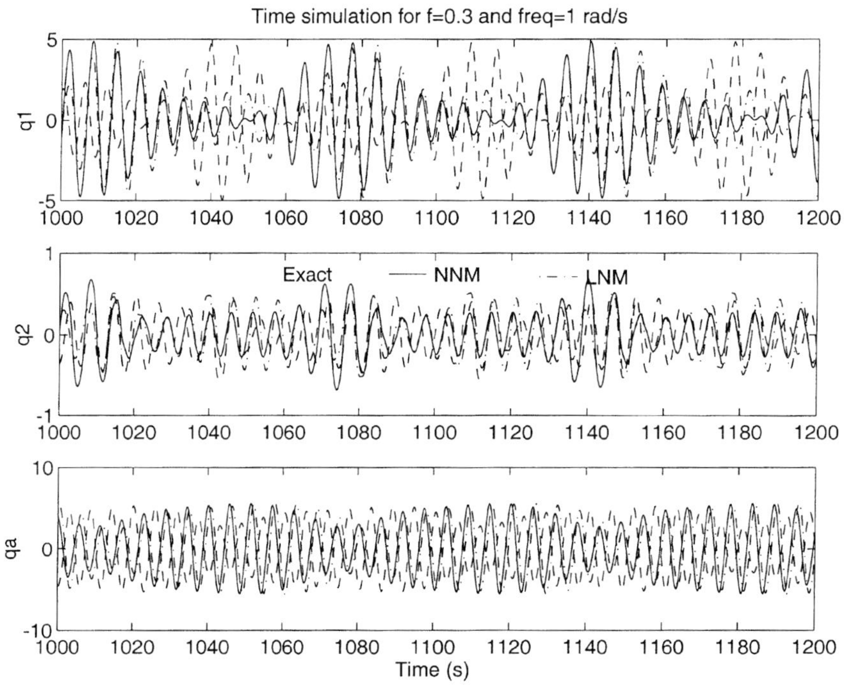

Figure 13. Steady-state response of the two-degrees-of-freedom system in modal coordinates for $f = 0.3$ and $\sigma = 0$.

observed, but is expected due to the approximations made by the method of multiple scales. After 1,000 seconds, a disagreement in the phase is expected as $O(\varepsilon^2)$ effects have had time to build up. This phase shift is visible if one compares the transient results and notes the phase shift begins to build by 180 seconds. For the case studied, ε is $O(0.1)$. Agreement in magnitude is all that can be expected. The steady state magnitudes are in general agreement. Some difference is noted and attributed to $O(\varepsilon^2)$ shifts in the suppression bandwidth. Thus qualitative agreement between the nonlinear normal mode solution and direct integration of the newtonian equations of motion is achieved. Quantitative agreement is lacking due to $O(\varepsilon^2)$ shifts in the location of the suppression bandwidth not captured by the $O(\varepsilon)$ approximations used.

5. Summary

In this paper, the equations of motion for a multi-degrees-of-freedom system with a vibration absorber were shown to reduce to those of the two-degrees-of-freedom absorber model through the application of nonlinear normal mode theory. The theory of nonlinear normal modes was briefly reviewed. Then, the theory of nonlinear normal modes for internally resonant systems was extended to model the forced and damped N-degrees-of-freedom system with an attached vibration absorber. Finally, the theory was then applied to an example commonly used in the nonlinear normal mode literature.

The example has verified the qualitative correctness of the nonlinear mode approximation to the dynamics of an N-degrees-of-freedom system with a vibration absorber. It has also, however, highlighted the limitations of the model near the suppression region. These limitations are due to $O(\varepsilon^2)$ shifts in the location of the suppression bandwidth not captured by the $O(\varepsilon)$ approximations in the model. Improvement should be possible via higher order approximations and is suggested for future research.

References

1. Sun, J. Q., Jolly, M. R., and Norris, M. A., 'Passive, adaptive, and active tuned vibration absorbers', *ASME Combined Anniversary Issue Journal of Mechanical Design and Journal of Vibration and Acoustics* **117**(B), 1995, 234–242.
2. Von Flotow, A. H., Beard, A., and Bailey, D., 'Adaptive tuned vibration absorbers: Tuning laws, tracking agility, sizing, and physical implementations', in *Proceedings Noise-Con 94*, 1994.
3. Roberson, R. E., 'Synthesis of a nonlinear vibration absorber', *Journal of the Franklin Institute* **254**, 1952, 205–220.
4. Arnold, F. R., 'Steady-state behavior of systems provided with nonlinear dynamic vibration absorbers', *Journal of Applied Mechanics* **22**, 1955, 487–492.
5. Pipes, L. A., 'Analysis of a nonlinear dynamic vibration absorber', *Journal of Applied Mechanics* **20**, 1953, 515–518.
6. Miller, H. M. and Gartner, J. R., 'Tunable, non-linear vibration absorber', ASME Paper No. 75-DET-9, 1975.
7. Hunt, J. B. and Nissen, J. C., 'The broadband dynamic vibration absorber', *Journal of Sound and Vibration* **83**, 1982, 573–578.
8. Kojima, H. and Saito, H., 'Forced vibrations of a beam with a non-linear dynamic vibration absorber, *Journal of Sound and Vibration* **88**(4), 1983, 559–568.
9. Jordanov, I. N. and Cheshankov, B. I., 'Optimal design of linear and non-linear dynamic vibration absorbers', *Journal of Sound and Vibration* **123**(1), 1988, 157–170.
10. Nissen, J.-C., Popp, K., and Schmalhorst, B., 'Optimization of a non-linear dynamic vibration absorber', *Journal of Sound and Vibration* **99**(1), 1985, 149–154.

11. Shaw, J., Shaw, S. W., and Haddow, A. G., 'On the response of the nonlinear vibration absorber', *International Journal of Non-Linear Mechanics* **24**, 1989, 281–293.
12. Rice, H. J. and McCraith, J. R., 'Practical non-linear vibration absorber design', *Journal of Sound and Vibration* **116**(3), 1987, 545–559.
13. Natsiavas, S., 'Steady state oscillations and stability of non-linear vibration absorbers', *Journal of Sound and Vibration* **156**(2), 1992, 227–245.
14. Agnes, G. S. and Inman, D. J., 'Nonlinear piezoelectric vibration absorbers', *Smart Materials and Structures* **5**, 1996, 704–714.
15. Agnes, G. S., 'Performance of nonlinear mechanical, resonant-shunted piezoelectric, and electronic vibration absorbers for multi-degree-of-freedom structures', Ph.D. Thesis, Department of Engineering Science and Mechanics, Virginia Polytechnic Institute and State University, Blacksburg, VA, 1997.
16. Agnes, G. S. and Inman, D. J., 'Nonlinear vibration absorbers for nonlinear structures, Part I: Theory', *Journal of Sound and Vibration*, to appear.
17. Agnes, G. S. and Inman, D. J., 'Nonlinear vibration absorbers for nonlinear structures, Part II: Experiments', *Journal of Sound and Vibration*, to appear.
18. Slater, J. C., 'Nonlinear modal control', Ph.D. Thesis, State University of New York at Buffalo, 1993.
19. Slater, J. C. and Inman, D. J., 'Nonlinear modal control method', *Journal of Guidance, Control and Dynamics* **18**(3), 1995, 433–440.
20. Shaw, S. W. and Pierre, C., 'Non-linear normal modes and invariant manifolds', *Journal of Sound and Vibration* **150**, 1991, 170–173.
21. Rosenberg, R. M., 'The normal modes of nonlinear N-degree-of-freedom systems', *Journal of Applied Mechanics* **30**, 1962, 7–14.
22. Rosenberg, R. M. and Kuo, J. K., 'Nonsimilar normal mode vibrations of nonlinear systems having two degrees of freedom', *Journal of Applied Mechanics* **31**, 1964, 283–290.
23. Rosenberg, R. M., 'On nonlinear vibrations of systems with many degrees of freedom', in *Advances in Applied Mechanics*, G. Kuerti, (ed.), Academic Press, New York, 1966, pp. 155–242.
24. Shaw, S. W. and Pierre, C., 'Normal modes for non-linear vibratory systems', *Journal of Sound and Vibration* **164**, 1993, 85–124.
25. Nayfeh, A. H. and Nayfeh, S. A., 'On nonlinear modes of continuous systems', *Journal of Vibrations and Acoustics* **116**, 1994, 129–136.
26. Nayfeh, A. H., Nayfeh, S. A., and Chin, C., 'On nonlinear normal modes of systems with internal resonance', *Journal of Vibration and Acoustics* **118**(3), 1996, 340–345.
27. Nayfeh, A. H. and Balachandran, B., *Applied Nonlinear Dynamics*, Wiley, New York, 1995.
28. Vakakis, A. F., Manevitch, L. I., Mikhlin, Y. V., Pilipchuk, V. N., and Zevin, A. A., *Normal Modes and Localization in Nonlinear Systems*, Wiley, New York, 1996.
29. Vakakis, A. F., 'Nonlinear normal modes (NNMs) and their applications in vibration theory: An overview', *Mechanical Systems and Signal Processing* **11**(1), 1997, 3–22.

A. F. Vakakis (Ed.), *Normal Modes and Localization in Nonlinear Systems*, 293, 2001.

Subject Index